Color Gamut Mapping

Wiley-IS&T Series in Imaging Science and Technology

Series Editor:
Michael A. Kriss

Consultant Editors:
Anthony C. Lowe
Lindsay W. MacDonald
Yoichi Miyake

Reproduction of Colour (6th Edition)
R. W. G. Hunt

Color Appearance Models (2nd Edition)
Mark D. Fairchild

Colorimetry: Fundamentals and Applications
Noburu Ohta and Alan R. Robertson

Color Constancy
Marc Ebner

Color Gamut Mapping
Ján Morovič

Published in Association with the Society for Imaging Science and Technology

Color Gamut Mapping

Ján Morovič

Hewlett-Packard Company, Barcelona, Spain

John Wiley & Sons, Ltd

Telephone (+44) 1243 779777

Email (for orders and customer service enquiries): cs-books@wiley.co.uk
Visit our Home Page on www.wiley.com

Other Wiley Editorial Offices

John Wiley & Sons Inc., 111 River Street, Hoboken, NJ 07030, USA

Jossey-Bass, 989 Market Street, San Francisco, CA 94103-1741, USA

Wiley-VCH Verlag GmbH, Boschstr. 12, D-69469 Weinheim, Germany

John Wiley & Sons Australia Ltd, 42 McDougall Street, Milton, Queensland 4064, Australia

John Wiley & Sons (Asia) Pte Ltd, 2 Clementi Loop #02-01, Jin Xing Distripark, Singapore 129809

John Wily & Sons Canada Ltd, 6045 Freemont Blvd, Mississauga, ONT, L5R 4J3

Wiley also publishes its books in a variety of electronic formats. Some content that appears in print may not be available in electronic books.

Library of Congress Cataloguing-in-Publication Data

Morovič, Ján.
Color gamut mapping/Ján Morovič.
p. cm.
Includes bibliographical references and index.
ISBN 978-0-470-03032-5
1. Color display systems. 2. Color separation–Data processing. I. Title.
TK7882.I6M66 2008
686.2'3042–dc22

2007050360

British Library Cataloguing in Publication Data

A catalogue record for this book is available from the British Library

ISBN 978-0-470-03032-5

Typeset in 10/12 pt. WarnockPro by Thomson Digital, Noida, India

Contents

Foreword		**ix**
Series Preface		**xi**
Preface		**xiii**
1.	**Introduction**	**1**
1.1	What is Color Gamut Mapping?	1
1.2	Historical Context of Gamut Mapping	5
1.3	Who is this Book for?	7
1.4	What is in the Rest of the Book?	8
2.	**Basics of Color Science**	**11**
2.1	What is Color?	11
2.2	Describing Color Experiences	12
2.3	The Color Ecosystem	16
2.4	The Human Visual System	17
2.5	Physical Color-related Properties	21
2.6	Colorimetry	25
2.7	Color Ecosystem Interactions	29
2.8	Digital Color Capture and Generation	35
2.9	Color Management	39
2.10	What can Affect the Appearance of a Pair of Stimuli?	41
2.11	Summary	42
3.	**Desired Color Reproduction Properties and their Evaluation**	**45**
3.1	Color Reproduction Framework	45
3.2	Desired Color Reproduction Properties	47
3.3	Evaluating Reproductions	53
3.4	Case Study: Evaluating Printed Reproductions of a Displayed Image	70
3.5	Summary	71
4.	**Color Reproduction Data Flows**	**73**
4.1	Device Color Spaces	73
4.2	Conceptual Stages of Color Reproduction	76
4.3	Closed-loop Color Management	81
4.4	sRGB Color Management	83
4.5	ICC Color Management	85
4.6	Windows Color System Color Management	88
4.7	Importance of Gamut Mapping in Color Reproduction	89
4.8	Summary	90

5. Overview of Gamut Mapping **91**
5.1 Definitions 91
5.2 Aims of Gamut Mapping 94
5.3 Gamut Mapping Algorithm Context 95
5.4 Types of Gamut Mapping 96
5.5 Building Blocks of Gamut Mapping Algorithms 97
5.6 Factors Affecting Gamut Mapping 105
5.7 Will Gamut Mapping become Redundant? 107

6. Color Spaces for Gamut Mapping **111**
6.1 Implications of Mapping Appearance Predictors 112
6.2 Which Appearance Attributes' Predictors to Map 114
6.3 Overview of Color Appearance Spaces 115
6.4 Mapping in Nonappearance Spaces 119
6.5 Choosing a Space for Gamut Mapping 121

7. Basic Computational Geometry for Gamut Mapping **123**
7.1 Spaces, Points, Lines and Planes 123
7.2 Intersections 126
7.3 Is a Point Inside or Not? 128
7.4 Normals 129
7.5 Triangulation 131
7.6 Summary 132

8. Color Gamuts and their Computation **133**
8.1 Challenges and Implications of Definition 134
8.2 Gamut Boundary Descriptor Algorithms 143
8.3 Evaluating and Operating on Gamut Boundary Descriptors 155
8.4 Examples of Salient Color Gamuts 160
8.5 Image Gamuts 167
8.6 Summary 169

9. A Case Study: Minimum Color Difference Gamut Clipping **171**
9.1 The Original 171
9.2 The Destination Gamut 175
9.3 Minimum Color Difference Gamut Mapping 176
9.4 The Destination Image 182
9.5 Effect of Alternatives 183
9.6 Summary 191

10. Survey of Gamut Mapping Algorithms **193**
10.1 Color-by-color Reduction 194
10.2 Color-by-color Expansion 220
10.3 Spatial Gamut Reduction 224
10.4 Spectral Gamut Mapping 234
10.5 Gamut Mapping for Special Applications 237
10.6 Summary 239

11. Gamut Mapping Algorithms and Color Management Systems 241
11.1 Gamut Mapping Algorithms for Closed-loop Color Management 241
11.2 Gamut Mapping Algorithms for sRGB Color Management 242
11.3 Gamut Mapping Algorithms for ICC Color Management 242
11.4 Gamut Mapping Algorithms for WCS Color Management 244
11.5 Types of Color Management Context 245
11.6 Spatial and Spectral Gamut Mapping 246
11.7 Conclusions 246

12. Evaluating Gamut Mapping Algorithms 247
12.1 Why is there no Reference Data Set in Gamut Mapping Algorithm Evaluation? 248
12.2 Characteristics of Published Gamut Mapping Algorithm Evaluation 250
12.3 What can be Said on the Basis of Existing Evaluation Results? 252
12.4 CIE Guidelines 253
12.5 Studying Specific Properties of Gamut Mapping Algorithms 255
12.6 Summary 258

13. Conclusions 261

References 265

Index 281

Foreword

Ján Morovič has been a regular lecturer at conferences on color science for some years now, and, like many others, I have been extremely impressed with his clarity of thought, and felicity of deliverance, on these occasions. It is therefore with much pleasure that I agreed to write the Foreword to his book on Color Gamut Mapping. This is a subject on which he has justifiably gained an international reputation as an expert, and he has served with distinction as the chairman of the CIE Technical Committee on gamut mapping.

Gamut mapping is not an easy subject. It is one of those endeavours for which there is no perfect solution. Ideally, we would like to be in a situation where gamut mapping is unnecessary because all reproduction media have the same gamut. The real world is very different: reflection prints, projected images, and self-luminous displays all have different ranges of tones and colors that they can reproduce. Furthermore, in each of these classes of display, various technologies are used. Thus, reflection prints can be made by silver-halide photography, by ink-jet, by electrophotography, by lithography, and by gravure printing; and the quality of the paper or other substrate used can have a large effect on the attainable gamut. Projected images can use film, liquid-crystal, and digital-mirror-device, projectors. Self-luminous images can be derived from cathode-ray tubes, liquid-crystal displays, plasma displays, and light-emitting diodes.

To add to the complexity, the conditions under which images are viewed greatly affect their appearance. The surround to the picture affects its contrast and colourfulness, and the ambient light can have a large effect on the color and tonal range that can be perceived. A superficial approach is therefore totally inadequate for gamut mapping, and, in this book, the relevant fundamentals are carefully explained before the practical applications are approached.

A treatment of this subject therefore requires a considerable breadth of knowledge, and a good grasp of what is important and what is of little consequence. Ján Morovič's experience in carrying out research in color science, and in being involved in developing commercial printing systems, has provided him with the necessary background to meet these exacting demands. These, coupled with his outstanding ability to present complex subjects with a remarkable clarity, have resulted in this excellent book. Its publication will give much pleasure to all those who have enjoyed being acquainted with Ján's delightful personality, and to other readers it will serve as a most valuable treatment of the subject.

Robert W. G. Hunt

Series Preface

"How do you want your steak cooked?" is a question we often get from the waiter when we dine out. You reply, "Well, what are my choices?" The waiter responds, "rare, medium rare, medium, medium well and well done." To which you might say, "just cook it so it is not red in the middle, but a little pink." The ambiguity of terms makes it necessary to express exactly what you want and not rely on the interpretation of the chef. In many ways the same ambiguity can be found in color reproduction. For the most part, color reproduction has become an open system where there is little control (by one person or one company) on what color you 'see' on the final print or display. In contrast, silver based photographic system were closed systems developed by integrated companies and when processed properly gave consistent results; the customers could argue over which company's color reproduction they like best. Color proofing systems have been developed over the years for the graphic arts and printing industry where a 'mistake' is very costly. These proofing systems allow the customer and supplier to determine exactly what the customer wants before the presses roll. These color printing presses may include 'spot color' presses that apply special inks to accurately depict the color of a very expensive car or just the corporate logo. However, what can be done to ensure 'correct color reproduction' in the far more common and now ubiquitous desk top color printers, be they color ink jet printers, color laser electrophotographic printers or color solid ink printers? For the sake of discussion only three or four color printers will be considered that use cyan, magenta, yellow and black inks or pigments. First what is meant by 'correct color reproduction?' For a digital photograph correct color reproduction might be an exact match to the color seen by the observer or something that was close enough to the 'original colors' to evoke a clear 'memory' of the experience associated with the image. For a business graphic or presentation, the correct color reproduction might be the one that best transmits the information to the audience. For an 'artistic' representation the correct color reproduction may have no relation to 'reality' but a color rendering that is able to express the author's inner feelings. No matter what the color rendering intent, the user is limited by the fundamental limitations of the inks, pigments, phosphors, color filters and light sources used in the printing or display systems. For the sake of discussion consider a digital camera system where the color information will be constrained by the spectral sensitivities of the sensor (including the color filter array use to encode the color), the taking illuminant, the color (white) balance used by the camera and the internal color image processing to produce a 'good' color space. All this information is then recorded in 8 to 12 bits per pixel in some color space like sRGB or its equivalent. Having done this, the ultimate color gamut has been established. Now consider using a three-color ink jet printer to print this digital image. The three inks, cyan, magenta and yellow, establish the printing primaries and the printer gamut is established as those colors that reside inside the hexagon formed by the three ink primaries and the red, green and blue colors formed by their binary addition (within the CIE x-y plot). Assuming that the digital image color gamut exceeds that of the printer, the concept of how to 'best map' the digital image colors into the printer color gamut is called Color Gamut Mapping and is the subject of this book. How this is done has more variations than how one might prepare the steak mentioned above.

Color Gamut Mapping by Ján Morovič is the fifth book in the Wiley-IS&T Series in Imaging Science and Technology. This text provides a color scientist or engineer a comprehensive study of the basics of color reproduction and an in-depth study of the many ways to implement color gamut mapping, detailing their individual methodologies, strong points and shortcomings. The reader can, with this text, learn the art and science of color gamut mapping and then create solution to his/her particular problem. Ján Morovič is an outstanding representative of a new generation of color scientists and engineers who have focused their collective efforts on turning complex and subtle aspects of color science into practical solutions for a rapidly expanding color enriched display and printing industry. Ján received his Ph.D. from the University of Derby, UK, under the guidance of Professor Ronnier Luo and is currently with Hewlett-Packard Española. He has chaired the CIE Division 8 Technical Committee 8-03 on Gamut Mapping and was the General Co-chair of the 15th Color Imaging Conference sponsored by IS&T and SID. Ján also received the Royal Photographic Society of Great Britain's 2003 Selwyn Award for his work in solving one or more technical problems in area of color photography.

MICHAEL A. KRISS

Formerly of the Eastman
Kodak Research Laboratories
and the University of Rochester

Preface

This book has been a long time in the making and I am as delighted that it is complete now as I was with writing it while that lasted.

I would first like to thank Professor Ronnier Luo, whose 1994 talk about color appearance modeling at the Institute of Physics in London inspired me to study color science. Ronnier's supervising of my PhD on gamut mapping is the source of my writing on this subject. I would also like to thank my PhD student, Dr Pei-Li Sun, for his insights into many of the topics covered here. Others working in the field of gamut mapping have also been a great inspiration to me and, at the same time as apologizing to those I may have omitted, I would like to thank the following: Raja Bala, Gus Braun, Mark Fairchild, Patrick Herzog, Tony Johnson, Naoya Katoh, Hideto Motomura and Todd Newman. They, and all my colleagues first at the University of Derby's Color and Imaging Institute and then at Hewlett-Packard Española S. L. in Barcelona, have greatly influenced my work in this area.

For supporting my writing effort and allowing me to balance it with other work assignments I would like to thank all my managers at HP during the last two and a half years: Ramon Pastor, Virginia Palacios, Jordi Arnabat, Albert Serra and Alan Lobban. Without them this book would have remained a missed opportunity. I would also like to thank the staff at John Wiley and Sons Ltd – Simone Taylor, Mary Lawrence, Nicky Skinner, Erica Peters, Jo Bucknall, Emily Bone and Wendy Hunter – for being patient during the period before the book writing started and for then supporting me throughout its duration.

The content of this book would have been noticeably compromised had my amazing wife Karen not read it all, word for word, and scrutinized it with her sharp legal mind, had my brother Peter not given me all that invaluable feedback, had my colleague Utpal Sarkar not taken great care to snoop out mistakes of various magnitudes of mathematical and physical gravity, had my colleague Huan Zeng not provided the great gamut visualization tool used in some of the figures, had my colleagues Joan Manel Garcia (who impressively spotted an inverted L:M ratio in Chapter 2), Joan Uroz, Marc Casaldaliga, Ana Maria Cardells and Pau Soler not volunteered their opinions and had John McCann and Alessandro Rizzi not provided an advance copy of their paper on glare in HDRI. To all of them, and to any others whom I forgot, I would like to thank from the bottom of my heart.

It would be unthinkable not to thank my parents, Ján and Juliana, for everything they have done and still are doing for me, and my siblings, Peter, Monika and Beátka, for being the best siblings I can imagine.

I am eternally indebted to my beautiful and patient wife Karen and to my lovely children, Ján and Thomas, for their love and support. They are my everything, and without them I could do nothing. In the truest sense of the word, Karen has been my co-author here and this book is as much her achievement as it is mine.

Finally, I would like to thank God, who is Love.

Ján Morovič

1 Introduction

1.1 WHAT IS COLOR GAMUT MAPPING?

On an average day most of us will come across a myriad of visual content generated using an ever-increasing variety of means. Already at the moment of waking up we may be presented with the flashing digits of an unwelcome alarm clock, accompanied by a sharp burst of light from a window or by a gradual emergence of shapes in the dark. During breakfast we may browse a newspaper or watch the news on a television. We may check our email using a personal computer or send a message, photo or video using our mobile phones and on the way out catch a glimpse of the latest artwork of our kids stuck to a fridge door or glance at some junk mail.

This scenario, which is so everyday as to be unremarkable, is an example of the ubiquity and variety of sources of visual content in our lives. We are regularly exposed to at least natural imagery, i.e. our homes themselves, whose reflective surfaces can be lit by natural or artificial light sources. Besides, we almost certainly view the output of traditional, analog imaging technologies such as drawings, print, photography and television and we regularly interact with digital imaging devices such as mobile phones and personal computers. In the process of work and leisure we also come across other imaging technologies, including digital projectors, cameras and printers. Many of us, therefore, are likely to have already used many, if not all, imaging technologies developed to date.

Reflecting on the diversity of visual content in our environment brings us to the observation that a variety of means can represent the same image and that these means, therefore, need to communicate among themselves. For example, we may wish to capture a moment from a birthday party (Figure 1.1) by taking a picture of it using a digital camera. Supposing we like the picture a lot we may also email it to our friends, send it to others via a mobile phone, print it out on a desktop printer, place it on a website, have a larger version of it printed in a copy-shop and include it in a presentation stored on a DVD and viewed on a television or projected onto a screen. Here, the same content (the scene from the birthday party) is present in at least 10 instances. Depending on how many friends we email and how many visitors come to the website, this number can be a lot larger.

Color Gamut Mapping Ján Morovič

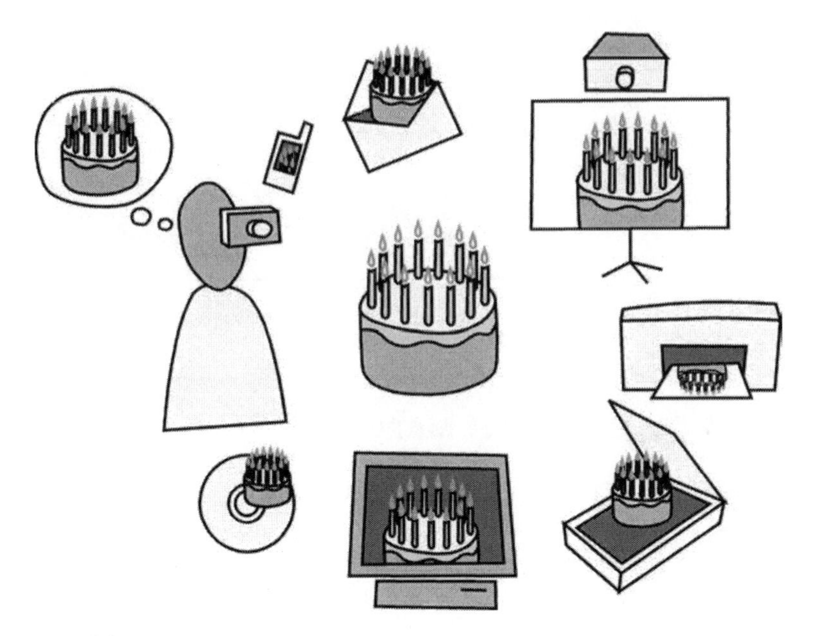

Figure 1.1 Birthday party images.

Next, let us think about how we would like the various instances of a given image to relate to each other. Clearly, the simplest and most immediate answer is that we want them all to look the same. In other words, we would like to see 'the same' when we look at the photo on the website and the print we made on a printer. Furthermore, we would like both of these to be 'the same' as when we looked at the actual scene during the birthday party. In some cases, though, we are less interested in an accurate record of an event and instead prefer to have an image that 'looks nice.' If the birthday party took place on an overcast day and everything looked a bit dull at the time, then we may still prefer for a more cheerful appearance to be represented by the images we took of the occasion.

Even though some differences between the various instances of visual content can be considered to be improvements, there are other differences that no one likes to see. Take, for example, the case of receiving the birthday party image and viewing it on your mobile phone in bright daylight. The image is likely to have much lower contrast than the original scene did, so much so that it could be difficult to see altogether. Viewing the image on your personal computer's display could result in a darker result than how you remember the party. Printing the image might give a less colorful appearance than what was shown on your display. So, even if we want all instances of some visual content to 'look the same' or to have changes that make it 'look nice,' there are in practice many other differences between the various instances of an image that are undesirable.

Why is it then that different instances of a given image can look so different? The answer is a rather complex network of interactions among many individual factors, including the following:

1. *The instances of an image are viewed by different people.* As each one of us responds slightly differently to the light entering our eyes, there will be differences between the

experiences two people have when viewing a single image. Furthermore, as soon as we communicate about what we see, our experiences, skills and habits also play a role. When two people talk about a single image and even experience it in the same way, they are likely to express it differently.

2. *The instances of an image are viewed under different viewing conditions.* If we view two physically identical instances of an image in different environments, then they will look different. For example, viewing a television in the dark can give rise to a greater range of colors (i.e. greater contrast in images, more colorful parts of images, e.g. grass looking more vibrant) than when it is viewed in bright daylight, when it looks a lot duller overall. Note that in this case the television outputs the same image in both cases, but the dark versus bright environment changes its appearance significantly. This is yet more dramatic for images that reflect light (e.g. prints), where in the dark they too are dark and only when more light is present to view them do they acquire a clear appearance. In addition to how much light is present when viewing images, the background of the image is also important (e.g. what color the wall is behind a television), as is the distance at which it is viewed. Finally, it also matters what else is seen when instances of an image are viewed (e.g. whether the paper of a newspaper looks 'white' depends on whether we also see other kinds of paper at the same time) and how the instances are arranged (e.g. whether they immediately next to each other or far apart).
3. *The instances of an image are created using different technologies.* When the digital data from a camera are displayed on two different displays they are likely to look different, as displays can differ in terms of the materials they use for outputting color as well as in their settings. The digital data – which describe an image as a series of red, green and blue (RGB) values for a grid of spatial locations – give only relative instructions to imaging devices, i.e. the instruction may be to use 100% of a display's red and green colorants and 50% of its blue colorant. However, if the displays have different colorants, then following the same relative instructions will give different results. This constraint can, however, be overcome. The key to the solution is that it is possible to understand the relationship between digital data input to a display and the color appearance of the corresponding output (e.g. we can know what sending RGB = [100%, 100%, 50%] will look like on a given display and we can also work out what RGBs to send to the display if we want a certain color output from it). Then, to get two displays to look the same, we can take the RGBs we send to the first display, work out from them what the corresponding output colors will look like and from these color appearances work out what other RGBs to send to the second display. In other words, sending the same RGBs to two displays gives different colors, but sending different, appropriately chosen RGBs to each one can give the same color. Furthermore, the approach also extends to getting colors to look the same on different types of imaging device.

The above is a highly simplistic and incomplete sketch of why different instances of a given image can look different; more detail will be provided in Chapters 2 to 4.

One factor, however, was not made explicit in the above list. Namely, that, all else being equal (i.e. same viewer and viewing conditions), each imaging technology is capable of accessing only a specific part of all possible colors. For example, when viewing a television in a brightly lit room, it is not capable of providing a viewer with the experience of a very

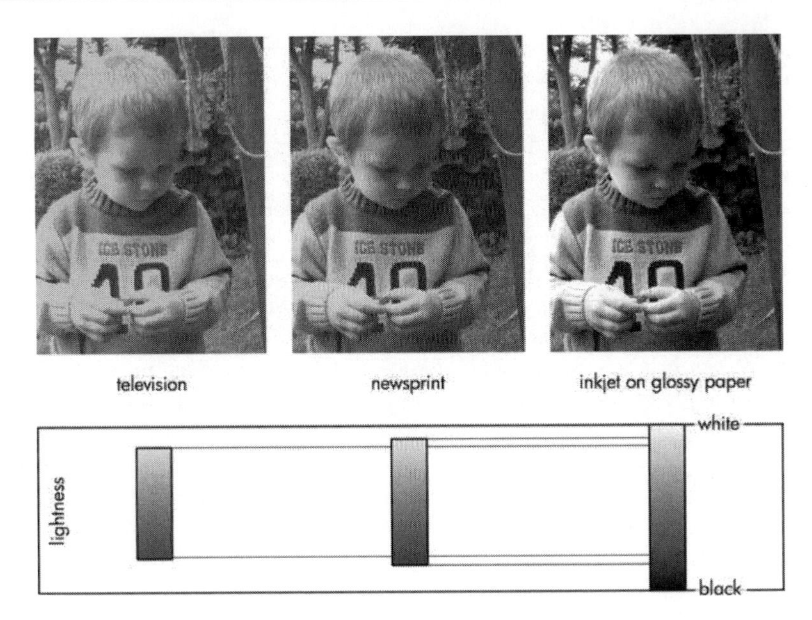

Figure 1.2 Getting dark shades in a brightly lit room. Top: a scene represented using different media; bottom: ranges of lightnesses possible on those media.

dark black. Looking at an image in a newspaper has such limitations too, whereas a glossy inkjet print does not and can give rise to the experience of dark blacks (Figure 1.2). A computer display is likely to be able to display more colorful, bright greens than are possible in print under the same viewing conditions, whereas print is more likely to be able to contain dark, colorful greenish-blues that cannot be obtained on a display.

As was proposed above, we tend either to look for colors being reproduced in the 'same' or a 'nice' way in all instances of an image. However, as individual imaging technologies provide different ranges of colors, our desire for sameness or niceness may in some cases be impossible. For example, start with an image that has the colors we want (e.g. a painting) and attempt to reproduce it by means that are unable to provide all of them (e.g. capture the painting using a digital camera and view the result on a television). While there will be parts of the painting whose colors we can replicate on a television, there are likely to be other parts that we cannot and decisions need to be made about how to deal with such colors.

Most important, these decisions can have a strong effect on the final image appearance and the same set of decisions is not likely to be appropriate in all cases. For example, when an image of a piece of clothing is to be reproduced in a printed catalogue, it is very important for that image to be as similar to the actual piece of clothing as possible. If there are some colors in the fabric that cannot be matched in the printed catalogue, then what matters is to represent them by colors that look most similar to the original. Alternatively, if a business presentation prepared on a computer display includes a pie-chart that has a segment in a bright, pure yellow, then it is preferred for such a yellow to be reproduced as bright and pure when printed, rather than as a yellow that is as similar to the original one as possible but which does not preserve the original's brightness and purity. Finally, when scanning and reproducing traditional holiday snapshots on a display, the color of the sky is typically preferred to be more vibrant than is possible in the photograph, rather than to be as close to it as possible.

What the above examples illustrate are both variety and, less directly, the potential for misapplication. Take, for example, the case of the printed clothes catalogue and consider the (mis)application of the solution from the holiday snap reproduction case. Clothes would appear more vibrant in the catalogue than they are in reality and this would be likely to result in customer disappointment, returned orders and reduced turnover. Conversely, if the catalogue solution were applied to holiday snaps, some displayed images would look dull and disappointing. Hence, careful decisions need to be made when dealing with the different color ranges available to different instances of an image.

More technically speaking: the color ranges mentioned above are called *color gamuts*, the changes made to images to deal with different instances of the image having access to different color gamuts are called *color gamut mapping* and the likelihood of gamut mapping being necessary in any given color communication is very high.

1.2 HISTORICAL CONTEXT OF GAMUT MAPPING

To counter the popular impression that gamut mapping is a modern phenomenon, it is worthwhile to consider its historical background before proceeding with a technical analysis. Such an insight into its origins will put gamut mapping into perspective and facilitate a better understanding of its expected effect. Since gamut mapping is the dealing with different color ranges being available to different instances of an image, its history coincides with that of man-made images. To create an image involves its reproduction either from another visible source or from the mind of its author and, hence, the addressing of differences between available color ranges.

As soon as humans created the first images, gamut mapping took place and it has been continuously practiced ever since (Figure 1.3). The earliest such images are Paleolithic cave paintings that date back 30 000 years and record hunting experiences on cave walls. The materials (i.e. colorants) used for creating these images, prime examples of which can be found in France (e.g. Chauvet-Pont-d'Arc, Lascaux) and Spain (e.g. Altamira), were charcoal and red and yellow ochres. When these images were created, there had to be at least an implicit consideration of how to represent the wide range of color experiences from a hunting event using the very limited range of available colors. Therefore, already these paintings are instances of the use of gamut mapping, albeit in a way that is implicit and certainly not like current approaches.

Following its Paleolithic beginnings, we can see a gradual but continuous expansion of the palette of color reproduction means, and a very brief look at their emergence will be taken next (Wikipedia, 2005a). Note that the development of the various means of creating color images often also provided access to greater color gamuts or to finer color variation within a given gamut. For example, the simple palette of the earliest cave paintings that consisted of three to five distinct colors became more extensive over time and was followed by the development of frescoes (i.e. 'painting in pigment in a water medium on wet or fresh lime mortar or plaster' (Wikipedia, 2005b)) in 15th century BC Greece. These frescoes were created using a much more varied palette of colorants and also allowed for

Figure 1.3 Examples of color reproduction from the Paleolithic age to the present. Digital original image reproduced by permission of Joe Kondrak.

an easier creation of transitions between them. Similar results were also achieved by painting onto dry walls, e.g. as seen in the tombs of Egyptian rulers from the 13th century BC onwards.

The 6th century BC saw the introduction of painted, enameled tiles in Babylon and the use of elaborate painted pottery in Greece. Mosaics and tapestries appeared around the 2nd century BC; and the earliest remaining carpet, which too is a means of reproducing color imagery, is considered to be the 5th century BC Middle Eastern *Pazyryk* rug excavated in Siberia. Frescoes from the 1st century AD can be seen to exhibit finer detail than earlier uses of the technique allowed, as do mosaics from the 5th and 6th centuries AD. In this period we also start seeing the first examples of illuminated manuscripts, which involve painting and the application of gold leaf onto parchment. The beginning of the second millennium AD heralds the use of elaborate, multicolored ceramics and stained glass windows, and imagery painted onto paper is preserved from subsequent centuries.

The earliest printed book that survives to this day is the *Diamond Sutra* from 9th century China and it is among the first examples of relatively large numbers of copies of visual content being produced. These beginnings of mass production are an important stage in the history of imaging, as they have two key implications. First, they lead to a process that is currently blossoming in the widespread availability and use of imaging technologies. Second, they introduce an additional source of imaging that is distinct from art and that has strong commercial and mass-consumption aspects. Had the creation of visual content remained exclusively in the hands of artists and craftsmen, the practice of gamut mapping might never have become explicit.

On the long road from the first use of book printing to the present and diverse variety of technologies, we also find the practice of painting onto silk, popular in 16th-century Japan, the invention of three- and four-color halftone printing by Le Blon in the 17th century (Pankow, 2005) and the widespread use of woodblock color printing in 19th-century Japan.

Around the beginning of the 20th century, color imaging experiences a rapid expansion thanks to the introduction of a variety of color photography techniques (starting with Lippmann's method, demonstrated in 1891, and followed by solutions that take us into the present day). The launch of color television in the late 1930s, of the first drum scanner in the 1950s, the first color computer display and inkjet printer in the 1970s and the first color digital cameras and color laser printers in the early 1990s continue the trend. The development of such imaging technologies, and especially the emergence of digital ones, is paramount to the explicit arrival of gamut mapping on the scene in 1978 (Buckley, 1978), as it is digital imaging that for the first time in history allowed for an explicit dealing with the differences of color ranges for the parts of an image, i.e. explicit gamut mapping.

The key point to take away from this overview is that gamut mapping has implicitly been present ever since man-made images were first painted on cave walls during the Paleolithic age and that its emergence in the 1970s is only the becoming-explicit of an ever-present process rather than its beginning. This realization is particularly helpful, in that it highlights the tight coupling between gamut mapping and the skills of artists and craftsmen who have practiced it for millennia, before its explicit form was taken over by mathematicians, scientists and engineers. An implication of this is also the need to confront scientifically developed solutions with a critique by those who create visual content professionally rather than only to focus on their scientific merits, which are in the foreground of contemporary gamut mapping research.

1.3 WHO IS THIS BOOK FOR?

The primary aim of this book is to help its reader to acquire the ability to choose from among existing, alternative ways of doing gamut mapping for an actual imaging task or product and then to implement them. More specifically, the focus will be on understanding and choosing existing gamut mapping solutions, as the other aspects of such an endeavor require familiarity with disciplines far beyond gamut mapping itself (e.g. color and imaging science, computer science, computational geometry, etc.). As regards these broader disciplines, a basic, and very much gamut-mapping-focused, overview will be provided of color and imaging science and of computational geometry. Experience with using at least some computer programming language and an introductory, university-level mathematics knowledge are highly recommended. Scientists, researchers, engineers, training instructors and students of undergraduate and graduate courses in color and imaging and related subjects are the primary audience for this book.

The following content will also address questions about gamut mapping that arise in the context of its use, rather than development and implementation. Hence, those who create color and imaging content (e.g. photographers, graphic designers, artists), those who provide technical services in the imaging industry (pre-press houses, print service providers, consultants), those who are involved in the preservation and distribution of color content, as well as undergraduate and graduate students of engineering and computer science courses will also benefit from the material presented here.

1.4 WHAT IS IN THE REST OF THE BOOK?

To aid both developers/implementers and users of color imaging technologies, Figure 1.4 provides a recommendation for how to use this book, depending on the reader's background.

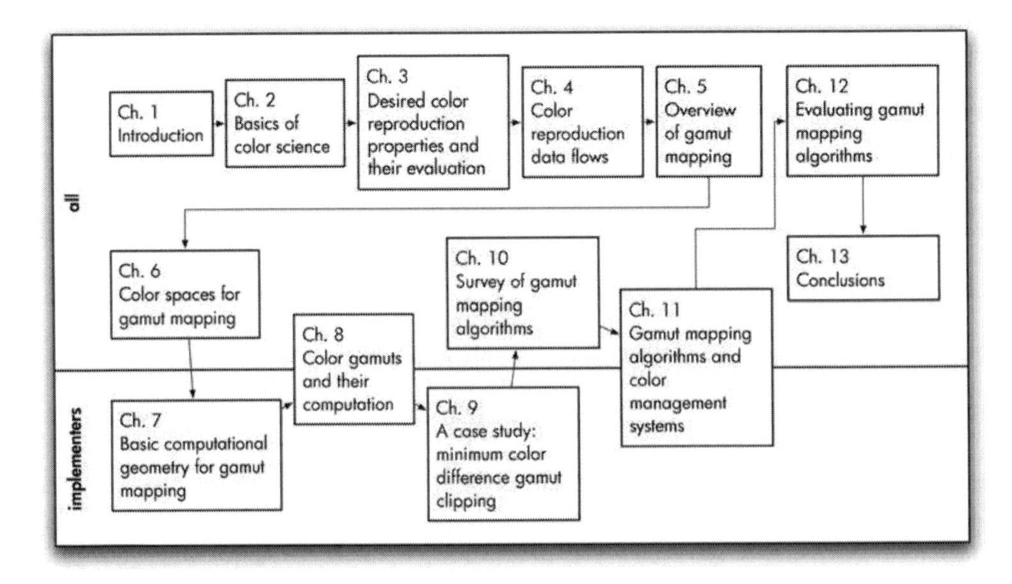

Figure 1.4 Book content overview.

Following this introduction, the basics of color science will be presented in Chapter 2, where the aim will be to give an overview of what needs to be taken into account in a system where gamut mapping is to be performed. This will include discussions of color appearance phenomena, viewing conditions, human vision, color-related physical properties, color measurement and quantification, color difference prediction and the basic principles of color imaging devices and media. The emphasis will be on providing a high-level outline and references to relevant literature. Finally, the chapter will conclude with applying the general overview to the question of what affects the perceived relationship of a pair of color stimuli (i.e. objects that evoke the experience of color perception).

Chapter 3 will take a look at various aims that color reproduction can have and at how reproductions can be assessed with reference to them. Psychophysical/psychometric versus measurement-based approaches will be introduced and an overview will be given of the *Commission Internationale de l'Eclairage's* (CIE's) *Guidelines for the Evaluation of Gamut Mapping Algorithms*. The concepts discussed in this chapter will be essential for a clear understanding of gamut mapping and will be referred to throughout the remainder of the book.

An 'under the hood' look at color reproduction workflows will follow in Chapter 4, where ways of representing original color information and the transformations it needs to undergo on the way from original to reproduction will be made explicit. Attention will also be paid to the importance of gamut mapping in different applications (e.g. what the role of gamut mapping is in newspaper printing versus press proofing on an inkjet printer).

Following the initial chapters that set the scene, Chapter 5 will take a detailed look at gamut mapping itself. The aims of gamut mapping, the types of gamut mapping algorithm (GMA – i.e. processes), their components and the factors affecting them will be introduced and different ways of looking at the role of gamut mapping will also be compared. Chapter 6 deals with the impact of the choice of color space in which gamut mapping is performed and Chapters 7 and 8 look at what color gamuts are and how they can be computed.

Armed with the content provided in the first seven chapters, a gamut mapping case study will be taken under the microscope in Chapter 9. For a specific color reproduction system, the journey of a color image will be traced from original to reproduction; also, how the image's colors change in the process due to the application of the minimum color difference gamut clipping algorithm will be looked at explicitly. This chapter will show the reader how the concepts introduced so far are applied to a concrete case and the effect they have will be scrutinized both in visual and underlying numerical terms.

Chapter 10 will then give a survey of gamut mapping work published to date, with an emphasis on considering the pros and cons of individual algorithms. Chapter 11 confronts gamut mapping research and development with its implementation in commercial color management systems, such as those based on the *International Color Consortium* (ICC) specification, the Windows Color System and sRGB workflows.

How GMAs can be evaluated, e.g. in terms of whether their performance differs for different kinds of images or in terms of how well they deal with smooth transitions in originals, will be covered in Chapter 12, and Chapter 13 considers current challenges and future trends in this area and the future role of gamut mapping in the wider context of imaging.

Finally, the book will not cover the following topics, apart from briefly mentioning their relationship to general gamut mapping:

- *First, the optimization of gamut mapping implementation.* As GMAs are defined in terms of the effect they are to have when applied to the colors of originals, there can often be considerable variation in the resources (e.g. processing, memory and time) that need to be used to achieve gamut mapping, depending on how it is implemented. Nonetheless, the focus here will be on looking at gamut mapping proper, in terms of the various ways in which sets of original colors can be adjusted to fit a reproduction color gamut, and the reader is advised to consult general texts on algorithm optimization where applicable.
- *Second, high dynamic range imaging (HDRI).* This refers to techniques that allow for the capture and subsequent processing of images containing large 'dynamic ranges,' i.e. ratios of the brightest and darkest parts of the image. Whereas, for example, computer displays have dynamic ranges of around 200:1, natural or computer-generated scenes can be of the order of magnitude of 100 000:1 (Hunt, 1995b: 787) and HDRI deals with such high dynamic ranges. HDRI addresses the capture, encoding and also the subsequent reproduction of high dynamic range (HDR) images and it is with regard to HDR image reproduction that there is a link to gamut mapping. However, as HDRI and gamut mapping currently deal with different domains (the former is applied to quantities of light in scenes/images and the latter to color appearance), there will be little discussion of HDRI in this book. Nonetheless, attention will be drawn to cases where there has been cross-fertilization between the two areas and also where there might be potential for it in the future. The reader interested in HDRI is advised to consult Reinhard *et al.* (2005).

Given the above sketch of this book's content, let us now proceed to discover in more detail the landscape at whose heart gamut mapping lives.

2
Basics of Color Science

Understanding the framework in which gamut mapping takes place requires some familiarity with the basics of color science, which will be very briefly introduced here. Please note that the following will be far from a comprehensive overview, but is instead intended to serve as a guide to topics that need to be considered before gamut mapping can be approached. As a consequence, only what is relevant to gamut mapping will be covered; other topics, which may be no less important in color science as such, will be omitted.

With the above caveats in mind, let us now revisit the 'birthday party' scenario from Chapter 1 and use it as a basis for mapping out the content of the present chapter. The example of a birthday party being captured using a digital camera and then reproduced in a variety of ways is a good basis for thinking about the questions that need to be dealt with to understand how various instances of color content relate to each other. This, in turn, is a key prerequisite to comprehending gamut mapping itself.

The first question clearly needs to be 'what is color?', as it is necessary to have an explicit understanding of what it is that is being communicated. Next, we need to understand the key elements that contribute to the coming about of color and even just from our own experience we are likely to arrive at the following two: a person looking and an object being looked at. Upon further reflection we may also recognize the difference made by the environment in which an object is viewed (e.g. the way in which given objects appear in a dim versus a brightly lit room) and the effect that time has on how an environment appears (e.g. when entering a dark room, our visual experience of it changes over time and detail emerges where we initially only saw darkness). Furthermore, it is also necessary to understand the basics of how color can be generated (e.g. using digital imaging devices such as displays or printers), and we will conclude this chapter by bringing its content together and applying it to the question: 'what can affect the appearance of a pair of colors?'

2.1 WHAT IS COLOR?

A useful starting point for considering the essence of color is Maund's (2002) 'color' entry in the *Stanford Encyclopedia of Philosophy*, which provides a clear overview of how color

Color Gamut Mapping Ján Morovič

has been understood over time. Before considering philosophical stances, the natural, everyday concept of color is illustrated in terms of two of its accepted truths: '(1) that colors are properties in the world (i.e., properties of physical objects) to which one's color vision is sensitive; (2) that colors are qualities that perceptual experience represent[s] (or presents) objects as having' and the fact that color, understood in this natural way, plays an important role in people's lives is emphasized.

Such a natural concept of color is most closely related to various *objectivist* accounts, which consider the color of an object to be in some way the object's intrinsic physical property (e.g. following from the object's microstructure or the way it changes light that is incident on it). However, a key weakness of such accounts is the difficulty of providing a specific explanation of what makes an object a given color without reference to how it affects someone who views the object. Alternatively, color can be considered to be a *subjective quality in experience* (i.e. a quality of the experience of an object) and, therefore, residing in the viewer rather than the viewed. Seen in this *subjectivist* way, objects in themselves do not have color in the everyday sense, and color attains the status of *illusion* or of being a *virtual property* of objects.

Next, the *ecological* theory of color accounts for it as being a relational property. It states that 'being colored a particular determinate color or shade is equivalent to having a particular spectral reflectance, illuminance, or emittance that looks that color to a particular perceiver in specific viewing conditions' (Thompson, 1995); even though it is arguable whether this is sufficiently different from the subjectivist approach, it is pretty much what (often implicitly) underlies contemporary color science.

Apart from considering the relative merits of alternative views on color, what such diversity demonstrates very clearly is the complexity of the concept and the difficulty of defining it one way or another. For a detailed discussion of the philosophy of color, see Westphal (1987) and Byrne and Hilbert (1997).

For a viewer who views an environment over time, color will here be considered to be a quality in the resulting sequence of visual experiences that relates to the physical properties of the viewed environment.

While this is still fairly vague, it is challenging to pin down the definition of color further. Nonetheless, the following artificial case may help to distinguish it from among other qualities in visual experience: Color is that quality in an observer's visual experience that results when all that is seen appears the same for all parts of what is seen. In such a case, there is no texture, shape or motion and the visual experience's only quality is color. Note, however, that this kind of artificial, visually uniform environment cannot lead to all possible color experiences, since some (e.g. colors that would be described as 'brown' or 'pink') are only possible in more complex environments.

2.2 DESCRIBING COLOR EXPERIENCES

Given the above definition of color, let us consider how it can be communicated and in what terms it can be described. At the simplest level a description of color involves words like 'red,' 'mauve,' 'salmon,' 'crimson,' etc., which are simply labels for various individual colors and which do not express structure (i.e. they do not lend themselves to organizing or ordering color experiences).

2.2.1 Color names

The seminal work on *color naming* published by Berlin and Kay (1969) defined basic color names as being simple (i.e. not composite terms), salient (i.e. being used consistently within a given culture), general (i.e. applicable to different kinds of objects), and context independent. Studying 78 languages, the authors found that there are up to 11 such basic names, which in English are three pairs of opponent colors (black–white, red–green and yellow–blue) and an additional five color names, i.e. gray, orange, pink, purple and brown. It is also worth noting here that corresponding color names from different cultures often refer to different color experiences (e.g. the English 'red,' 'green' and 'blue' terms are used to refer to lighter colors than their Chinese linguistic equivalents (Lin *et al.*, 2001)), which limits their use for accurate color communication. Nonetheless, there is growing evidence that color categories are common to all humans and that they derive from how our visual system is made rather than from social convention (Kuehni, 2005). Of interest is also a new web-based method of understanding the color experiences that correspond to various color names (Moroney, 2003), and the importance of such understanding in color reproduction is also evident in the work of Motomura (2002), which will be looked at in more detail in Chapter 10.

2.2.2 Color order systems

A complementary approach to linguistic analysis is to think about how colors could be described in terms of basic properties that would allow for:

1. their organization;
2. for a clearer, less ambiguous description of any given color; and
3. for the arrangement of color sets that have pleasing properties.

This, in fact, is an effort that has a very long history and includes work by philosophers like Plato and Aristotle, artists like Goethe and da Vinci, and scientists like Newton and Maxwell (Silvestrini and Fischer, 2005). Significant milestones here are the work of Robert Grosseteste in the 13th century AD and Leon Battista Alberti in the 15th century AD, the latter of which came up with a spatial representation of colors in three dimensions: one defined by the colors black and white and a further pair formed by the four colors red, green, blue and yellow (Figure 2.1a).

The basic structure of Alberti's color space can be found under numerous guises in the *color order systems* that have been put forward since (Rhodes, 2002; Kuehni, 2003) and it is also consistent with the attributes of color appearance used in color science.

2.2.3 Color appearance attributes

To describe color appearance, the following three basic attributes (shown in Figure 2.1b) are used (Hunt, 1995a: 29–33; Fairchild, 2005: 83–93):

1. *Hue* – the 'attribute of visual sensation according to which an area appears to be similar to one, or to proportions of two, of the perceived colors red, yellow, green, and blue.'

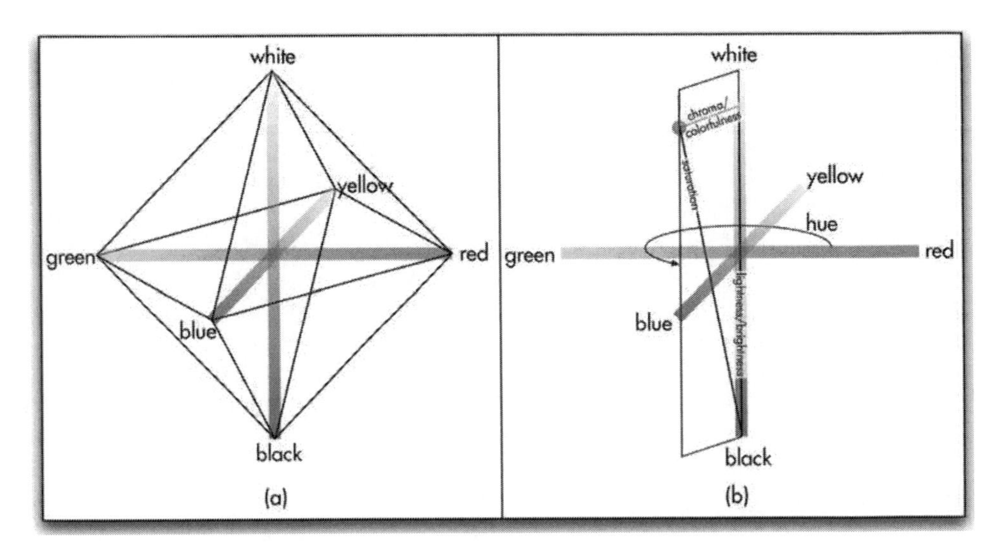

Figure 2.1 (a) Alberti's 1435 color order system; (b) contemporary color appearance attributes.

Looking at an orange, for example, we can see that its color is similar to both red and yellow and we could express its hue as some percentage of each of the two (e.g. 60% red and 40% yellow).

2. *Brightness* – the 'attribute of a visual sensation according to which an area appears to exhibit more or less light.' Returning to the same orange, we can also express this attribute of its color appearance and we would judge it to be higher for parts of it that are directly lit and lower for parts that are in the shade.
3. *Colorfulness* – the 'attribute of a visual sensation according to which an area appears to exhibit more or less of its hue.' Another way of looking at colorfulness is that it expresses how different a color is from a gray color of the same brightness (with gray colors having zero colorfulness). In terms of the orange, colorfulness too would be greater for its directly lit parts than for those in the shade, but lesser for any parts that reflect the light source under which it is viewed. An example of colorfulness decreasing is also what happens when a fabric fades as a result of prolonged exposure to light, repeated washing or abrasion.

While hue, brightness and colorfulness are sufficient for describing any color experience, it is useful to also introduce the terms *lightness* and *chroma.* These are derived from brightness and colorfulness respectively by being 'judged relative to the brightness of a similarly illuminated area that appears to be white or highly transmitting' (Fairchild, 2005: 86–87). The key implication of making the attributes relative to the illumination conditions is that it allows for expressing color experiences in a way that disregards whether the object described was brightly or dimly lit. This allows for the communication of what an object looked like compared with an entire scene and opens the possibility of reproducing that relationship in another scene that may be lit very differently.

In the context of color reproduction, using lightness and chroma allows for the communication of relative appearance across very different levels of illumination while focusing on brightness and colorfulness could, for example, result in very dark reproductions when going from a dim to a bright environment and very light ones when going in the other direction (Figure 2.2).

In Figure 2.2 we can see an example of starting with an image under some reference level of illumination and wanting to make reproductions of it under a higher and a lower illumination level. When moving to a higher level of illumination, the choice of maintaining the original's brightness is possible; however, the same brightnesses that corresponded to a pleasing image under the reference conditions will give a duller result under the higher level of illumination, where pure white has a higher brightness. Conversely,

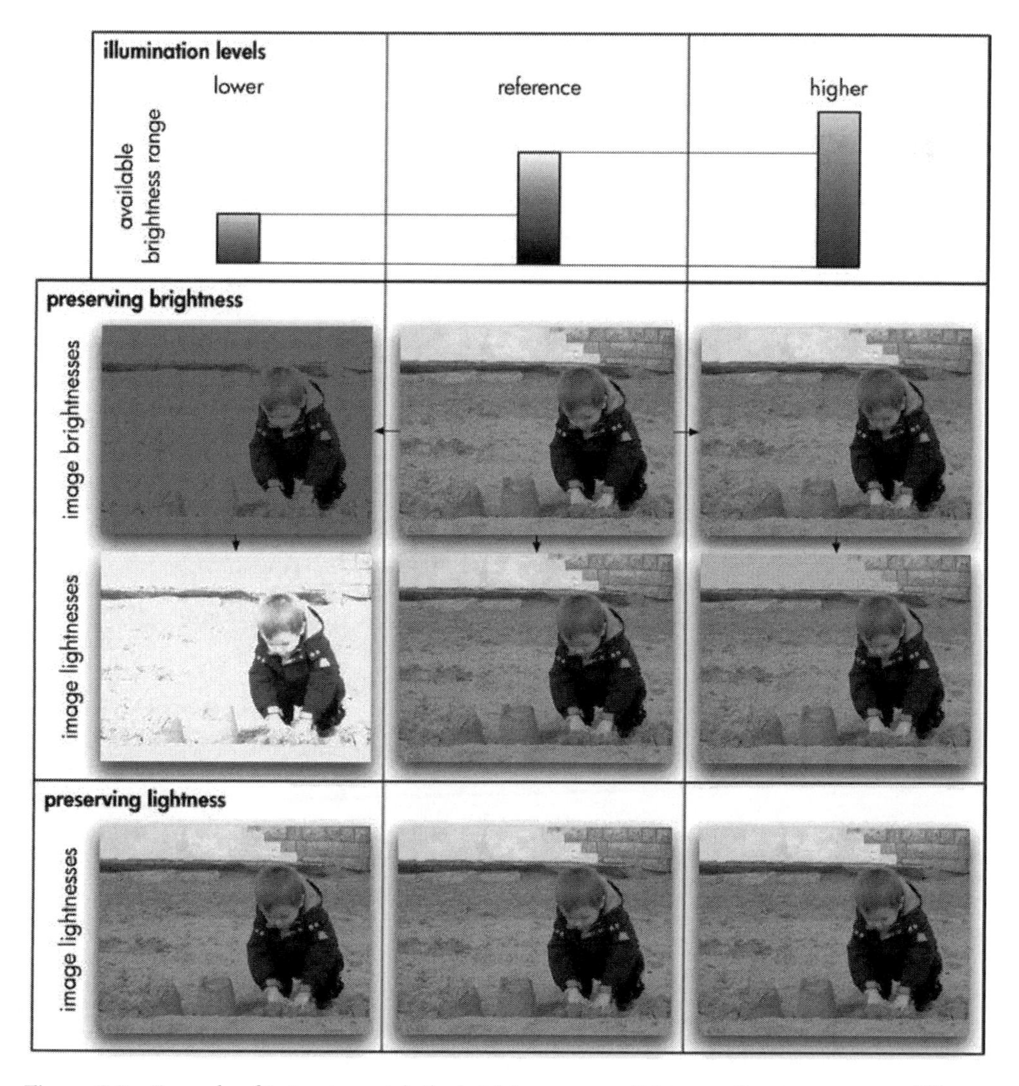

Figure 2.2 Example of trying to match the brightness versus lightness of an image across different levels of illumination.

trying to maintain brightness when moving to a lower level of illumination would result in loss of difference among the higher brightnesses from the original, which would be all above the highest brightness possible under the lower level of illumination (e.g. notice how the differences between white and light gray parts of the wall in the reference's background get lost – clipped). Looking at such a reproduction under the lower level of illumination would make the result appear overexposed and neither of the brightness-preserving reproductions would, therefore, be liked.

On the other hand, approaching the problem in terms of lightness allows for the full range of the original image to be represented in a way that adapts itself to the available range and it and chroma are therefore, the predominant color appearance attributes used in color reproduction.

Finally, it is also useful to consider another color appearance attribute, i.e. *saturation.* This attribute is the 'colorfulness of an area judged in proportion to its brightness' (Hunt, 1995a: 33) and its use has a number of distinct advantages (Hunt, 2001). The key to these advantages is that saturation relates well to the color appearances that a reflective object can have under different illumination conditions. For example, an orange's surface is lit differently, with some areas being brighter and more colorful while others are darker and less colorful. However, the ratio of brightness and colorfulness (i.e. saturation) remains unchanged.

2.3 THE COLOR ECOSYSTEM

Before going into more detail about the individual elements that affect color experiences (Figure 2.3), i.e. the experiences a viewer has when viewing an environment over time, and

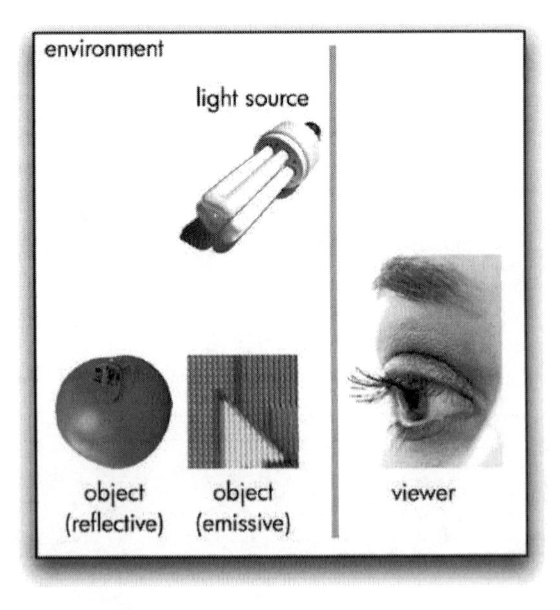

Figure 2.3 The color ecosystem.

that can be considered to form an ecosystem due to their interrelationships, we will first look at their key high-level properties.

Starting with the viewer, we find that color experiences are related to brain activity responding to signals received from the eyes. These signals arrive to the brain via the optic nerve, which collects the output of two kinds of light-sensitive cells lining the back of the eye in a way where the responses to any one part of the environment are also influenced by responses to other parts (e.g. changing the background against which an object is viewed will also change the response to the projection of the object itself and not only to that of the changed background). The first kind of light-sensitive cell (rods) provides vision when little light (i.e. electromagnetic radiation from a specific wavelength range) is present in an environment and only distinguishes between levels of that light's intensity. The second kind of light-sensitive cell (cones) operates when more light is present and there are three types, each of which is particularly sensitive to a different part of a wavelength range of electromagnetic radiation that enters the eye. Furthermore, the responses of the visual system (i.e. the eyes and the brain) change over time in a way where they can result in varying color experiences even when there is no change in the environment (e.g. when moving from a dark into a bright environment, color experience changes even after the move is complete).

The light a viewer's eyes respond to is a projection of light present in their environment and its properties are the result of the interaction of light-emitting and light-modifying objects. Examples of light-emitting objects include the sun, candles and light bulbs, which are also referred to as light sources and whose output is instrumental in bringing out the color properties of other objects, but also televisions, whose light output is viewed directly. Objects that modify light can do so by selectively absorbing or refracting parts of the electromagnetic radiation that is incident on them and reflecting or transmitting other parts. The majority of objects surrounding us are light-modifying, and their impact on our eyes and subsequently on our brain includes a combination of their properties and the properties of the light illuminating them.

The most important point to take away from the above is that if any of the elements that contribute to color experiences change then so can the resulting color experiences. If one viewer is substituted by another in the same environment, if any part of an environment changes or if an environment is viewed for varying amounts of time, then the resulting color experiences can vary too, even if they are for the same, unchanged object in the environment.

Next, let us look at each of the above elements of a color ecosystem in more detail.

2.4 THE HUMAN VISUAL SYSTEM

Visual experiences in humans can be related to the chemical and electrical activity of nerve cells in a part of the brain's crust (*cerebral cortex*) located at the back of the head. This part of the brain, called the *occipital lobe*, is where the visual cortex resides; this can be further subdivided into areas relating to different aspects of vision (Zeki, 1993).

The above brain activity is a response to signals received from another part of the brain, i.e. the *lateral geniculate nucleus* (LGN); this, in turn, is stimulated via the *optic nerve*, which collects the outputs of cells in the eye's *retina*. This channel from eyes to brain is

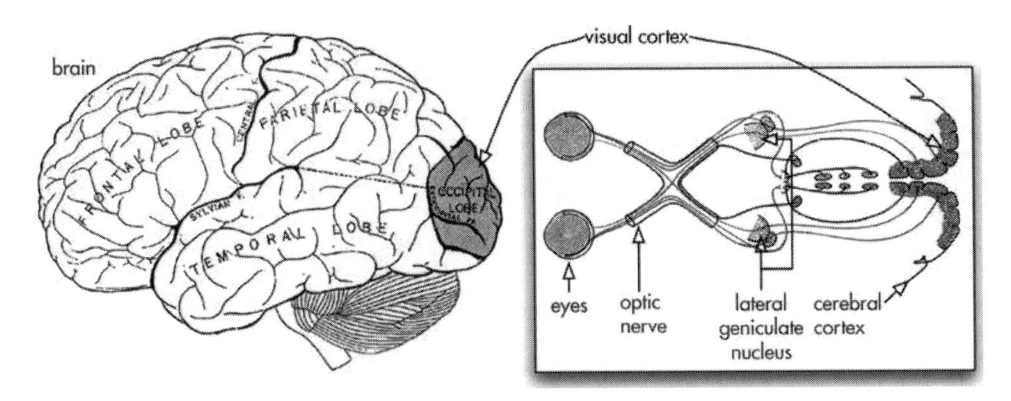

Figure 2.4 The visual pathway: from the eyes to the brain's visual cortex (adapted from Gray 1918).

referred to as the *visual pathway* (Figure 2.4), and we will next take a closer look at some of its details.

2.4.1 The eye and its photoreceptors

The light that enters the eye via the *pupil* (Figure 2.5) is focused onto the *retina* that lines its interior using the *lens* and on its way there passes through the *vitreous gel* filling the eye. The retina is made up of a series of interconnected cell layers, the penultimate one of which (as seen in the direction of light entering the eyes) consists of light-sensitive *photoreceptor* cells. These cells contain molecules of photopigment that change their

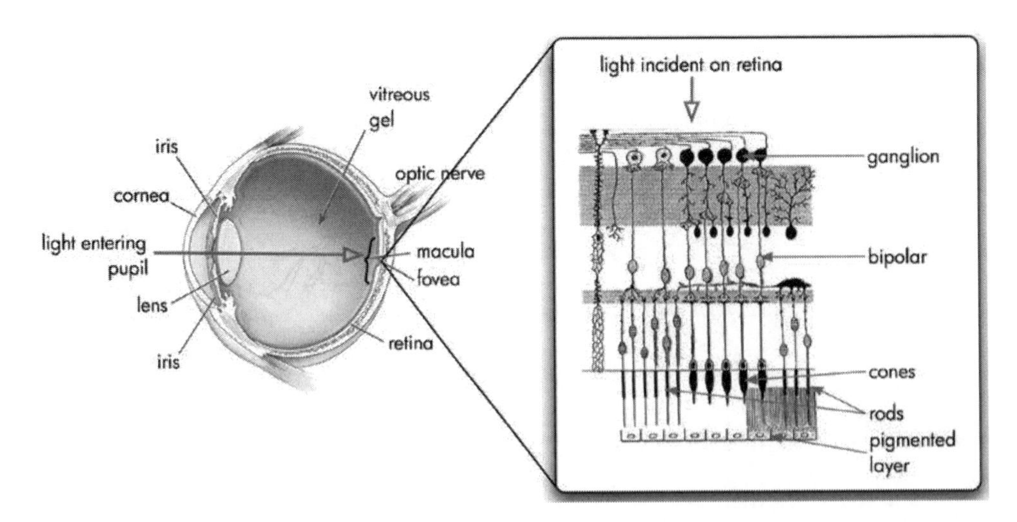

Figure 2.5 Left: cross-section of a human eye (courtesy of National Eye Institute, National Institutes of Health, USA). Right: cells in the human retina (adapted from Gray, 1918).

structure when light is incident on them, and a process involving their bleaching (i.e. oxidation) results in the activation of a neural signal that then gets modified along the visual pathway (Kolb *et al.*, 2006).

There are two kinds of photoreceptors: *rods*, which operate at low levels of light, and *cones*, which are active when more light is present and which result in signals leading to color vision. Vision at low light levels, when only rods are active, is called *scotopic*; at higher light levels, when cones are active, it is *photopic*; and at intermediate light levels, during which the transition between rods and cones takes place, vision is referred to as *mesopic*. The focus in the remainder of this chapter will be on photopic vision, as it is most relevant to scenarios where gamut mapping is used.

Although rods are present in abundance over a large area of the retina, there is a particularly high density of cones only in a very small retinal region, namely the fovea. As a consequence, the central part of the *field of view* (i.e. the part of our environment that we view at any one time) is also the one where we have highest *visual acuity* (i.e. ability to resolve fine detail). In terms of photoreceptor quantity, there is also a significant disparity between the two types: whereas there are around 100 million rods, there are only 6 million cones in a typical eye (Hunt, 1995a: 25).

The cones further come in three varieties, each sensitive to a different part of the spectrum of electromagnetic radiation. The L cones are most sensitive to long wavelengths and peak around 580 *nanometers* ($1\,\text{nm} = 10^{-9}\,\text{m}$), the M cones are most sensitive to medium wavelengths with a peak around 540 nm and the *S* cones are most sensitive to short wavelengths with a peak around 440 nm. How each cone variety responds to light incident on it can be seen, for example, by measuring how much light they absorb from different parts of the visible spectrum (Figure 2.6). As for their distribution, only about 1 in 100 are S cones, and the ratio of the number of M to L cones varies dramatically from person to person (i.e. between 1.1:1 and 16.5:1 (Hofer *et al.*, 2005), with a mean of around 1.86:1 (Carroll *et al.*, 2002)).

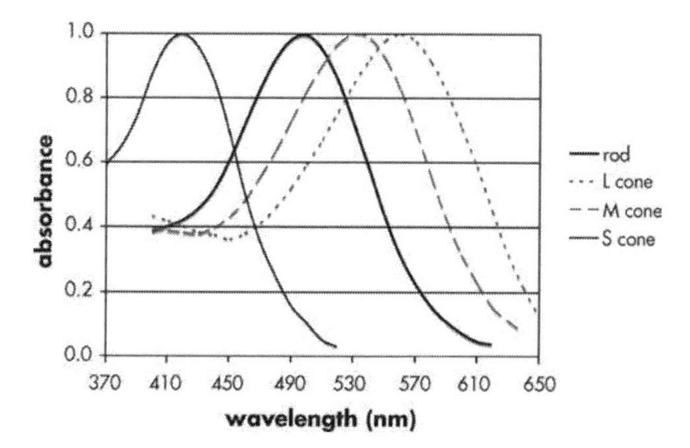

Figure 2.6 Normalized microspectrophotometrically measured absorbances of rods and the three cone types across the visible spectrum (Dartnall *et al.*, 1983). The larger these values are, the more light is absorbed in a particular wavelength band; the results for each photoreceptor are scaled to make their maximum equal one.

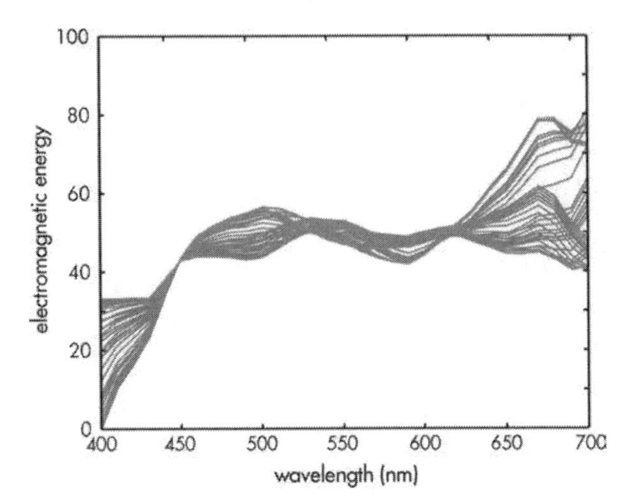

Figure 2.7 Spectra of electromagnetic radiation resulting in identical cone responses (Morovič P. M., personal communication, 31 January 2006).

An important consequence of this makeup of the retina is that, as far as color vision is concerned, the variation of electromagnetic properties of radiation across the visible spectrum results in only three varieties of response per unit area of the retina. This *trichromacy* of the human visual system means that even though the light stimulating the eye can vary freely across many bands into which the visible spectrum can be divided, the result is only three types of response – one each for the short, medium and long wavelengths. A consequence of trichromacy is also that the visual system will have identical responses to very different spectra, a phenomenon that is called *metamerism* (Figure 2.7).

2.4.2 From photoreceptors to the visual cortex

The signals arriving at the visual cortex are, however, very far from having a simple relationship with those emitted by individual retinal photoreceptors. Already the signal leaving the retina along the optic nerve is the result of outputs from several individual photoreceptors being combined by the *bipolar* cells they connect to, which in turn provide inputs to several *ganglion* cells that combine them (Figure 2.5). In addition to connections between cells of different layers in the retina, the cells in the individual layers also provide inputs to each other; hence, the signal that leaves the retina along the optic nerve is the result of a series of complex combinations of the outputs of photoreceptors via the layers of neural cells they connect to (Martin, 1998).

The result is one where the signal, in the form of electrical pulses, carried by the optic nerve does not have a point-to-point relationship with photoreceptor signals, but instead provides information about their local relationships. The signals that a ganglion cell generates have a *centre–surround* organization (Figure 2.8) and the signals they generate depend on the relationship between the signals from the central and surrounding areas

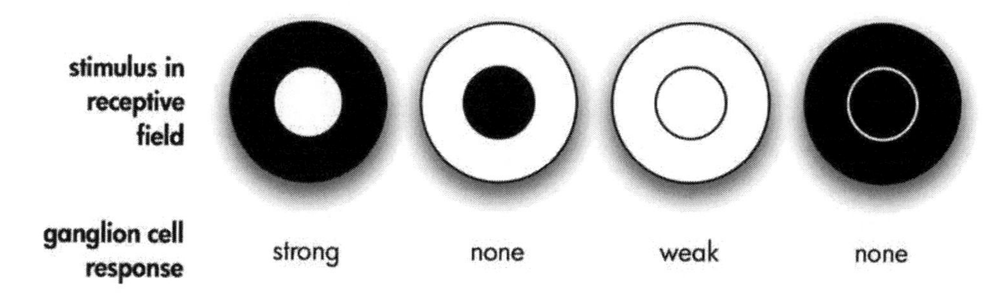

Figure 2.8 Centre–surround behavior of an on-center ganglion cell for different receptive fields (white represents a significant presence of light; black represents its absence in a stimulus).

within the part of the field of view to which photoreceptors that connect to it respond, i.e. the ganglion cell's *receptive field* (Wandell, 1995). As it is the difference between the signals from the centre and the surround that heavily influences the ganglion cell's output, edges generate stronger responses than uniform areas, and this leads to an improved sensitivity to them.

In the LGN, signals originating from different varieties of receptors are combined again and are thought to output three kinds of *opponent color* signals (de Valois *et al.*, 1966). The first is based on the difference of the L and M cones, the second on the difference of the S cones and the sum of L and M cones, and the third on the sum of the output from all three cones. These three signals also relate to a color space defined by three pairs of opponent colors: red–green, blue–yellow and black–white, which also underlies color appearance attributes. Note that both color and spatial opponency are combined, e.g. a ganglion cell can have a positive L response from the centre of its receptive field and a negative M response from its surround.

Finally, the signal originating from the retinal photoreceptors, which has already gone through a complex sequence of both color and spatially opponent combinations, arrives in the visual cortex and is further processed in its various subdivisions with even greater complexity (Livingstone and Hubel, 1988). Amongst other aspects, the signal is analyzed in terms of the orientation, shape, form, depth and motion of its content, and interaction with memory and other cerebral faculties takes place. For more detailed treatments of the human visual system, see Marr (1982), Hubel (1989), Wandell (1995), Zeki (1993), Hill (1997) and Fairchild (2005).

2.5 PHYSICAL COLOR-RELATED PROPERTIES

Having taken a broad look at the response of the human visual system to electromagnetic radiation entering the eye, in this section we will consider what determines the properties of such radiation, how it can be measured and how its impact on the visual system can be quantified, at least in some way.

Wikipedia (2006a) defines electromagnetic radiation as 'a propagating wave in space with electric and magnetic components [that] oscillate at right angles to each other and to

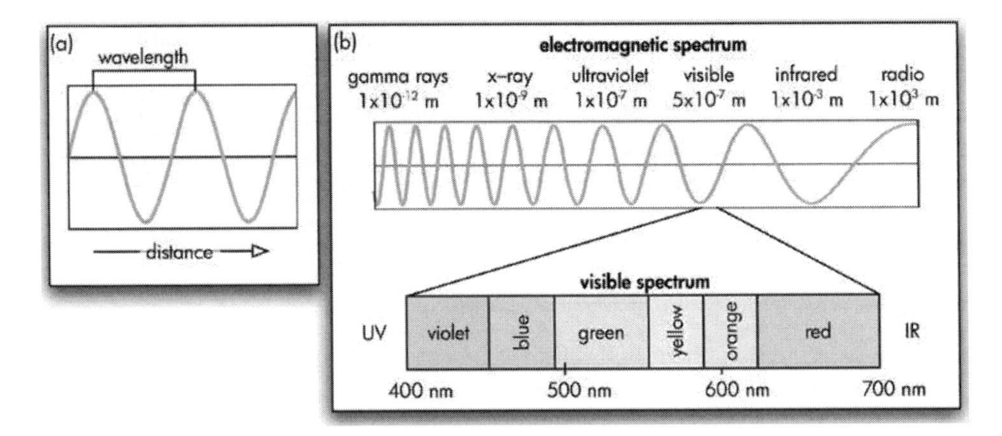

Figure 2.9 (a) A wave; (b) the electromagnetic spectrum.

the direction of propagation'. Depending on the wavelength of such radiation (i.e. the distance between the wave's peaks – Figure 2.9a), it then has different effects and is classified accordingly (Figure 2.9b). Near one extreme, radio signals are transmitted by radiation with very long wavelengths that are of the order of magnitude of kilometers (10^3 m) and near the other extreme of the *electromagnetic spectrum* are X-rays, whose wavelengths are around 1 nm. Within this range we can also locate radiation with wavelengths between about 400 and 700 nm that results in responses from the human visual system's photoreceptors, and such radiation is called *light.* Figure 2.9b also shows approximate descriptions of the color experiences that would result from viewing stimuli with energy only in individual parts of the spectrum (Hunt 1995a: 22). In other words, these are the experiences that a viewer would have if light from their entire field of view had energy only in a specific wavelength interval (e.g. energy only in the 650–700 nm interval would result in the experience of a red color).

The light that the visual system responds to has either been emitted or altered (e.g. refracted, absorbed, scattered, reflected, transmitted or fluoresced – absorbed at one wavelength and emitted at another, longer one) by the objects that are in its field of view. Here, the properties of light emitted by an object can be characterized by how much energy it sends out in different wavelength intervals across the visible spectrum (Figure 2.10a); this is referred to as the light's *spectral power distribution* (SPD).

In terms of light-emitting objects, two functional categories can be considered in imaging. The first are *displays* and *projectors,* which can emit light of different spectral properties depending on digital inputs to them and thereby give rise to color images (Section 2.8.3). The second are *light sources* (e.g. the sun, lamps using fluorescent tubes or incandescent light bulbs, etc.), whose role is to realize the color-altering potential of objects in our environment, without which they would all appear black. As light sources have a strong effect on the color of reflective and transmissive objects, the CIE has standardized the SPDs of some of them (CIE, 2004a). These SPDs, which are defined numerically and not as physical realizations, are referred to as *illuminants.* In the context of imaging, the most relevant illuminants are *CIE Standard Illuminant A* (tungsten-filament lamp), two variants

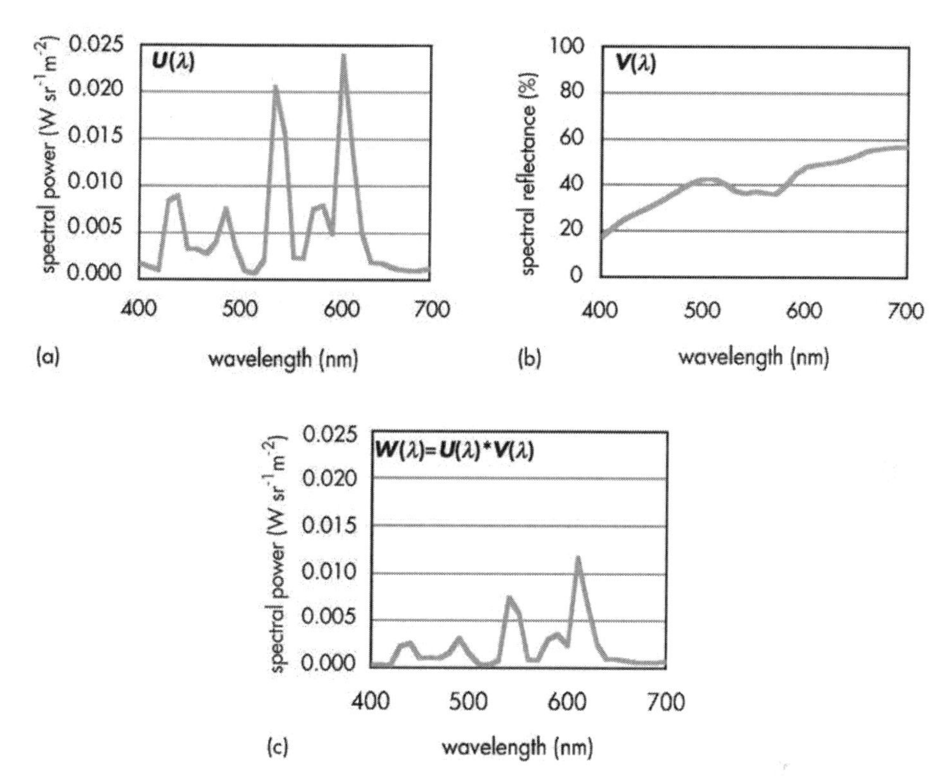

Figure 2.10 (a) SPD of light source U, (b) spectral reflectance function of surface V and (c) SPD of light W resulting from light U reflected by surface V.

of *CIE Standard Illuminant D* (D65 and D50; the former represents average daylight and the latter is a compromise between A and D65) and F11 (representing a type of fluorescent lamp also known as TL84). For further information on light sources and illuminants, see (Hunt, 1995a: chapter 4) and Luo (1998).

The light-altering properties of objects can be characterized by the percentages of light that they reflect or transmit at different wavelengths and the result is called a *spectral reflectance* or *transmittance function* (Figure 2.10b). In addition to spectral reflectance or transmittance, an object's shape, refractive index and surface texture, microstructure and topography also affect the properties of light that arrive from it at a viewer's retina. In surface property terms, an object can be:

- *matte*, in which case light that strikes it from a single direction is reflected with the same intensity in all directions (e.g. chalk has a matte surface);
- *glossy*, when light arriving from one direction is also reflected only (or predominantly) in one direction whereby the angles of the incoming and outgoing light, with respect to the normal of the surface (i.e. the line orthogonal to the surface), are equal (e.g. polished metal has a glossy surface).

For further details on light–object interactions, see Sinclair (1997).

For a light source illuminating a flat, diffuse surface, the light reflected by the surface can be predicted by multiplying the light source's SPD with the surface's spectral reflectance function at each wavelength λ (Figure 2.10), which in turn also allows for the prediction of how a given surface would interact with other illuminating light sources.

As spectral properties characterize the light that interacts with the human visual system to bring about experiences containing color, a first step towards predicting color experiences is to measure and quantify the spectral properties of light entering the eye. The measurement of the amount of power across the visible spectrum is called *radiometry* and its result is a set of *radiance* values for wavelength intervals (e.g. of 1, 5, 10 or 20 nm width) in the visible range. The unit of radiance is watts per steradian per square meter (i.e. $W\ sr^{-1}\ m^{-2}$), where *watt* is the unit of power and *steradian* is solid angle, as a result of which radiance expresses power per unit projected area. To measure radiance (Miller, 2000), what is needed are two key components:

1. a *photodetector*, which measures the energy of electromagnetic radiation incident on it;
2. a *monochromator*, which is a device that isolates a specific wavelength interval from the incoming light that can then be measured.

In addition to being necessary for the measurement of light source SPDs, radiometry is also valuable for measuring the light reflected by surfaces in a specific environment that is lit either by a nonstandard light source or by a mixture of light sources. The instrument used for making radiometric measurements is called a *telespectroradiometer* (TSR).

If only spectral reflectance or transmittance needs to be measured, then a *spectrophotometer* can be used instead. Here, the instrument is placed in contact with the sample surface that is to be measured and spectral characteristics are obtained by illuminating it using a light source internal to the instrument. What is then measured across the visible spectrum is the ratio of power detected from the sample surface and the power detected from measuring a (near) perfect diffuser supplied with the instrument. A perfect diffuser, then, is a matte object that reflects all light in each interval across the visible spectrum (i.e. its spectral reflectance is 100% regardless of wavelength). The result is spectral data about an object that can be used for predicting how it interacts with light from various light sources.

Even though spectral reflectance data can be obtained for wavelength intervals of varying widths and, therefore, with varying degrees of spectral resolution, it is worth noting that the spectral reflectances of many types of surface can be accurately characterized as weighted sums of a small number of component reflectances (i.e. basis vectors, since reflectances can be seen as vectors in n-dimensional space). Even though reflectance may be measured using 31 or 81 samples across the visible spectrum, this data can often be expressed with a high degree of accuracy by combining three to eight basis functions (Krinov, 1947; Vrhel *et al.*, 1994). The significance of this is that, even at a physical level, spectral properties are of much lower dimensionality than that of their measurements and are often close to the three dimensions of the human visual system's photoreceptors.

Both in radiometry and photometry it is important to consider and specify the geometric arrangement of the measuring apparatus with respect to the light illuminating a surface and the surface itself and there are a number of alternative geometries that the

CIE (2004a) have defined (e.g. light incident on surface at 45° and measurement taken at 0° – both with respect to the measured surface).

Finally, it is important to bear in mind that none of the measurement methods described here, or any of the other methods that are typically referred to as 'color measurement,' result in a measurement of color. Instead, they only measure physical properties that can then be the basis, alongside data about other aspects of a color ecosystem, for predicting what color experiences the corresponding objects give rise to. To truly measure color, one would have to measure such brain activity that can be related to the descriptions a viewer would make of their color experiences.

For more detail on physical color-related properties, see Wyszecki and Stiles (2000: chapter 1) and Nassau (2001).

2.6 COLORIMETRY

As the ability to perform gamut mapping in a controlled way relies on a quantitative description of color, it is necessary to be able to predict color experiences from physical measurements of the environment that is being viewed. Given the elements of the human visual system (Section 2.4) and the nature of physical color-related properties (Section 2.5), a first step in quantifying color experiences is to predict the response of the cones to electromagnetic radiation entering the eye. To do this, the SPD of light entering the eye needs to be integrated with the spectral responsivities of each of the three cone types; the result is a triplet of values expressing the degree of response by each cone type to light with a given SPD.

2.6.1 Color matching functions

As direct measurements of cone spectral responsivities have only been made very recently, an alternative approach to quantifying color in a way that relates to the cone response stage is to use *trichromatic matching* (Hunt, 1995a: chapter 2). The idea here is that the effect of a stimulus (i.e. something that elicits a response) on the cones can be expressed by the quantities of three lights that, when added, match that stimulus. As the visual system only has three types of cone, they can equally be stimulated using the properties of light coming from some object as by an appropriate mixture of three lights, each of which ideally stimulates only one cone type. A choice of three lights that is close to having this property is one where each has energy only in a narrow wavelength interval around one of 700 nm (red, R), 546.1 nm (green, G) and 435.8 nm (blue, B). Given these lights, Figure 2.11a shows the amount of each of them that is needed to match a stimulus that has energy only in a specific wavelength interval of the visible spectrum. For example, light with energy only around 600 nm can be matched by adding 0.3 units of R and 0.1 units of G light. These amounts of RGB lights, needed to match single-wavelength (i.e. *monochromatic*) stimuli across the spectrum, are called *color matching functions* (CMFs). The results of several observers setting up such matches (Wright, 1928; Guild, 1931) were then combined into a single set of CMFs, which the CIE declared to be the *Standard Colorimetric Observer* in 1931 (CIE, 2004a) and they are used as representatives of typical

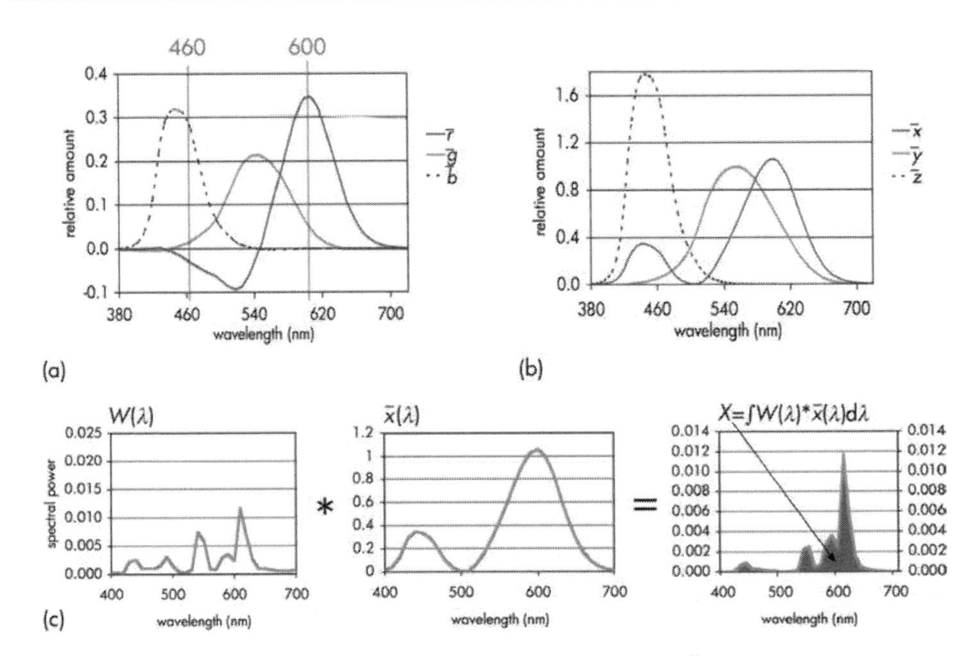

Figure 2.11 CIE 1931 Standard Colorimetric Observer: (a) $\bar{r}$, $\bar{g}$, $\bar{b}$ CMFs and (b) their linear transformation – the x, $\bar{y}$, $\bar{z}$ CMFs.

color matching behavior. Note that these RGB color matching functions (CMFs) also have negative parts, which is equivalent to adding light to the stimulus, rather than to the trichromatic mixture (e.g. a stimulus at 460 nm that has 0.03 R added to it, which is shown by a negative value in the $\bar{r}$ function, can then be matched by 0.01 G plus 0.3 of B). For a more detailed look at why some stimuli cannot be matched by a mixture of three lights, see Section 2.8, where the principles of color reproduction are introduced.

By transforming the CIE $\bar{r}$, $\bar{g}$, $\bar{b}$ CMFs into a linear combination that has no negative parts, the CIE defined another set of CMFs, i.e. $\bar{x}$, $\bar{y}$, $\bar{z}$ (Figure 2.11b), which are a ubiquitous means of colorimetric quantification. A further key feature of this transformation is that the $\bar{y}(\lambda)$ CMF matches the $V(\lambda)$ *luminous efficiency* function (Hunt, 1995a: 40–45) and thereby predicts the intensity of light seen as coming from a stimulus (albeit not in a perceptually uniform way). Integrating the SPD of a stimulus with the $V(\lambda)$ (or $\bar{y}(\lambda)$) function and choosing an appropriate scaling factor gives its luminance, expressed in units of *candelas per square meter* (cd m^{-2}). Tristimulus values scaled in this way are denoted as $X_L Y_L Z_L$, whereas those scaled so that the Y value for the perfect diffuser under given viewing conditions is 100 are referred to simply as XYZ (Hunt, 1995a: 54).

A frequently used transformation of XYZ tristimulus values is also the computation of *xy chromaticity* coordinates from them, which allow for a representation of stimulus colors without reference to luminance. Here, $x = X/(X + Y + Z)$ and $y = Y/(X + Y + Z)$; therefore, the full XYZ values can also be obtained when xy and Y are known. The use of chromaticity values is most appropriate for the specification of light source colors and should certainly be avoided when, for example, describing color gamuts.

All else being equal (i.e. observer, viewing conditions, spatial properties, etc.) CMFs allow one to determine whether two stimuli match or not. For a pair of stimuli with known SPDs, a pair of RGB (or *XYZ*) value triplets (*tristimulus values*) can be computed by integrating the product of each SPD in turn with each of the CMFs. Here, multiplying the SPD of a stimulus with a CMF results in a new function whose integral is then the area under it (Figure 2.11c), and this area is the tristimulus value. If the tristimulus values of the stimulus pair match, then so do the corresponding color experiences, if – and it is important to stress this constraint – all else is equal.

2.6.2 Color difference and uniform color spaces

While being able to tell whether two stimuli will match is a useful first step, it is even more important to predict how different stimuli are when they do not match. This is precisely what *color difference equations*, typically denoted by ΔE, set out to do with two key aims:

1. that a color difference equal to one (i.e. $\Delta E = 1$) should represent a *just noticeable difference* (JND), i.e. the smallest stimulus difference that results in a different response by a viewer, for all color pairs (Figure 2.12);
2. that color difference predictions should relate uniformly to judgments made by observers (i.e. a color pair with $\Delta E = 10$ should be seen as 10 times as different as another pair with $\Delta E = 1$).

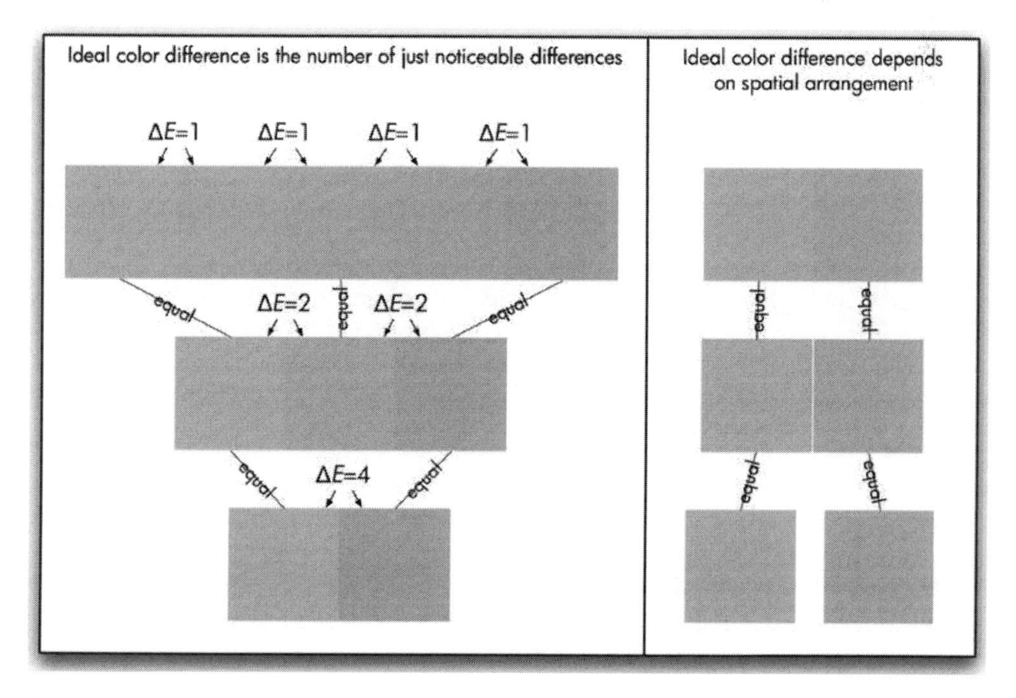

Figure 2.12 Expressing color difference.

As the use of CMFs assigns a set of three values to each stimulus, which in turn can be taken as coordinates in a three-dimensional space, the most immediate solution that presents itself for color difference prediction is to compute the Euclidean distance between the tristimulus coordinates of a pair of stimuli, i.e. $[(X_1 - X_2)^2 + (Y_1 - Y_2)^2 + (Z_1 - Z_2)^2]^{0.5}$. If the CIE *XYZ* (or CIE RGB) color space were uniform, then a given distance would always correspond to the same perceived difference regardless of where in color space the two stimuli were. In other words, two green stimuli and two red stimuli with the same distance between them would look equally different. However, it turns out that CIE *XYZ* is highly nonuniform and, in the case of our example, the green pair would look much more similar than the red one, even though their distances in *XYZ* would be the same (MacAdam, 1942).

Such nonuniformity means that distances between the *XYZ*s of colors do not express their perceived differences. Therefore, *XYZ* cannot be used for determining whether two stimuli can be distinguished or for setting thresholds below which differences are acceptable. In addition to nonuniformity, the CIE *XYZ* space also has the limitation that its dimensions do not predict how color experiences are described. As a consequence of nonuniformity and the absence of color attribute prediction, two lines of research (that are still active today) were triggered, one with the aims of defining a perceptually uniform space with appearance attributes as its dimensions and the other with the aim of defining a metric in a given space that makes uniform difference predictions.

The first attempts at uniform color spaces were the definitions of *CIELAB* and *CIELUV* by the CIE in 1976. These 'uniform' color spaces are nonlinear transformations of CIEXYZ (CIE, 2004a). The CIELAB space has three orthogonal dimensions: L^*, predicting lightness; a^*, predicting redness–greenness; and b^*, predicting yellowness–blueness. Furthermore a prediction of hue h^*_{ab} is also made using the counterclockwise angle between a vector in a^*b^* and the a^* axis. Chroma is predicted as $C^* = (a^{*2} + b^{*2})^{0.5}$, i.e. the distance from the L^* axis. CIELAB also comes with a color difference equation, which is that of Euclidean distance in the space: $\Delta E^*_{ab} = [(L^*_1 - L^*_2)^2 + (a^*_1 - a^*_2)^2 + (b^*_1 - b^*_2)^2]^{0.5}$. This color difference equation (sometimes referred to also as ΔE_{76}, after the year in which it was introduced) and the CIELAB space are still in extensive use today, in spite of their limitations.

A key point to bear in mind regarding these color spaces is that (just like color spaces in general) they only *predict* color appearance attributes instead of defining them. Given a stimulus, its L^* value is not its lightness, but only a prediction thereof. Lightness is how a viewer judges that color attribute in the given viewing environment. The importance of this distinction lies in the fact that the quality of a color space's attribute predictors is derived from the extent to which they match judgments made by viewers and from the fact that color spaces predict appearance attributes imperfectly. For example, a notorious weakness of CIELAB is the nonuniformity of its h^*_{ab} predictions relative to perceived hue. Namely, given a high chroma pure blue, maintaining h^*_{ab} and reducing C^*(the chroma predictor) gives a less chromatic purple color – instead of a less chromatic pure blue (Hung and Berns, 1995; Ebner and Fairchild, 1998a). If color space's predictions and the attributes themselves that they predict are confused, and this often happens, then it becomes difficult to talk about the imperfections of a color space and to work towards overcoming them. For more detail on this topic and its impact on gamut mapping, see Chapter 6.

In terms of color difference prediction, a significant effort has gone into the development of equations that make more accurate predictions of color difference judgments, with the latest and most accurate one being CIE ΔE2000 (CIE, 2001). Two properties of color difference and its prediction are also worth noting: First, that the perception of color difference between a pair of stimuli depends strongly on their spatial arrangement. For example, when the stimuli are immediately adjacent their difference looks greater than if there is even a hairline gap between them, and it looks smaller still when they are separated by a significant distance (Guan and Luo, 1999; Figure 2.12 right). Second, that color difference equations are based on data about how observers judge color differences and a large proportion of such data is collected under light sources simulating the CIE D65 illuminant. As a consequence, the majority of ΔE equations are tuned to D65 and their predictions under other light sources may be less accurate (Cui *et al.*, 2005). As color pairs for which judgments are available span a specific color range (gamut), the equations too are defined for that range. Using ΔE (or any other model for that matter) outside the conditions for which it was derived is unreliable and can result in incorrect predictions. For more detail on color difference equations, see Luo (1998; Luo *et al.*, 2001).

While predicting the perceived magnitude of color difference is a key step, it is important to realize that preferences also apply when a color needs to be substituted by another and zero ΔE is not possible. Given an original color and a number of alternatives that are each removed by the same number of JNDs, there will be some that will look more similar than others (Morovič *et al.*, 2007). For example, there is likely to be a preference for alternatives that are closer in lightness than in chroma (Ebner and Fairchild, 1997); and, depending on the original's hue, some hue change directions will be preferred over others (Taplin and Johnson, 2004).

For more on colorimetry, see Hunt (1995b) and Ohta and Robertson (2005).

2.7 COLOR ECOSYSTEM INTERACTIONS

With a glimpse of color vision, color-related physical properties and basic colorimetry, we can now turn to where the true complexity of dealing with color lies – namely the interactions among a viewer, an object that is being viewed, its environment and the temporal sequence of its viewing. The understanding of these interactions is essential for making reliable judgments about color and for obtaining successful gamut-mapped color reproductions.

In the following paragraphs we will look at the consequences of these key properties of the human visual system:

1. that it *adapts* its response to what is in its field of view;
2. that such adaptation is not instantaneous but happens *over time*;
3. that it responds to *changes* rather than to absolute physical magnitudes of stimulus properties (and, therefore, varies spatially); and
4. the nature of how we *remember* what we have seen.

Considering our experiences of the colors around us, we can notice that they have a certain degree of variation depending on environmental conditions. Colors at sunset are different from those at midday; there are differences between overcast and clear days; colors in dimly lit rooms look different from those when illumination is brighter.

2.7.1 Adaptation to intensity

Measuring physical properties related to color (e.g. spectral power or luminance) under such a range of conditions shows that differences are of many orders of magnitude. For example, a white surface can have the following luminances under different conditions: 60 cd m^{-2} in a typically lit living room, 3000 cd m^{-2} in an operating theatre and 30 000 cd m^{-2} in bright sunlight (Hunt, 1995b: 787). Even just across this range, the physical properties of white-looking stimuli alone have a 500:1 range. Thinking about the corresponding differences among our experiences, we can see, though, that a white surface in bright sunlight does not look 500 times brighter than when seen in a typically lit living room. The reason for this is that our visual system has a *dynamic response* and adapts to the level of illumination by adjusting its sensitivity (involving changes in pupil size, change of photoreceptor pigment concentration and higher level processes in the visual pathway, such as opponency).

An example of trying to predict adaptation can be found already in the CIELAB color space, which takes the ratio of the tristimulus values of a stimulus and the tristimulus values of a similarly lit white surface (the *reference white*) as its basis rather than only the *XYZ*s of the stimulus alone. As a consequence, for example, the lightness of a surface that reflects 50% of light across the spectrum will be predicted as being the same regardless of the level of illumination, since the ratio of its *XYZ*s to those of a white surface will not change.

An important property of adaptation to illumination level is its temporal aspect, which, furthermore, is asymmetrical. Whereas adaptation to an illumination level increase is relatively fast (i.e. it stabilizes after about 5 min), the inverse is significantly slower (i.e. around 30 min) (Fairchild, 2005: 19–23). The significance of this is that critical comparisons between, for example, an original and a gamut-mapped reproduction should be made only once an observer is adapted to the environment in which the viewing takes place. Not waiting for complete adaptation can result in a judgment that is done under conditions other than those for which the gamut mapping was performed and in different observers performing the viewing under different conditions.

2.7.2 Adaptation to chromaticity

In addition to adaptation to illumination level, the visual system also adapts to the color – more specifically to the chromaticity – of illumination (called *chromatic adaptation*) and this process, too, has a temporal aspect. As a result of such adaptation (which is a consequence of the largely independent adaptation of each of the three channels of the visual system), the appearance of objects is more or less preserved under light sources of different colors. Even though tungsten-filament light bulbs emit very little light in the short wavelength part of the spectrum and significantly more light in the long wavelength part, objects under such illumination do not exhibit the strong red cast that would be seen if there was no chromatic adaptation. The degree of adaptation to a scene also depends on the color of the illumination, and for light sources that differ significantly from daylight (e.g. a tungsten-filament lamp), adaptation can be incomplete and surfaces will retain some of the cast of the light source's color.

The reason for objects not preserving their appearance completely with illumination color change is due to the fact that adaptation occurs in terms of adjusting the responses of three photoreceptors, whereas stimulus change occurs with much higher dimensionality (i.e. the combined dimensionality of the surface and light-source spectral properties). The phenomenon whereby the color of surfaces can change with light-source changes even after adaptation has taken place is referred to as *color inconstancy*, which can be quantified using a *color inconstancy index* (CII; Luo *et al.*, 2003). The higher the CII value of an object is, the more different are the color experiences resulting from viewing it under different light sources. While surfaces resulting in some colors (e.g. grays) can be perfectly color constant, other (more colorful) colors are by nature related to color-inconstant surface reflectances (Morovič and Morovič, 2005).

An important aspect of chromatic adaptation is also that it involves *cognitive mechanisms* in addition to physiological ones. This results in the ability to discount the color of a light source even beyond the adaptation provided by physiological mechanisms (Fairchild, 2005: chapter 8). The degree of adaptation is also influenced by what is present in a scene. For example, when familiar objects (e.g. one's hands) are seen under some illumination, the degree of adaptation increases and this is a purely cognitive feature (Fairchild, 1993). In terms of temporal aspects, chromatic adaptation stabilizes after about 1 min (when not accompanied by a change in luminance level) (Fairchild, 2005: chapter 8).

2.7.3 Elastic versus plastic adaptation

Beyond such adjustments of the visual system relative to a base state, that base state itself is subject to change depending on broad properties of a viewer's environment. Short-term adaptation can, therefore, be seen as *elastic* (i.e. compensating for changes in an environment to provide stable visual responses – a base state). There is then also a *plastic* process in place that alters the base state of visual responses. For example, spending a significant amount of time (i.e. many hours) in an environment lit with a strongly green light will have long-term effects on how 'normal' environments look (e.g. as can be established by asking a viewer to chose what stimulus looks perfectly yellow). Instead of the base state being a consequence of the physiological makeup of a viewer (i.e. the proportions of L:M:S cones, which have recently been found to vary significantly between viewers), it is strongly conditioned by a viewer's environment (Neitz *et al.*, 2002).

2.7.4 Predicting adaptation and degrees of color constancy

A simple prediction of chromatic adaptation is made already by the CIELAB color space, as it scales the *XYZ*s of a stimulus independently for each of *X*, *Y* and *Z*, which adjusts for differences between the relative outputs of light sources in the three parts of the visible spectrum. However, as such predictions are agnostic of more complex mechanisms, they are of limited accuracy and more complex models have been developed for making them.

At present, the most advanced and accurate color appearance model is *CIECAM02* (CIE, 2004b), which makes reliable predictions of a number of color appearance phenomena, like chromatic adaptation, and also of appearance attributes (i.e. brightness, colorfulness, hue, etc.).

The effect of a dynamic response to illumination levels and chromatic adaptation is that our experiences vary much less than the physics of stimuli we see and this phenomenon is referred to as *color constancy.* Besides these color-related mechanisms, the visual system also adapts to the spatial properties of what it views. For example, viewing an image after a more blurred image makes it appear sharper than viewing it after a sharper image and the same also holds for noise (Fairchild and Johnson, 2005; Wandell, 1995: chapter 7). This fact of seeing a stimulus in a way that depends on what other stimuli were seen before it also needs to be borne in mind when viewing originals and their reproductions.

2.7.5 The spatial dimension of adaptation

One aspect of these types of adaptation is that they result in an enhanced ability to discern differences – both in terms of color and in the spatial structure of a stimulus. This focus on differences, which is already indicated by the nature of ganglion cell receptive fields, can further be seen in the phenomena of *simultaneous contrast* and *crispening* (Fairchild, 2005: 113–115). What *simultaneous contrast* refers to is the color of a stimulus depending on the color surrounding it. Figure 2.13a shows a pair of gray patches connected with a line and surrounded in one case by white and in the other case by black. However, even though the patches and line match in their physical properties, their appearance is different: the patch on the black background looks lighter than the other patch and this is a consequence of the visual system's focus on and amplification of differences. When the size of the central stimulus decreases significantly, simultaneous contrast no longer takes place and is supplanted by its opposite – i.e. *spreading,* whereby the central stimulus becomes more similar to its surround.

The *crispening* effect refers to the amplification of differences when seen against a similar background. Figure 2.13b shows an identical pair of different gray patches against a background that is similar to them (in the middle) and against two backgrounds that are significantly different from the patch pair. What we can see here is that the middle pair looks more different than the others even though they are all physically the same.

Figure 2.13 (a) Simultaneous contrast: identical gray patches shown against black and white backgrounds. (b) Crispening: identical differences against different backgrounds.

In addition to having an effect on the color of uniform stimuli and on the perceived differences between such stimuli, the background also affects the appearance of complex images (e.g. photographs). Given an image, viewing it against a white background will make it appear less chromatic, but having more contrast, than viewing it against a black background (Bartleson and Breneman, 1967; MacDonald *et al.*, 1990).

Differences between a stimulus and its surround, like the ones discussed above and like those between one part of a visual stimulus and an adjacent part (e.g. different parts of an image), are by definition spatial, and it is valuable to be able to express the sensitivity of the visual system to them. This sensitivity, the *contrast sensitivity*, is the inverse of the amount of difference (*contrast threshold*) between two parts of a stimulus necessary for eliciting a response in a viewer (Wandell, 1995: 135–137). In other words, if two adjacent parts of a stimulus have a difference below the contrast threshold then they will be seen as the same; and if the difference is above it, then it will be perceived.

Instead of being fixed, it turns out that the visual system's contrast sensitivity depends on the spatial frequency of the stimulus differences as well as on the temporal frequency of stimulus change. Here, spatial frequency relates to the distance (or angular subtens) in a stimulus across which difference occurs, e.g. if there is a given magnitude of difference between two parts of a stimulus that are 1° apart then it has a higher spatial frequency than if that difference occurred across a 2° separation (Figure 2.14a).

What can be seen from Figure 2.14c is that the visual system can best distinguish spatial variation that is in a specific interval (around between 1 and 20 cycles per degree) and that its ability to perceive differences both in lower and higher spatial frequencies is reduced. The significance of this for gamut mapping is that color reproduction can exploit these limitations and focus on that original image content, which can be distinguished by the visual system.

In addition to local differences being enhanced and perceived in a certain way, the absolute perceived brightness of any part of a stimulus also depends on its neighborhood. A striking example of this is the experiment reported by McCann and Rizzi (2007) where observers were asked to look at different parts of a stimulus and make a judgment of its brightness on a scale of 0 to 100 (where 100 was assigned to the brightest part of the stimulus and 0 to the darkest). What made this experiment special was its dynamic range of 18 000:1 (i.e. the ratio of the largest and smallest measured luminances (cd m^{-2}) in the stimulus) and its containing four sets of gray scales where the lowest luminance in each set

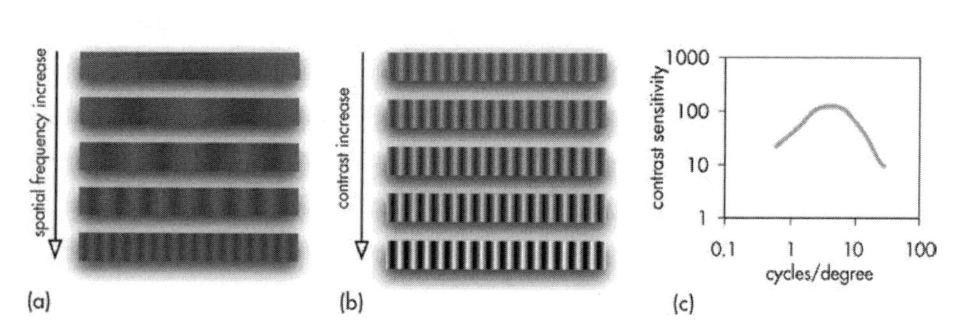

Figure 2.14 Stimuli with increasing (a) spatial frequency and (b) contrast; (c) a contrast sensitivity function (after De Valois and De Valois (2002)).

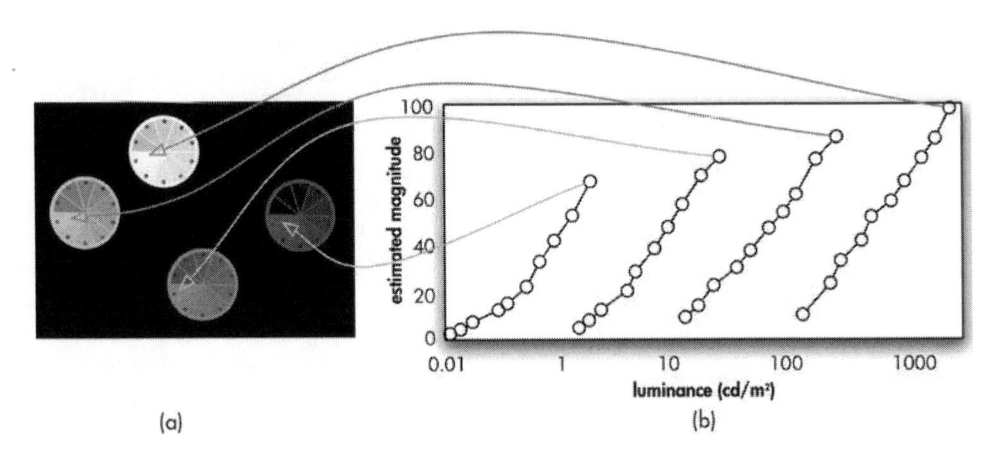

Figure 2.15 (a) A high dynamic range stimulus and (b) its brightness judgments. Note that the stimulus you are looking at in this book is likely to have a dynamic range of as little as 20:1, which is illumination-level dependent. Reproduced by permission of John McCann.

matched the highest one in the next (Figure 2.15a). If luminance related in a unique way to brightness in the entire stimulus, then a monotonic (i.e. always growing) relationship would be expected between brightnesses reported for patches with increasing luminance. Instead, what observers reported with great consistency was that it was the most luminous parts of each of the four subsets of the stimulus that had high brightnesses rather than the most luminous parts of the stimulus as a whole (Figure 2.15b). Another way of looking at the magnitude estimates is to say that, within a single scene, stimuli with luminances of <1, around 10, around 100 as well as several hundred candelas per square meter can result in the same perceived brightness of 50 units. These results are also consistent with Land's (1964) *Retinex* theory that places particular emphasis at local luminance ratios.

2.7.6 Mixed illumination, viewing modes and color memory

The interactions dealt with so far have all had either a single viewing environment with a single light source or a sequence of single light-source environments in mind. However, in practice this is very often not the case. Instead, a series of light sources tend to interact – e.g. a room can be lit with a mixture of artificial light and sunlight or a reflective, illuminated object can be viewed alongside a self-luminous one (e.g. a printed and a computer monitor displayed image). Even outdoors our environment is effectively lit not only by the sun but also with light scattered by the sky, which illuminates those parts of a scene that are in a shadow (Hubel, 1999). The relationship between stimuli and resulting color experiences under such mixed illumination conditions is in general beyond the predictive capabilities of contemporary models, although there are solutions for specific cases, like the viewing of displays and print with different white points (Katoh *et al.*, 1998).

In the context of gamut mapping it is important to consider not only the simultaneous viewing of stimuli in a single environment, but also their viewing in different environments and at different times. This aspect of how stimuli are seen will be referred to as their *viewing mode*. For example, a photographer may take photographs of landscapes using a digital camera; when the captured images are then viewed and adjusted in the studio, the original color experiences are no longer available and comparisons are made with reference to memory and under very different viewing conditions.

What is of value to note here are the limitations of color memory and the cognitive phenomenon of memory colors. Testing the accuracy of *color memory* suggests that a time difference of even just 15 s between viewing a color and then trying to match it from memory introduces a color difference of around five ΔE^*_{ab} units, and this difference slowly increases over time (de Fez *et al.*, 1998). For familiar objects (e.g. sky, grass, skin) a further phenomenon takes place, whereby our memories of their colors are distorted towards the prototypical colors of the corresponding objects. As a result, we remember grass to be greener, the sky to be bluer and skin tones to look more healthy and pleasing than what they were when we saw them. Objects to whose memories this applies are said to have *memory colors* (Hunt *et al.*, 1974; Yendrikhovskij, 1998; Fairchild, 2005). In fact, this effect is so strong that even black and white images of objects with familiar colors (e.g. bananas) have a color cast in the direction of the objects' colors (Hansen *et al.*, 2006).

Returning to the photographer example, let us imagine that the photographs being worked on are for a gallery exhibition of their prints. What matters then is not the relationship between what one sees on the display and the prints as seen in the studio. Instead, it is the print appearance in the gallery that needs to relate to the display; and as there is likely to be both a difference in viewing conditions and in the time at which the display and exhibited prints are viewed, both the effects of adaptation to different environments and the properties of color memory and memory colors will be involved in making a judgment. If the prints are of a different size compared with their display originals, then another phenomenon will be observable, i.e. that the perceived color of a stimulus depends also on its size. For example, taking a stimulus and scaling it up will result in the perception of increased lightness compared with a smaller stimulus with the same XYZ values (Xiao *et al.*, 2004; Nezamabadi and Berns, 2005).

2.8 DIGITAL COLOR CAPTURE AND GENERATION

As gamut mapping takes place between ranges of colors generated using different means, we will next take a look at the basics of how those that are digitally controlled operate. First, however, let us take a look at the two underlying color reproduction principles, i.e. additive and subtractive, that these means employ. Note that the following overview is based on Hunt (1995b: chapters 2 and 4), where more detail is available on this subject.

2.8.1 Additive color reproduction

Additive color reproduction exploits the fact that the human visual system has only three types of chromatic photoreceptor. It is sufficient, therefore, to match their responses to give a reproduction the same appearance as that of an original (when seen under the same viewing conditions). First, it is necessary to know the extent to which the different types of cone are stimulated by a given original stimulus and a means of reproducing it is needed. The level of stimulation resulting from various parts of a stimulus can be recorded by capturing it through filters that have spectral transmissions like the spectral responsivities of the three cone types. The capturing medium also needs to have a response to light intensity like that of the cones. Once the extent of stimulation is captured, the ideal way of reproducing a stimulus would be to use three image projectors with filters that result in each of the projections stimulating only one of the cone types. Reproductions made using a system like this would result in the same stimulations of the eye as their originals, and the entire range of possible color experiences could be reproduced if the projectors could output the same range of intensities as present in the original.

While such capture is physically achievable, the above kind of projection is not. In particular, it is impossible to have light sources that only stimulate one cone type at a time, and there will always be *unwanted stimulations* (Hunt, 1995b: 40–43). This follows directly from the spectral sensitivities of the cones, where, for the vast majority of wavelengths, at least two of the cone types are sensitive (Figure 2.6). Even though it is theoretically possible for light sources to stimulate only L or S cones, for M cones this is impossible. The best that can be done (i.e. giving least unwanted stimulation) is to have narrow band lights at 450, 510 and 650 nm. Even in this case, the 510 nm (green-looking) light would stimulate the L and S cones in an unwanted way. The problem would be worst when the M response to a color is large, as matching it would increase the extent of unwanted stimulation of the other two types of cone. As a consequence, reproductions of greens would look paler than their originals and neutrals would have a magenta cast.

The consequence of the overlap of photoreceptor sensitivities is that adding the output of three lights cannot reproduce the full range of color experiences and that such color reproductions are, by definition, limited.

2.8.2 Subtractive color reproduction

To overcome some of the problems of additive color reproduction, an alternative trichromatic approach was developed on the basis of subtraction. Here, the visible spectrum is first divided into three regions (above about 580 nm, between 490 and 580 nm and below 490 nm) corresponding approximately to red, green and blue respectively. A beam of white light shining onto a white surface can then be considered to be an additive mixture of red, green and blue light from the three regions of the spectrum. If the amount of light can then be controlled in each of the three regions, then a wide range of colors can be reproduced. One way of doing this is by applying varying concentrations of three different materials, each of which absorbs light predominantly in one of the regions and reflects light in the other two, and then illuminating it with white light.

Figure 2.16 shows the spectral reflectance curves of a set of three inks, i.e. cyan, magenta and yellow, at a various levels of concentration. When, for example, the yellow ink changes concentration, its spectral reflectance in the blue region changes while spectral reflectances in the green and red regions are unchanged. Therefore, the yellow ink controls the amount of blue light that will be reflected from a surface coated with it. Similarly, the magenta and cyan inks primarily control the green and red regions of the spectrum respectively (even though they vary in other regions as well – which is unwanted). A scene

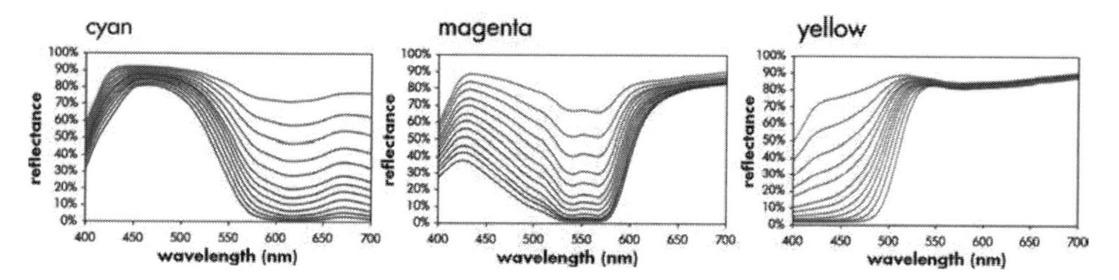

Figure 2.16 Spectral reflectances of cyan, magenta and yellow inks.

could, therefore, be reproduced using the subtractive method by first capturing it through red, green and blue filters and then applying amounts of the inks that are inversely proportional to the amounts captured through the corresponding filters. For example, an area where large amounts of red and small amounts of green and blue were captured will contain small amounts of cyan (because little red needs to be absorbed) and large amounts of magenta and yellow (because large amounts of green and blue need to be absorbed).

As the subtractive system also uses the additive principle – i.e. after subtracting parts of the light in the different regions of the spectrum the result is added together to form the stimulus that the visual system responds to – it is also susceptible to the limitations of the additive system. The difficulty again is that the inks (or other colorants) should absorb only in regions of the spectrum to which a single cone type is sensitive. As has already been discussed, this is impossible and *unwanted absorptions* necessarily take place in subtractive systems. Just as was the case with additive systems, subtractive trichromatic reproduction is also incapable of reproducing the full range of color experiences.

2.8.3 Color reproduction media and devices

These means of color reproduction (using the additive and/or subtractive principles) are typically referred to as *color reproduction media* or *devices* (where devices, sometimes in addition to other components, e.g. a screen for a projector, bring about media – i.e. means of color reproduction) and their two highest level categories distinguish between input and output devices. Input color reproduction media are those that capture digital color information, given a visual stimulus, and output media are stimuli that were created from digital inputs.

Input media come in two types, i.e. scanners and cameras, both of which use the additive principle and share a lot of the underlying functionality. In the case of a *scanner*, an original stimulus is illuminated using the scanner's built-in light source and the light reflected or transmitted by the original is then incident (via appropriate optics) on a photodetector that outputs a signal, which is finally digitized. Typically, the illumination and photodetector sense one strip of the original at a time and by scanning across it obtain samples from its entirety. Key contributors to the quality of the resulting digital data are the details of the photodetector in color, spatial and signal noise terms. As human color vision is trichromatic, scanners also typically sense an original via a set of three filters – each of which is most transmissive in one of the three spectral bands. How well the colorimetric data can be predicted from scanner output depends on how its filters relate to the CMFs. In spatial terms, a scanner will record a given number of picture elements (*pixels*) per unit distance in the original and the number of these (in conjunction with the quality of the scanner's optics) will determine what original spatial frequencies the scanned data will encode.

While scanners contain a built-in light source, *digital cameras* do not; as a result, their responses depend both on the objects they capture and the light sources that illuminate them. As a consequence, digital cameras have a much more adaptive response than scanners and, in that sense, are more like the human visual system. Cameras need to adjust their responses both to the illumination intensity and color to produce pleasing images. However, as a result of this adaptation, it is difficult to predict the appearance of the scene that their data encodes if information about the camera's internal settings (e.g. aperture, exposure time, filters, etc.) is not available.

Owing to their operation, both scanners and digital cameras can be thought of as combined spatial and color filters to the two-dimensional electromagnetic stimulus projected onto them. Also, in both cases an input device will report a set of RGB values for each of the pixels it records and these values represent the amount of light detected at a given stimulus location through the device's RGB filters. As a consequence, the RGB values recorded by an input device have a colorimetric meaning that is a function of the device's photodetector properties, and the RGBs of each individual camera end up having somewhat different color meanings. Since virtually all cameras and scanners have a small number of filters (i.e. three), they are also subject to metamerism; consequently, the degree to which it will be possible to predict from their RGBs what a viewer would see given a certain stimulus depends on how metameric they are relative to the human visual system.

Output media, on the other hand, take two-dimensional arrays of digital value n-tuplets as inputs and generate a stimulus as a result. Examples of such n-tuplets are RGB, CMYK (cyan, magenta, yellow, black), CMYKOV (CMYK, orange, violet), etc. and such n-tuplets define *device color spaces* (sometimes also called *native* color spaces – see Section 4.1). The key property of device color spaces is that they do not in themselves have colorimetric meaning. For example, a device RGB (dRGB) of [0%, 50%, 100%] does not give information about colorimetry or color appearance, and whether it corresponds to one hue or another depends on the device that it is sent to (or the device that has recorded it, for input media). Device color spaces also map onto the color range of the device that they are used with, which makes them particularly suitable for controlling device outputs.

The three main types of output color reproduction media are displayed, projected and printed. *Displayed media* (i.e. generated using displays) are brought about by devices that take digital inputs (typically RGB) and as a result emit, and in some cases filter, light to give a corresponding balance of energy in the three bands of the visible spectrum. Examples of displays are cathode-ray tubes (CRTs), liquid-crystal displays (LCDs), organic light-emitting diode displays (OLEDs) and plasma displays. Note that in this case (unlike for input media or the other types of output media) the device used for bringing about the medium is also the medium itself, i.e. the device itself is the stimulus that is viewed. For other media, such as projected or printed ones, it is not the projector or printer that are viewed but the stimuli that they create.

Projected media are similar to displayed media, in that RGB inputs control light output in the three spectral bands. The key difference is that in this case it is not the projector devices that are viewed. Instead, it is their output, as reflected or transmitted by a screen, that constitutes the final stimulus. Key technologies used in projectors are CRTs, LCDs and digital micro-mirror devices (DMDs). The color appearance of both displayed and projected media, both of which use additive color reproduction, is strongly dependent on the ambient viewing conditions. For example, an image projected in a dark room can give rise to a very large range of color experiences, whereas its parts may become hardly distinguishable from each other in a brightly lit room.

Finally, printed media employ subtractive color reproduction and are created using printers that deposit colorants on a substrate (i.e. a carrier such as paper) based on digital inputs. The colorants can be dye- or pigment-based inks, dry or liquid toners, dyes in a wax carrier, etc., and substrates can range from paper, via plastics, to textiles, and more specialist choices like wood, glass, ceramics, etc. The technologies for depositing colorants on substrates include ink-jet, laser, thermal wax and dye sublimation, and the digital data can come in a variety of device color spaces. In addition to RGB, CMYK is popular, and color spaces of higher dimensions can also be used for printers that use a larger number of colorants. The end result here is a reflective or transmissive object that modulates the light illuminating it and in that way results in a specific stimulus. The colorants filter (i.e. selectively subtract from) light by absorbing parts of its energy in specific spectral bands, the combined result of a set of colorants being a sequence of subtractions from the light that illuminates them.

For more information about the principles of operation of various color reproduction devices and media, see Hunt (1995b) and Sharma G. (2003).

2.9 COLOR MANAGEMENT

Given color imaging devices, it is frequently necessary to provide interfaces between them. For example, a color image captured using a digital camera can be uploaded to a computer for display on its screen and then sent to a printer for output. The processes that translate and communicate color information at such interfaces are referred to as *color management.*

In practice, color management can be addressed in a number of alternative ways, of which the ICC's color management framework is currently the *de facto* standard, at least as far as the reproduction of still images is concerned. The approach proposed by the ICC is one where the color management process is divided into two transformations: first, a forward one that takes device color data (e.g. a display's RGB or a printer's CMYK) and transforms it into a colorimetric description for specific viewing conditions (called the *profile connection space* – PCS); second, an inverse one that takes such a colorimetric description and transforms it back into device color space data. Note that these transformations also include color gamut mapping, as will be discussed in more detail in Chapter 4.

Color interchange between devices is then achieved by being able to perform one or both parts of the transformation for each of the color reproduction media among which color is to be managed (Figure 2.17a). The parameters, based on which the forward and

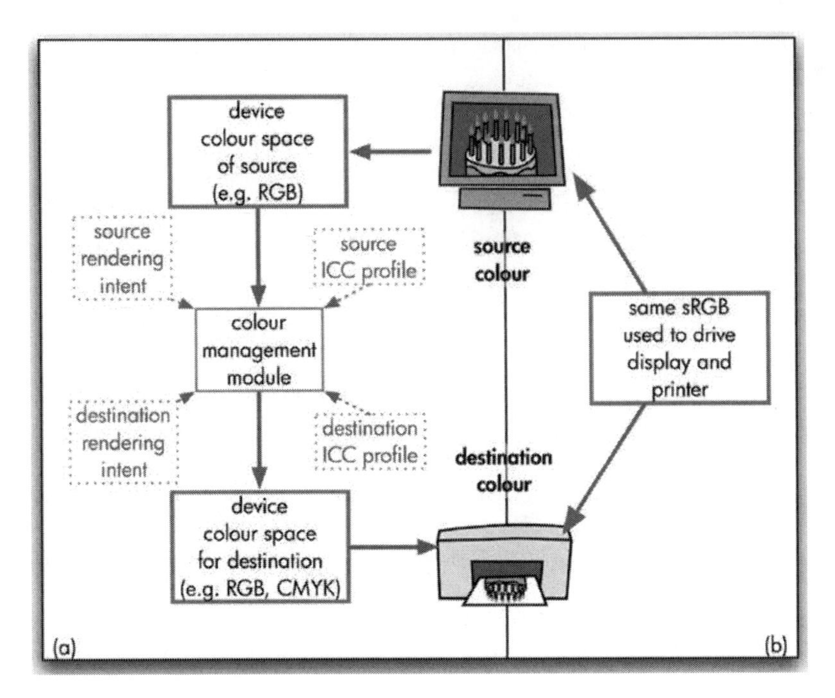

Figure 2.17 Color management frameworks: (a) ICC and (b) sRGB.

inverse transformations are performed for a given color reproduction medium, are stored in a data file called the *ICC device profile*; its detailed specification and the specification of the entire architecture can be found at http://www.color.org.

An important, complementary approach to the ICC color management architecture, where the color imaging behavior of each device is characterized with reference to colorimetry, is to base all color communication on a single device-related, but colorimetrically defined, color encoding. Specifically, color communication workflows can also provide good results by taking two decisions: first, that RGB will be used to communicate color information between devices; second, that RGB will be given a unique colorimetric interpretation, e.g. sRGB (IEC, 1999). Each device then does the best it can either to encode its native color information in sRGB so that the result is pleasing (e.g. scanners, digital cameras) or to provide pleasing color output given sRGB input (e.g. printers, displays, projectors) (Figure 2.17b). Note again that the transformation that will need to be performed by devices in response to sRGB inputs, or to generate sRGB outputs, will also involve color gamut mapping.

A further alternative is the *Windows Color System* (WCS) recently introduced by Microsoft, which involves a more complex, but flexible and modular, sequence of transformations to interface color information between devices. For more on this approach to color management, see Microsoft (2006) and Chapter 4. More detailed discussions of color management in general can be found in Giorgianni and Madden (1998) and Morovič and Lammens (2006).

2.10 WHAT CAN AFFECT THE APPEARANCE OF A PAIR OF STIMULI?

To conclude this chapter on the basics of color science, let us consider the viewing of the output of two color reproduction devices and think about the factors that can affect the resulting experience (i.e. whether they match or what the nature is of their difference).

The first thing to notice in Figure 2.18, which should already be apparent from the preceding content of this chapter, is that there are many factors involved in determining the outcome of comparing the color appearances of two stimuli and that the outcome of the comparison can change if any of these factors change. More specifically, it can be seen that, in addition to the pair of color stimuli (which are the only parts of the setup that can be directly controlled by the inputs to color reproduction devices), there can also be differences in at least the following factors:

1. The presence of other stimuli generated using the two devices or simply viewed alongside them (e.g. Are the white points of the two media visible? Is there a neutral

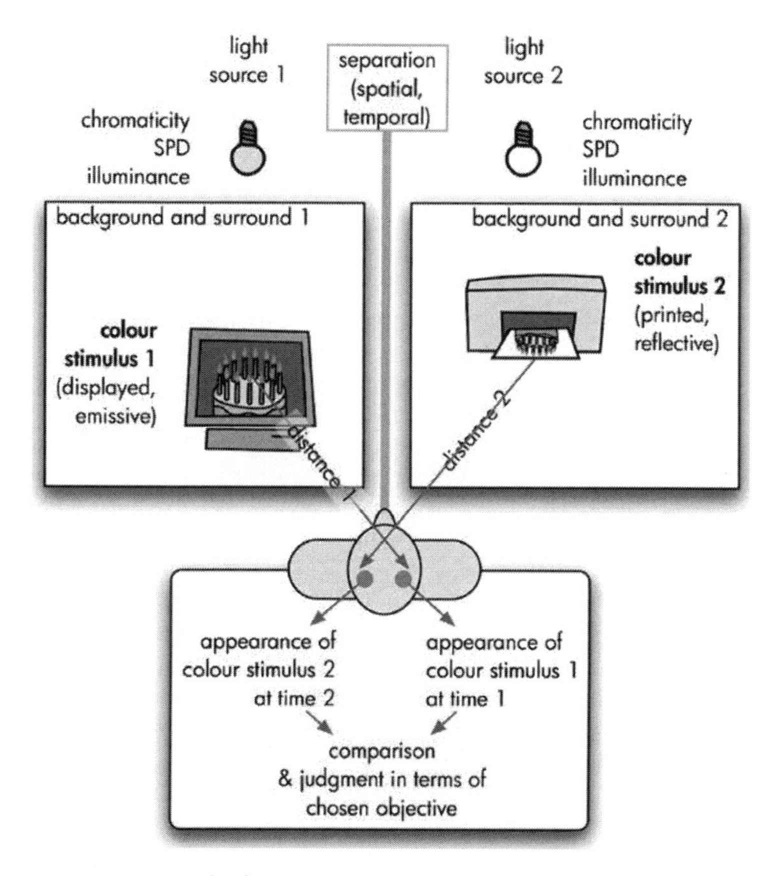

Figure 2.18 Viewing a pair of color images generated using different color reproduction devices.

more luminous than the two white points present when the color stimuli are viewed? This affects adopted white (i.e. what looks white in a scene) and hence perceived appearance attributes (i.e. lightness, chroma and hue)).
2. The background and surround against which the stimuli are viewed (affecting adaptation, involving simultaneous contrast and crispening effects; influencing perceived image contrast and colorfulness).
3. The SPDs, chromaticities, and illuminances of the light sources under which the stimuli are viewed (strongly affecting adaptation, setting limits to possible colors perceptible under them, determining the *Hunt effect* (Fairchild, 2005), i.e. surface colors appearing more chromatic as illumination level increases).
4. The distances from which the stimuli are seen (affecting perceived lightness and chroma).
5. The nature of separation between the two stimuli's viewing environments (i.e. stimuli can be seen simultaneously under mixed viewing conditions, where the state of adaptation can be complex to determine (Katoh, 1994; CIE, 2004c), or at different times, in which case the judgment about the relationship of stimulus appearances also involves memory).
6. The specific physiology of the individual observer (determining the particular CMFs, affecting perceptibility thresholds).
7. The experience the observer has with making the requested observations, comparisons and judgments (affecting perceptibility thresholds, judgment tolerances and repeatability and potentially involving personal preferences and cultural background).

The key points to take away from this analysis are: first, that the relationship between the ways in which two stimuli appear to an observer is not a property solely of those two stimuli, but rather of a complex constellation of numerous states; second, that it is not color stimuli that are compared and judged, but the mental representations of their appearances, and this underlines the inherently subjective nature of any color comparison task.

2.11 SUMMARY

Before moving to Chapter 3, where we will consider what aims color reproduction can have and how the degree to which a given reproduction fulfils such aims can be evaluated, let us first reflect on the key points that this chapter focused on:

1. Color is difficult to define in specific terms, and in this book it will be considered to be a *quality in a sequence of a viewer's visual experiences* (Section 2.1).
2. In everyday language, color can be described using a small number of terms; and more specific descriptions of color experiences can also be had by judging their appearance attributes (e.g. brightness, colorfulness and hue) (Section 2.2).
3. Activity in the brain's crust relates to color experiences and is the result of responding to a signal carried and transformed along the visual pathway from three types of light-sensitive cells in the eyes. This signal is based on spatial and temporal differences between different parts of the scene and is then processed in a highly complex way in the visual cortex (Section 2.4).

4. The eyes' light-sensitive cells respond to electromagnetic radiation from a specific wavelength range, whose properties are determined either by objects that emit the radiation themselves or also by objects that modify it via reflection or transmission (Section 2.5).
5. The response of the eye's light-sensitive cells can be predicted from measurements of electromagnetic radiation and expressed as CIE *XYZ* tristimulus values. *XYZ*s are then the starting point for making predictions about color appearance attributes and for predicting the effect of the visual system's processing (Section 2.6).
6. The visual system adapts to what it sees, and there are complex interactions between the properties of various parts of a scene that affect the resulting experiences (Section 2.7).
7. Color stimuli can be generated using a range of media that typically involve additive and/or subtractive color reproduction, and color-related information can also be captured using digital means (Section 2.8).

In a sense, even more important than the content of this chapter is for it to become clear that the subjects covered in the preceding pages are of significant complexity and that their treatment here is nothing but a whirlwind tour. While this may have been a useful refresher for those already aware of the material presented here, the less-experienced reader is highly encouraged to read up on the topics discussed in this chapter and to follow the references given before proceeding with the remainder of the book.

3

Desired Color Reproduction Properties and their Evaluation

The result of using gamut mapping will always be judged by someone asking themselves, or at least implicitly acting on the outcome of the question: 'Is this a good reproduction?' It is essential, therefore, to be clear about how this can be answered from the start. Furthermore, such a question quickly leads to two others: 'What does "good reproduction" mean?' and 'How can it be established whether a given reproduction is "good"?'

In terms of what 'good' means, G. K. Chesterton (1874–1936) provides a vivid example:

> The word 'good' has many meanings. For example, if a man were to shoot his grandmother at a range of five hundred yards, I should call him a good shot, but not necessarily a good man.

A variety of meanings can apply to goodness in the color reproduction context too, and a choice of interpretation, therefore, needs to be made and stuck to throughout the development and evaluation of a gamut mapping solution. The second part of this chapter will present an enquiry into robust methods of evaluating color reproduction goodness, given a choice of interpretation.

3.1 COLOR REPRODUCTION FRAMEWORK

Before going into the details of what goodness of color reproduction can mean, the term *color reproduction* (Figure 3.1) will be defined first as 'the process by which color

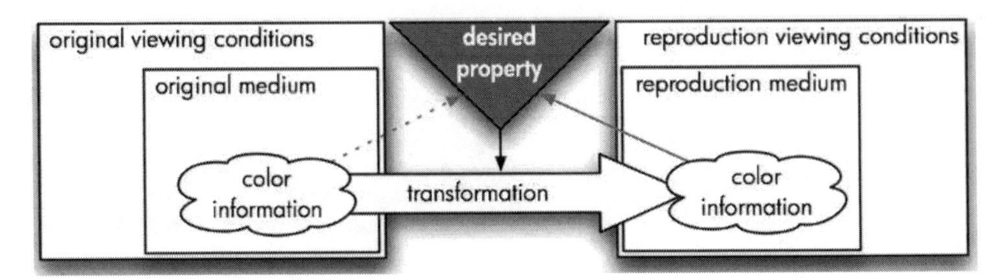

Figure 3.1 Elements of color reproduction.

information from an original medium, under the original's viewing conditions, is transferred to a reproduction medium, under its viewing conditions, with the intention of the reproduction having a predetermined property' (Morovič, 1999). Note that in some contexts the pair *source–destination* is used instead of *original–reproduction* and that the meaning of these pairs is the same.

Color *information* can be understood as having two properties: type and context. *Type* can be physical, color-related properties (e.g. spectral power, reflectance or transmission characteristics), the extent to which a viewer's photoreceptors are stimulated (e.g. cone responses, *XYZ* tristimulus values) or the attributes of a viewer's color experiences (e.g. appearance attributes like brightness, colorfulness and hue). *Context* can mean colors being considered by themselves, against a simple background or as part of an image.

The term *media* applies to anything that can contain or exhibit color information, which includes a person's imagination, nature and man-made media (e.g. textiles, metallic or plastic surfaces, print, television, projection, displays, etc.). Note that a medium is often facilitated by a device and that the two can be distinct. For example, a print is a color reproduction medium and a printer is the device that brings it about. The printer is not a medium in itself, though (i.e. it is not the printer that is viewed).

As should be plain from Chapter 2, stimuli (e.g. color reproduction media) are heavily influenced by *viewing conditions*, including viewers, illumination and spatial and temporal arrangements of a given stimulus versus others. For example, viewing a printed image against a black or a white background results in different color information being conveyed.

The process of transferring or *transforming* color information from original to reproduction media will be looked at in more detail in Chapter 4, and the *desired property* of the reproduction, which is the aim of the transformation, will be considered in more detail here.

Finally, since the majority of this book will focus on the details of color reproduction, it is useful to bear in mind the bigger picture: namely, that the foremost purpose of reproducing color information is beyond itself. Instead of being about color information being reproduced, the aim is to convey an idea, a message, a certain meaning, some feelings; to allure, please, sell, inform, enlighten, shock, warn or even to disgust. If a color reproduction system fails to deliver the intended effect, then its inner workings, however scientifically solid or well engineered, will have failed too.

3.2 DESIRED COLOR REPRODUCTION PROPERTIES

A number of dimensions can be identified from looking at how the desired properties of color reproduction have been defined in the literature, whereby they can be:

1. defined with reference to a *reproduction alone* or with reference also to the original and, therefore, to the *relationship* between the two;
2. defined by an *application* to which they relate (e.g. the reproduction of artwork) or be *independent of application*;
3. *potentially descriptive* (i.e. a reproduction can be known to have such a property) or expressing an *ideal state*;
4. *quantitatively* measurable by *physical* means or primarily expressible in *qualitative* terms and measurable *psychovisually*.

To explain what these nominal dimensions mean, let us next survey the desired reproduction properties put forward by various authors and categorize them using the taxonomy listed in Table 3.1.

3.2.1 Yule's reproduction properties

In his *Principles of Color Reproduction*, Yule (1967: 258) describes two variants of a desired reproduction property that he calls *accurate reproduction*:

> *Duplication*. All combinations of the coloring materials of the reproduction process ... must be accurately reproduced.
> *Colorimetric Quality*. All visually identical colors ... must be similarly reproduced.

The first of these, i.e. duplication, applies to a very particular type of reproduction that is made using the same medium as the corresponding original. Accuracy then means a match in how the medium's colorants are used. This reproduction property defines a desired relationship between original and reproduction (S = R), is made in an application-independent way (A = I), can be descriptive (N = D) and be expressed by physical measurement (E = H) – hence it can be classified as (S, A, N, E) = (R, I, D, H). The second type of accurate reproduction refers to *visual identity* as the desired relationship between original and reproduction. Therefore, it is a relaxed version of duplication. Whether it can be measured physically and whether it can be descriptive or not depends

Table 3.1 Taxonomy of desired reproduction properties.

Dimension	States
Scope (S)	Reproduction alone (S = A) or relationship of reproduction and original (S = R)
Application (A)	Dependent (A = D) or independent (A = I)
Nature (N)	Descriptive (N = D) or ideal (N = I)
Expression (E)	Physical (E = H) or psychovisual (E = S)

on how original and reproduction are viewed. If viewing conditions are the same, then this desired property is (R, I, D, H); and if not, then nature and expression depend on the complexity of the viewing condition mismatch.

3.2.2 Hunt's reproduction objectives

The most popular set of aims by far are those proposed by Hunt (1970). His six desired reproduction properties, i.e. *color reproduction objectives*, form a hierarchy of degrees of match between original and reproduction (Hunt, 1970; Hunt, 1995b: chapter 11):

1. *Spectral* (R, I, D, H). Spectral power distributions of original and reproduction are identical. This is physically the strictest form of match between an original and its reproduction and is in practice rarely achieved.
2. *Exact* (R, I, D, H). Chromaticities and absolute luminances are identical. As a consequence, if the spatial surround and temporal viewing sequence of original and reproduction are the same, then this also means a match in brightness and colorfulness.
3. *Colorimetric* (R, I, D, H). Chromaticities and relative luminances (i.e. relative to the luminances of corresponding reference whites) match. All else being equal, this translates to a match in lightness and chroma, but not in brightness and colorfulness. In other words, if original and reproduction are both reflective (or transmissive), then that member of the pair that is more strongly illuminated will have greater brightness and colorfulness.
4. *Equivalent* (R, I, D, S). Chromaticities and relative and absolute luminances of reproduction have the same *appearance* as those of original. This objective takes into account the human visual system's adaptation mechanisms and says that original and reproduction should look the same under their respective viewing conditions. Unlike previous reproduction properties, this one can no longer be measured physically but needs to be judged psychovisually.
5. *Corresponding* (R, I, D, S). Chromaticities and relative luminances in the reproduction *appear to be* the same as in the original when both have the same luminance levels. This is a further relaxation of the equivalent reproduction objective that allows for different appearances in terms of brightness and colorfulness as a result of different levels of illumination.
6. *Preferred* (A, I, I, S). Equality of appearance is sacrificed in order to achieve a more pleasing result in the reproduction. Unlike the previous reproduction properties, preferred color reproduction focuses on the reproduction itself, rather than on its relationship with the original. This is because reproductions are often seen without reference to the original. Therefore, it is often their specific qualities that need to be liked by their viewers rather than the nature of how they represent some original. The production of photographic prints (e.g. of images taken during a holiday) is a good example of when preference is desired. In this case the original scenes are no longer available and their viewers have a tendency to prefer colors being reproduced in an idealized way (Hunt *et al.*, 1974; Yendrikhovskij, 1998). The sky in photographs is preferred to look bluer and more colorful than it might have looked in a photographed scene, grass to look greener and skin to look healthier. Furthermore, there is also some

evidence that color reproduction preferences vary with culture (Fernandez and Fairchild, 2002).

An important observation to make about this system of desired reproduction properties is that no reference is made to potential differences in color gamuts that are the result of factors other than viewing conditions.

3.2.3 CIE subjective accuracy

However, as different color reproduction media, obtained using different imaging devices, can have very different color gamuts even under the same viewing conditions, the definition of a further desired reproduction property is useful (Morovič, 1998; CIE, 2004d):

> [The] *subjective accuracy* intent is one which aims at reproducing a given colour image in a way where the reproduction is as close to the original as possible, this similarity is determined psychophysically and the process has no image enhancing aims. (CIE, 2004d: 1, italics added)

Such *subjective accuracy* (R, I, I, S) is related to Hunt's *equivalent* and *corresponding* color reproduction properties, but with some key differences. First, there is no link to potential viewing condition relationships. Second, a specific allowance is made for gamut differences by aiming at the reproduction being 'as close ... as possible' to the original. Third, the desire for the reproduction to be as close as possible to the original also means that the property is an *ideal* one, since it is currently not possible to know whether a given reproduction is the one that will look closest to the original when there are gamut differences.

Subjective accuracy also has some important implications: first, that an unpleasant-looking original (e.g. having some defects) will result in a reproduction that also looks unpleasant; second, that a reproduction medium's color capabilities might not be used to its full potential. For example, when reproducing an original printed on uncoated paper with a reproduction on coated, glossy paper that has a larger color gamut, the subjectively most accurate reproduction would not make use of the entire available color range.

3.2.4 ICC rendering intents

The ICC is the source of another set of widely used desired color reproduction properties, referred to as *rendering intents* (ICC, 1994: 14). The *ICC-absolute colorimetric* and *media-relative colorimetric* rendering intents (ICC, 2001) are variants of Hunt's *corresponding* reproduction objective and have also been described by Viggiano and Moroney (1995). An *ICC-absolute colorimetric* reproduction is defined as one where the original's CIE *XYZ* tristimulus values are chromatically adapted to the D50 illuminant and then matched in the reproduction under that same illuminant. This also means a match in D50 CIELAB values if the perfect diffuser is the reference white. In the *media-relative colorimetric* case, original and reproduction tristimulus values match when scaled by the *Y* values of their respective media white points (e.g. unprinted paper, a displayed color that results from maximum output in all of the display's RGB channels), which is equivalent to a D50

CIELAB match when the original and reproduction use their respective medium white points as reference whites. Note that both of these intents would be classified as (R, I, D, S). Even though they are defined in terms of physical relationships (i.e. *XYZ*s), the definition is not of a relationship between original and reproduction directly, but of a relationship of representations of each in an abstract intermediate space – the ICC *profile connection space* (PCS; Section 4.5) – to which original and reproduction are transformed using models of chromatic adaptation.

Note also that these rendering intents only define how to reproduce original colors that are inside the reproduction's color gamut. The treatment of original colors that are outside the reproduction gamut is not addressed.

The ICC (2004: 9–10, italics added) also provides the *perceptual* and *saturation* rendering intents and describes them in the following ways:

> The *perceptual* intent is useful when it is not required to exactly maintain image colorimetry (such as with natural images), and the input and output media are substantially different. . . . The . . . *saturation* intent . . . involves compromises such as trading off preservation of hue in order to preserve the vividness of pure colours.

Both of these intents are related to Hunt's preferred reproduction objective, in that they express a desire for departure from the original's properties in a preferred way. Viggiano and Moroney (1995) proposed these under the *raster* and *business graphics* labels respectively. It can be seen that these definitions are much vaguer than those of other desired reproduction properties and are more like hints at what is wanted. Such vagueness makes these two rendering intents more difficult to classify, since they could be seen as either (R, I, I, S) or (A, D, I, S) (the latter being more clearly applicable to the proposal by Viggiano and Moroney (1995)).

3.2.5 Re-purposing and variants of re-targeting

An alternative way of looking at desired reproduction properties is to say that a reproduction is to be suitable for a particular application. Here, the ICC describes two intended uses for a reproduction (ICC, 2006):

> *Re-purposing* starts with color content that has been color rendered to one output color encoding and then applies a color re-rendering to achieve another output color encoding. . . . [I]f the media are different there will likely be intentional differences in the colorimetry . . . because the objective is to make the "best possible" reproduction in each case, which will depend on the reproduction media and on the use case preferences.
> *Re-targeting,* e.g., proofing, can be thought of as an alternative to re-purposing . . . because the reproduction goal is to produce not the best reproduction possible, but rather the closest possible match to some [original] . . . [T]he intent is always to preserve rather than to re-shape.

Of the two, *re-purposing* is equivalent to Hunt's *preferred* color reproduction and would be classified equally (A, I, I, S), whereas *re-targeting* can be interpreted in two ways, depending on the domain in which the closest possible match is sought. If the domain is that of colorimetry, then it matches Hunt's *colorimetric* reproduction objective (R, I, D, H);

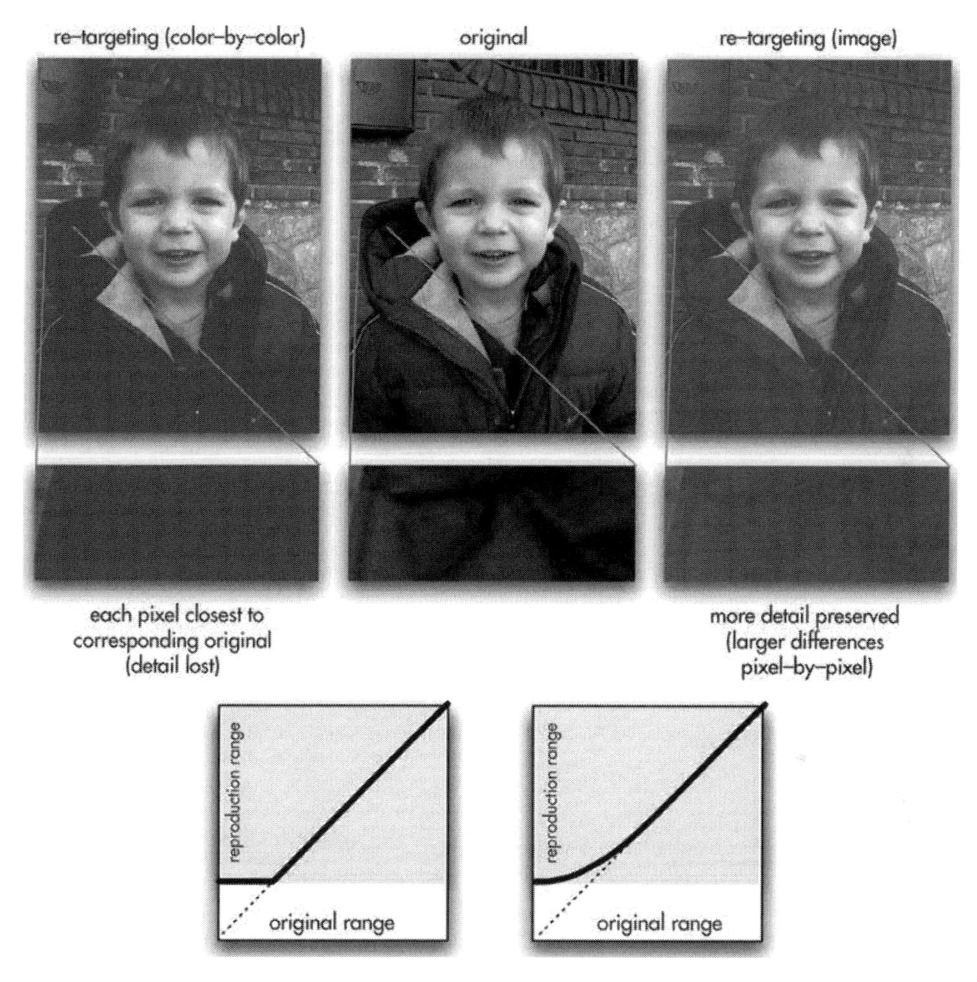

Figure 3.2 Original and reproduction that is (a) colorimetrically closest and (b) aims at being visually closest.

however, if the closest match is sought visually, then it matches the CIE's *subjective accuracy* objective (R, I, I, S).

The key impact of this difference in interpretation of re-targeting is that in the former case a physically measurable relationship is the aim on a color-by-color basis in an image, whereas in the latter it is about similarity of overall image appearance. Figure 3.2 illustrates this point by showing a grayscale original (middle) and two alternative reproductions in a more limited dynamic range. The reproduction on the left has pixels that are individually closest to the corresponding original pixel, and as the reproduction has a more limited range, the original pixels that are darker than the reproduction's black point are all mapped to it with any differences between them being lost. The reproduction on the right, however, is an attempt at representing the original in the more limited dynamic range in a way where it is considered as an image and preserving the original's detail is part of obtaining a similar reproduction. A consequence of this is that original colors that are just above the reproduction's black point are not matched individually but made lighter than the original to accommodate differences among the darker original colors.

3.2.6 Field's reproduction objectives

Finally, in his *Color and Its Reproduction*, Field (1997) divides reproduction objectives into three groups: exact, optimum and creative. Field's *exact* desired reproduction property is equivalent to Hunt's *corresponding* color reproduction (R, I, D, S) and he further divides *optimum* reproduction (A, I, I, S) into three subgroups: *preferred* (matches Hunt's reproduction objective of the same name), *corrective* (preferred reproduction applied to originals that have defects, such as color casts) and *compromise* (preferred reproduction that acknowledges difference in color ranges between original and reproduction). The most interesting of Field's reproduction objective groups is the *creative* one, which refers to cases when the impact and appeal of a reproduction are paramount and the examples of fashion, entertainment and advertising are given. In terms of classification Field's creative desired reproduction property would be (A, D, I, S).

The above range of desired reproduction properties (Figure 3.3), which can broadly be divided into variants of accuracy and preference, should be a clear indication of the degree of choice and variety of potential outcomes that are possible when one is given an original

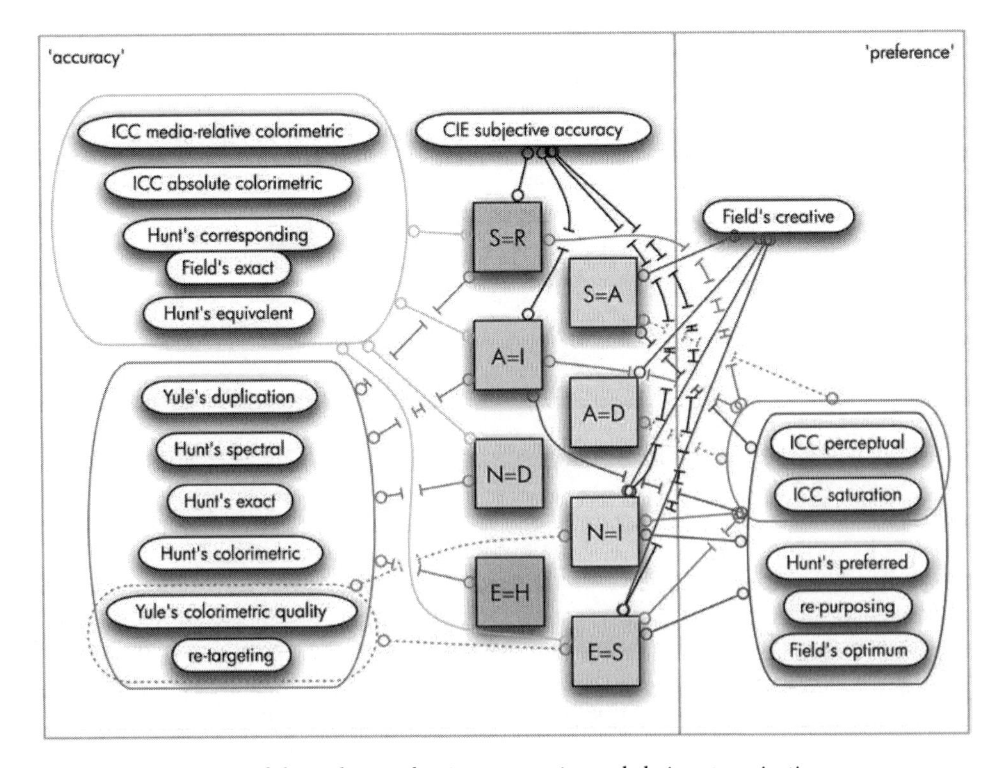

Figure 3.3 Overview of desired reproduction properties and their categorization.

and asked to reproduce it using a given reproduction medium. Instead of being able to proceed immediately with creating a reproduction, it is first necessary to determine which reproduction property is sought. The chosen property will then be a constant criterion for choosing from among alternatives in a color reproduction system (including GMAs) and ultimately be what determines whether the resulting reproduction is considered a success or failure.

3.3 EVALUATING REPRODUCTIONS

Regardless of the chosen desired reproduction property, the viewing conditions under which original and reproduction are to be viewed need to be defined too. When reproductions are made for a specific location (e.g. gallery, shop, etc.) this is conceptually problem free (although not so in practical terms); however, when the aim is for reproductions to be viewed under 'typical' (e.g. typical of an office or a photographic studio) or any viewing conditions, then this step can be demanding too. What is often done in practice when reproductions are meant for viewing under unknown conditions is to evaluate them under some standard ones (e.g. ISO 3664, defined for the graphic arts; ISO, 2000).

Once viewing conditions are known or chosen, two different methods need to be chosen between, depending on whether the desired reproduction property is expressed quantitatively or qualitatively.

A key point to note, before looking at details of evaluation methods, is that the results of an evaluation procedure are never simple statements, like: 'Reproduction A exhibits the desired reproduction property twice as much as reproduction B.' Instead, they need to be prefixed with the details of the method used to obtain them and are a function not only of the reproductions that were evaluated, but also of the details of the evaluation method used. As a consequence, evaluation results for different reproductions are directly comparable if and only if the same evaluation method was used to obtain them.

3.3.1 Measurement-based methods

The simpler case of evaluating desired reproduction properties is when they are expressed in physical terms (e.g. Hunt's *spectral* or *colorimetric* reproduction). Given a physical expression, both original and reproduction can be measured in its terms, the measurements can be compared and the resulting distribution of differences can be either given in full or some of its descriptive statistics can be reported. Therefore, the key questions here are how to measure chosen physical properties, what to measure in the given stimuli, how to express their differences and how to decide whether the differences are acceptable or not. As it is the colorimetric desired reproduction property that is most commonly used from among those that are physically expressed, let us take a closer look at it in terms of the four aspects of measurement, sampling, expression of differences and establishment of acceptability.

Since all of the existing desired reproduction properties that are expressed physically also have a scope that refers to the relationship between original and reproduction, their evaluation will also be based on measurements of the two. The starting point, therefore, is a pair of spatially uniform or complex (e.g. photographic image) color stimuli in their respective viewing environments and the first task is to obtain their physical measurements.

Measurement

The instrument of choice for measuring originals and reproductions, when accuracy is the aim, is a TSR because it measures spectral power as present in an actual viewing environment and, therefore, relates most directly to what a viewer would respond to. If a TSR is placed where a viewer's eyes would be when looking at a stimulus, then the resulting measurements express the properties of light that would enter the viewer's eyes and, therefore, have the greatest potential for correlating with visual experience. Spectral power distributions measured in this way can then be transformed into tristimulus values and from them to CIELAB or CIECAM02 coordinates (given further information about the adapting white and other viewing condition parameters).

Alternatively, other means of measurement can be (and often are) used for the sake of resource constraints or greater repeatability (setting up TSR measurements has its challenges). In such cases, it is important to make suitable adjustments to the measurements to relate them to the viewing conditions they are meant to represent (e.g. adding flare to contact measurements of a display; measuring print reflectance using a spectrophotometer but computing tristimulus values using a TSR measurement of the actual illuminating light source, etc.). In cases where this is not possible, it is at least important to bring attention to the mismatch between measurement and viewing conditions when making decisions on the basis of their results.

A typical example of such a mismatch is between display and print colorimetry as predicted using ICC profiles on the one hand and visual experiences of viewing the two media under typical office viewing conditions on the other hand. Here, the ICC profile colorimetry refers to what those media would be like if their white points were D50 and if the display was viewed in the dark and the print illuminated with a light source with an illuminance of 500 lx. As this is never the case in an actual viewing environment (never mind an office), such colorimetric data will not relate well to viewing the two media under a mixture of daylight and light emitted by fluorescent tubes. For a start the darkest color on the display will have a lightness significantly greater than zero and the two media will be seen in a mixed state of adaptation. For more on the effect of viewing conditions on the color gamuts of color reproduction media, see Section 8.4.9.

Sampling

Sampling is simplest when applied to spatially uniform stimuli where only a single measurement, or a number of measurements that are later averaged, can be taken for the original and reproduction and subsequently compared. If, as is more common, the original is a complex image, then sampling becomes more challenging and the following alternatives can be considered (Figure 3.4):

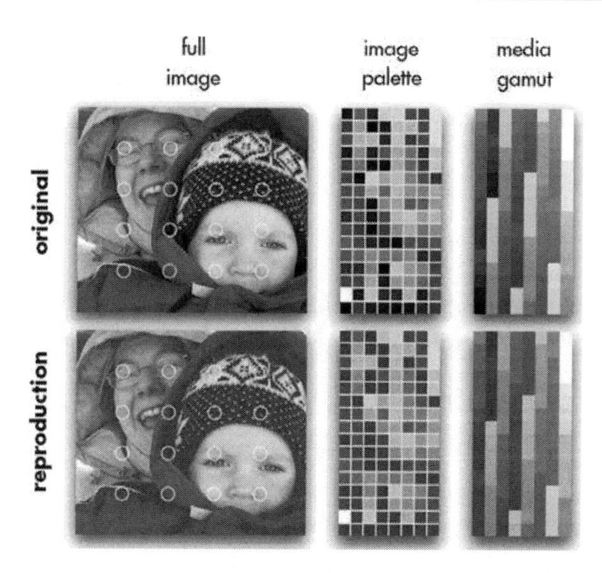

Figure 3.4 Alternative sampling approaches. Left: direct measurement of original and reproduction (e.g. at locations indicated by circles); centre: measurement of image palettes; right: measurement of media gamut samples.

1. *Measuring stimuli at specific locations.* If only the original and reproduction are available and there is no access to the devices that brought them about, then the only option (when using a TSR) is to select corresponding locations in the original and reproduction and measure them. Such an approach, however, has serious weaknesses, as even slight differences in the placement of the measured area on the stimuli can have a large effect on the measured result and as the choice of locations to measure is also problematic (e.g. resolution of uniform sampling; choice of nonuniform sampling).
2. *Generating new, image-dependent stimuli on original and reproduction media.* If there is access to the devices that generated the original and reproduction (and these are in similar states to when the pair were made), then it is possible to analyze image content and generate a palette of the original image (e.g. using Foray clustering of its pixel colors (Gose *et al.*, 1996)). The palette can then be rendered using the original medium and from there reproduced in the reproduction medium (note that the reproduction palette is a reproduction of the original palette rather than the palette of the reproduction image – to ensure correspondence). The result is a number of spatially uniform patches that represent the original's image content and their reproductions. The measurement of such patches then does not have the shortcomings of the first alternative.
3. *Generating new, image-independent stimuli on original and reproduction media.* This is a variant of the second alternative and consists of sampling the original medium's color gamut, rendering patches with the resulting colors, reproducing them in the reproduction medium and then measuring them.
4. *Obtaining colorimetric representation of entire images.* If both original and reproduction were rendered using digital devices, then the digital data that was used to render them can be used to obtain colorimetric data for all of their pixels, if characterization models

(or ICC profiles) of the two devices are available. Such models and profiles take digital data sent to a device and predict from it the resulting stimulus properties (more on these in Chapter 4). While this can give access to entire originals and reproductions in colorimetric terms it has drawbacks, since models tend to involve some error and also since they tend to be set up for default viewing conditions rather than the specific ones in question.

In addition to the challenge of sampling the colors of a given original–reproduction pair, there is a further question about how well results from a single pair (or even a larger number of original–reproduction pairs) generalize; as will be shown in Chapter 12, such generalization is indeed very difficult if not impossible.

Difference

Given pairs of TSR measurements for each sample (one each from the original and the reproduction), the next step is to compute their differences. While a simple, Euclidean distance between the pair in the desired reproduction property's expression space may be the most obvious choice, it often is not the most useful one.

The choice of difference computation needs to be made on the basis of how its results will be interpreted. If all that is needed is to distinguish between no difference and some difference, then the choice is not of great importance, since a zero difference is reported for identity by all the alternatives. If, however, one wants to be able to answer questions like: 'will the difference be perceptible?,' 'will I be able to tell how much larger one difference is than another?,' or 'will I be able to tell whether the difference is acceptable?,' then the perceptual uniformity of the difference computation plays an important role and the choice will again depend on whether the original is a spatially uniform or complex stimulus.

Currently, the most perceptually uniform way of predicting differences between spatially uniform stimuli is CIEDE2000 (CIE, 2001; Section 2.6). Applying it to the original and reproduction measurements will result in a distribution of distances, and appropriate descriptive statistics need to be chosen for communicating its properties in a succinct and easily comparable way.

The key question to ask when choosing from among alternative statistics is whether the population from which the sample was obtained is normally distributed (i.e. Gaussian, bell shaped) or not. To answer this question, there are two alternatives. The first is most accurate, but impractical, and consists in taking the entire population of the original's colors and its reproductions and testing its properties. The second alternative consists in testing whether the available sample (i.e. the set of CIEDE2000 color differences obtained for the chosen sampling) conforms to the properties of the normal distribution. If the sample is large enough (e.g. contains in excess of around 100 samples), then this second alternative is reliable too. In either case, the test to perform is the *Kolmogorov–Smirnov* test (Weisstein, 2006a) with respect to the normal distribution.

The CIEDE2000 distribution in Figure 3.5 shows a sample that does not come from a normally distributed population, and computing normal statistics (i.e. mean μ and

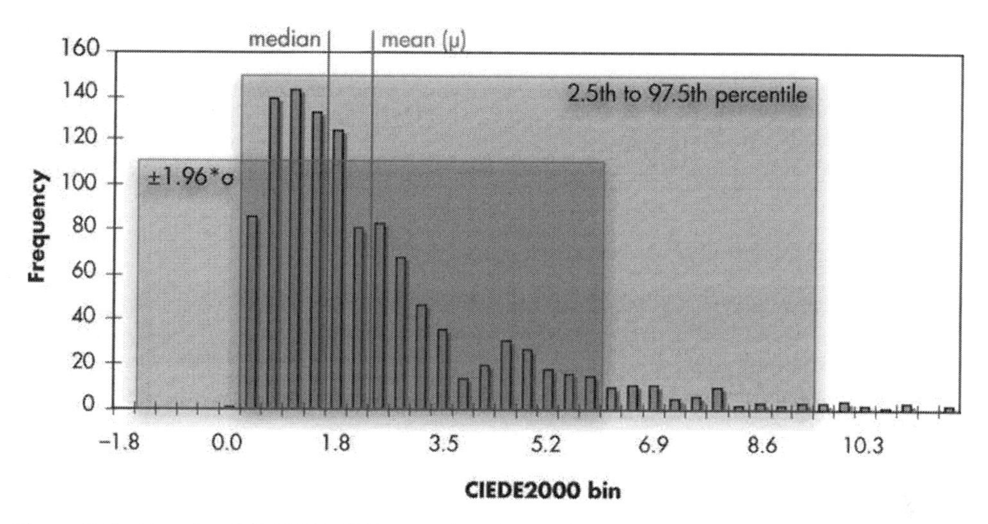

Figure 3.5 Sample CIEDE2000 difference distribution between an original and its reproduction. Also shown are values of alternative descriptive statistics.

standard deviation σ) from it gives results that do not describe it adequately. Instead, the more appropriate choice is the use of the *median* and a pair of *percentiles* (i.e. values below which a certain percentage of a distribution lies). In any case, the maximum difference in a distribution should also be reported.

When the original is a complex image and available digitally, then further consideration needs to be given to taking into account the spatial sensitivity properties of the human visual system (Section 2.7). The reason is that there may be some detail in the original and reproduced images that is invisible to the visual system or to which there is lesser sensitivity than to other detail. As a result, it is advisable to pre-filter the original and reproduction images spatially in a way that mimics human response before extracting colorimetric data from them. Examples of such spatial filtering are those used in the S-CIELAB metric (Zhang and Wandell, 1996), the iCAM model (Fairchild and Johnson, 2004) and the work of the CIE's technical committee 8-02 on *Colour difference evaluation in images* (http://www.colour.org/tc8-02/). In all these cases an image is taken and transformed so that local detail in it that is not visible (for given viewing conditions) is removed and the magnitude of detail that is left in the image is proportional to its perceptibility. The colors of pixels in the resulting spatially filtered versions of original and reproduction can then be used more reliably for color difference computation.

Acceptability

Given descriptive statistics of the differences between original and reproduction, the question often is whether they are acceptable. The answer to that question, however, depends very much on the intended use of the reproduction. While in some cases no visible differences are acceptable, in others the expectations are less stringent and even visible differences will not inhibit the reproduction's usefulness.

The first of these cases, where the criterion of acceptability is perceptibility, is the one that can be defined more clearly, and there are two groups of answers for what the perceptibility threshold is, depending on whether the original is spatially simple or complex.

For simple originals, the perceptibility threshold (for 50% of the population of observers) is one ΔE, since ΔE metrics are set up so that their unit is a JND (Section 2.6) when applied to spatially uniform stimuli. Note, however, that this is the intention for all ΔE metrics and that it is, therefore, important to use as accurate a ΔE metric as is available, which currently is CIEDE2000 (CIE, 2001). It is also worth bearing in mind that perceptibility varies from individual to individual; and as metrics aim to represent the human population's central tendency, there are observers who can distinguish between stimuli with $\Delta E < 1$. In some industries, where even very small color differences are not acceptable, thresholds can even be set below one (e.g. around 0.7 in some textile manufacturing contexts).

For complex originals (e.g. photographic images) the answer to the question about perceptibility comes from studies that have looked at ΔE distributions between the (processed) pixels of image pairs and tried to predict from them whether a viewer looking at them would see a difference. While each of these studies reports slightly different results, the general consensus is that it is some high percentile (e.g. 99th) that correlates better with perceptibility than the mean or median of the difference distribution and that images whose pixel-by-pixel ΔE distribution has a 99th percentile of below about $2.5\Delta E_{94}$ (CIE, 1995) are not distinguishable (Uroz *et al.*, 2002). For further studies into the perceptibility of color differences between complex images, see Stokes (1991), Farnand (1995), Fedorovskaya *et al.* (1997) and Sano *et al.* (2003).

In the second case, where even perceptible differences are acceptable, the acceptability threshold will depend greatly on what reproduction is being judged. In some applications (e.g. fine art reproductions), acceptability thresholds are much lower than in others (e.g. printing of consumer photographs), but there is no reliable published data about what such thresholds are for individual applications. In cases where some perceived differences are acceptable, therefore, it is necessary to perform psychovisual tests to determine acceptability thresholds.

3.3.2 Psychovisual methods

When the desired reproduction property is expressed qualitatively (e.g. Hunt's *preferred* reproduction and the CIE's *subjective accuracy*) its evaluation becomes significantly more complex. Instead of collecting data about the original and reproduction using measuring instruments, it is necessary to get observers to view and judge them. In effect this becomes an attempt to measure either observers' *attitudes* towards originals and reproductions (*psychometrics* – Wikipedia, 2006b), or their *perceptions* of them (*psychophysics* – Wikipedia, 2006c). While psychometrics is likely to be a more correct label for the methods used for visually evaluating the desired properties of reproductions, they are often also referred to as psychophysics in the literature. In the end, the kinds of experiments performed to evaluate gamut mapping visually are somewhere in the overlapping regions of these two disciplines, and here they will collectively be referred to as *psychovisual.*

A key point worth noting here is that while the intention may indeed be to *measure* attitudes or perceptions, what actually happens in these experiments is something far less direct (Yendrikhovskij, 1999). When an observer is presented with a stimulus and asked a question about it, it is not the observer's perceptions that are recorded in an experiment, but their response made on its basis and the basis of the question they were asked. Such a response is not even a direct representation of the percept formed in the observer's mind but a response decided by the observer in a process of interpreting the question they were asked, judging the percept, taking decisions and finally formulating a verbal (or numerical, gestural, etc.) response (Figure 3.6). The resulting relationship between the verbal response and the percept is one like that of *ekphrasis*, i.e. the writing of a poem about a painting or sculpture (Eco, 2003: 158), which means that the percept is transmuted or adapted from one system of representation into another – the response. Hence, instead of being a measurement of perceptions or attitudes, a psychovisual experiment reports the decisions an observer expresses in response to viewing a stimulus and being asked a question. Even though this may be seen as either obvious or facetious, it is important to bear it in mind and not interpret experiments simplistically.

In addition to the evaluation process being much lengthier and more indirect, it also presents a number of challenges that are more dramatic than those of their measurement-based equivalents:

1. *Dependence on specific observers.* Even though measurement results, too, depend on specific instruments, the choice of specific observers can have a much more marked effect here. For example, when asking a group of observers to express their judgments about preferences among alternative reproductions of an original, the results that would be obtained from professional photographers would very likely be quite different from those that a group of estate agents would give. This is a simple consequence of preference being a function of what an observer's needs are: the photographer is likely to want to represent the image they edited on a display in print, whereas the estate

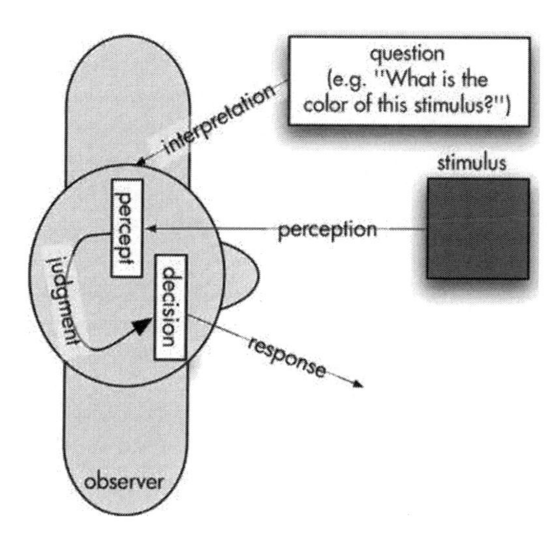

Figure 3.6 Some conceptual stages of a stimulus–response process.

agent wants the print to help in the sales process. What is most preferred in the two cases is often different reproduction properties.

2. *Dependence on specific experimental procedure.* To evaluate psychovisually the extent to which a reproduction has chosen properties, it is possible to design and conduct a myriad of different experiments. The environment in which original and reproduction are shown, the spatial and temporal arrangement of the images, the specific instructions given, the duration of the experiment, the choice of experimental method and the rigor with which the design is followed are only some of the aspects that can be different between experiments that aim to evaluate the same original and reproduction pair from the point of view of the same desired property. If both stimuli and desired reproduction property are the same, then it will be apparent that different results are the consequence of experimental differences. However, the real problem is when different results are reported for different stimuli and different experimental procedures. In this case it is then not possible to know whether the stimuli were different in terms of the chosen property or whether the results mainly reflect differences in procedure. This has plagued gamut mapping research for a long time and it has been addressed only recently by the CIE's publication of *Guidelines for the Evaluation of Gamut Mapping Algorithms* (CIE, 2004d), which we will look at more closely in Chapter 12.

Given the above caveats, the key elements of psychovisually evaluating the desired properties of color reproductions will be looked at next in terms of the following aspects: presentation of stimuli, evaluation method, instructions given to observers and data analysis of raw experimental results. As psychovisual evaluation of color reproductions is a complex subject, the reader is strongly advised to become familiar with more detailed treatments of this topic before attempting the use of this method (Bartleson, 1984; Engeldrum, 2000; Fairchild, 2005).

Presenting stimuli

As should be clear from Chapter 2, the color experiences that an observer will have for given originals and reproductions depend very much on how these are presented. What the viewing conditions are, whether a stimulus is seen in isolation or as part of a set of stimuli, in what order stimuli are presented, whether the observer is adapted to the viewing conditions, whether stimuli have content with which the observer is familiar, whether stimuli that are compared are seen side by side or sequentially and whether they match in angular subtens or not can all play important roles in the resulting experiences and, therefore, affect the responses observers will give. As a minimum, it is necessary to keep these variables constant and to report their states when evaluation results are given, and ideally the choices made for these various aspects should follow a standard (i.e. CIE (2004d) when evaluating GMAs).

Evaluation methods

Given an original, a reproduction (or set of alternative reproductions) and a chosen way of presenting them, there are a number of methods that can be applied to obtain judgments

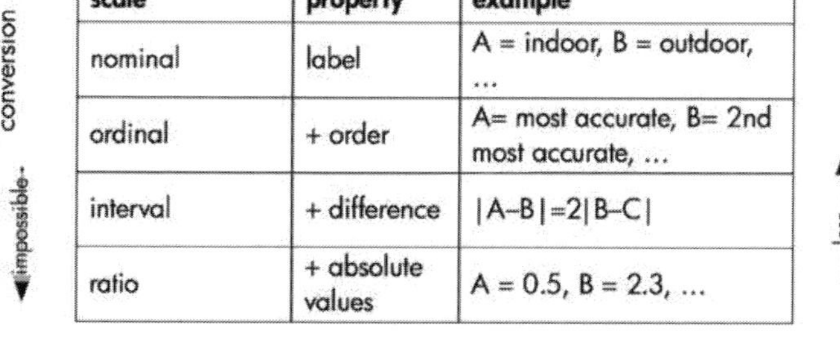

scale	property	example
nominal	label	A = indoor, B = outdoor, ...
ordinal	+ order	A= most accurate, B= 2nd most accurate, ...
interval	+ difference	|A–B|=2|B–C|
ratio	+ absolute values	A = 0.5, B = 2.3, ...

Figure 3.7 Hierarchy of scales.

about reproduction properties. The three most important ones are ranking, category judgment and pair comparison.

Before looking at their details, let us briefly review the hierarchy of scales on which experiments can yield results (Stevens, 1946). Note that in this hierarchy each level also has the properties of all lower levels. At the lowest level, a *nominal* scale is one where stimuli are only labeled (e.g. as 'portrait,' 'business graphic,' 'outdoor,' 'indoor'). As a consequence, they allow for discrimination but there is no concept of order among the labels. An *ordinal* scale then introduces this concept of order, which allows for the expression of one stimulus having more of some property than another. Also, to show the magnitudes of differences requires an *interval* scale, and when there is also a meaningful (i.e. nonarbitrary) origin, the result is a *ratio* scale (Figure 3.7).

Ranking involves presenting several alternative reproductions simultaneously and instructing an observer to indicate their order in terms of the degree to which they exhibit the chosen reproduction property. The result is the assignment of numbers from 1 to n to the reproductions (where n is the number of reproductions that are being evaluated). A value of 1 is assigned to the reproduction that is judged to have most of the chosen property, a value of 2 is given to the reproduction coming second, etc. The data then provide ordinal information about the set of evaluated reproductions, and while it shows which reproduction performs best, second best, etc., it does not express by how much the reproductions differ or how well they do in absolute terms. In other words, the reproductions ranked first and second could be very similar or very different and they could both have a high degree of the chosen property or very little of it in absolute terms. Neither of these aspects would be reflected in the ranks judged in the experiment though. On the positive side, ranking experiments tend to be relatively simple to perform for observers, which contributes to their greater repeatability (i.e. due to their simplicity, observers are more likely to give similar answers when they repeat the experiment at different times).

In *pair comparison*, an observer is always presented with a pair of reproductions and the task is to choose the reproduction that exhibits more of the desired reproduction property. When reproductions are judged to be the same in terms of the chosen property, observers are either forced to choose one anyway (*forced choice*) or are allowed to indicate a match. The result of such pair-wise judgments can then be analyzed to give interval-scale results about a set of reproductions. However, while such data express the relative magnitudes of differences, they cannot give information about their absolute states (i.e. all reproductions could be very inaccurate or very accurate and this would not be shown in the data). A further constraint of the pair comparison approach is the large number of judgments that need to be made to obtain choices made for all pair combinations. For n stimuli, the number of pairs that need to be compared is $n(n-1)/2$, which for $n = 10$ is 45. To address such a requirement of judging relatively large numbers of pairs, methods have also been proposed for analyzing the result of only judging some of the possible pair combinations (e.g. Clark, 1977).

Finally, *category judgment* involves the definition of equally spaced (*equi-interval*) categories along a scale and observers assigning each of a set of reproductions (or even just a single reproduction) to one of the defined categories. For example, a scale of nine *subjective accuracy* categories can be defined by instructing an observer as follows (Sun, 2002: 239):

> Please, give your opinion on a scale of numbers from one to nine where one represent the most accurate image and nine represent the least accurate image you can think of. Use numbers between one and nine to represent equal intervals of accuracy so that the difference between any neighboring categories should be the same.

Here, categories are defined in an absolute way (i.e. they will be interpreted as absolute by an observer rather than having an absolute application to stimuli, which is not possible with psychovisual methods) and the raw experimental data, therefore, can be analyzed to yield ratio-scale results. It is then possible to see whether a reproduction has a high or a low degree of subjective accuracy.

Observer instructions

The instructions given to observers who participate in a reproduction evaluation experiment play a critical role in shaping the experiment's results. What is most important is to ensure that all observers are given the same instructions, which in practice means that instructions need to be read out verbatim to each observer and any deviations initiated by observers need to be noted (e.g. when an observer asks for clarification).

The first kind of instruction that can be given to observers is background information about the task ahead. For example, the desired reproduction property can be explained either using examples of where it would be used or by showing stimuli that embody it in specific ways. How to (or *whether* to!) provide such background training should be considered carefully, as it can easily lead to one of two undesirable extremes. On the one hand, excessive and restrictive training can lead to results that are a consequence of it and that do not express the observers' own views. For example, showing observers stimuli and telling them what degree of pleasantness each has can force on them an interpretation of pleasantness that is not their own but that of the experiment's designer. On the other hand, insufficient explanation of the background can lead to different interpretations by individual observers and, therefore, 'noisy' overall results. A careful balance needs to be struck here, and it is advisable to perform a pilot study where alternative instruction approaches are tested.

Following an introduction, observers need to be given instructions on what task to perform and the wording of such instructions, too, needs to be given careful consideration. For example, a category judgment experiment asking observers to judge accuracy so that the most accurate of the set of reproductions they are shown is given a 1 and the least accurate a 9 will yield interval scale results. Instructing them to interpret category 1 as containing the most accurate reproductions they can imagine and category 9 to contain the least accurate ones then results in data that can be analyzed to give ratio-scale values. For a good example of analyzing the effect of instructions on experimental results, see Fairchild's (1993: 52) discussion of the issue.

Data analysis

The direct, raw results of an experiment are a record of the responses each observer gave when presented with each of the stimuli (or sets of stimuli) that were evaluated. Next, statistical techniques are used to analyze the raw data to understand how a group of observers judged each of the stimuli.

Ranking In ranking, the raw ordinal data can be analyzed by computing the mode of the ranks (i.e. the most frequent rank) assigned to each reproduction by the group of observers. To get a more detailed view (which is highly advisable), the percentages of each rank assigned to a reproduction can also be computed and visualized. Looking at these percentages per rank shows whether or not the response is *unimodal*, i.e. whether the ranks assigned to a given reproduction concentrate around a single rank. If they do not, and the data are, for example, *bimodal* (i.e. having two peaks; Figure 3.8), then the group of observers could be divided into two subgroups with different takes on what makes a reproduction have a certain property. This, in turn, could be the consequence of genuine alternative interpretations or preferences, or an indication of an issue with the experimental procedure (e.g. differences in presentation, instructions or observer profiles). The mode (or any single value) would, therefore, be an inadequate summary for bimodal (or multimodal) distributions. Finally, note that even the results of ranking can, given certain assumptions, be transformed to an interval scale (Handley *et al.*, 2004).

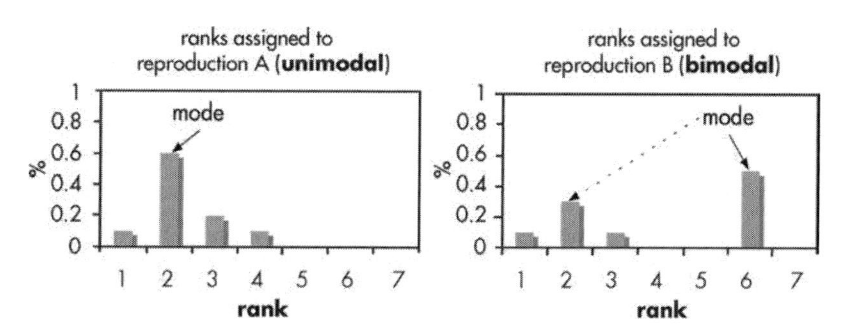

Figure 3.8 Uni- versus bi-modal distributions of rank percentages (rank distributions of two of a set of seven reproductions shown).

Pair comparison Raw pair comparison data show which stimulus was chosen for each judged pair by each observer. Assigning 1 to choice and 0 to nonchoice (and, if a tie is allowed, 0.5 to both of a pair of stimuli) and subsequently summing up the judgments made by each observer then gives a frequency matrix, like that shown in Table 3.2a. Here, the values in each cell show how many times the stimulus indicated by the column was chosen over the stimulus of the row (e.g. B was chosen nine times when compared with A and seven times when compared with C here). Dividing these frequencies by the number of observations then gives a percentage matrix (Table 3.2b); while this is interesting information, it does not express how different the various stimuli were judged to be in the experiment.

To obtain a quantification of differences between stimuli, it is necessary to have a model of what happens in such an experiment, and the first such model is *Thurstone's law of comparative judgment* (Thurstone, 1927). This law (i.e. model) is 'based on the notion that the proportion of times stimulus A will be judged greater than stimulus B is determined by the degree to which sensation A and sensation B differ' (Gescheider, 1985: 152). Furthermore, Thurstone suggests that an observer's response to a given stimulus will result in a range of responses following a *normal distribution* (Weisstein, 2006b; Figure 3.8) on the *psychological continuum*. Therefore, the difference between two stimuli is determined by the distance between the means of their response distributions, which can in general be computed as

$$\bar{\psi}_{\mathrm{B}} - \bar{\psi}_{\mathrm{A}} = z_{\mathrm{BA}}\sqrt{\sigma_{\psi_{\mathrm{A}}}^2 + \sigma_{\psi_{\mathrm{B}}}^2 - 2r_{\psi_{\mathrm{A}}\psi_{\mathrm{B}}}\sigma_{\psi_{\mathrm{B}}}\sigma_{\psi_{\mathrm{A}}}} \tag{3.1}$$

Here, $\bar{\psi}_{\mathrm{A}}$ and $\bar{\psi}_{\mathrm{A}}$ are the means of the response distributions to stimuli A and B respectively, r is the *Pearson correlation coefficient* (Weisstein, 2006c) between the two distributions, σ is the standard deviation and z is the *normal deviate*, or *z-score*, corresponding to the proportion of times stimulus B is chosen over stimulus A. Given a percentage of times that stimulus A was chosen over stimulus B, the corresponding z-score is the distance from the mean (on a scale whose units are the distribution's standard deviation) that corresponds to an area under the normal distribution's curve equaling the given percentage. For example, Figure 3.9 shows a normal distribution with zero mean and unit standard deviation (i.e. the *standard normal distribution*). Here, we find that for an area equaling 0.82 (i.e. 82%), which is the percentage of times that stimulus B was chosen over stimulus A in Table 3.2b, the corresponding z-score is 0.91 (Table 3.2c), and this is also the distance between the two stimuli on the ratio scale obtained using the law of comparative judgment.

As the standard deviations and correlations of response distributions are almost always unknown and cannot even be reliably estimated from experimental data, Thurstone defined five special cases (I–V) that make different assumptions about their states. Of these, Case V is most frequently used and assumes that $\sigma_{\psi_{\mathrm{A}}} = \sigma_{\psi_{\mathrm{B}}}$ (i.e. that responses to both of a pair of stimuli vary equally), and that $r_{\psi_{\mathrm{A}}\psi_{\mathrm{B}}} = 0$ (i.e. there is no correlation between the responses to different stimuli). Given these Case V assumptions, the difference between mean responses can be computed as

$$\bar{\psi}_{\mathrm{B}} - \bar{\psi}_{\mathrm{A}} = z_{\mathrm{BA}}\sigma\sqrt{2} \tag{3.2}$$

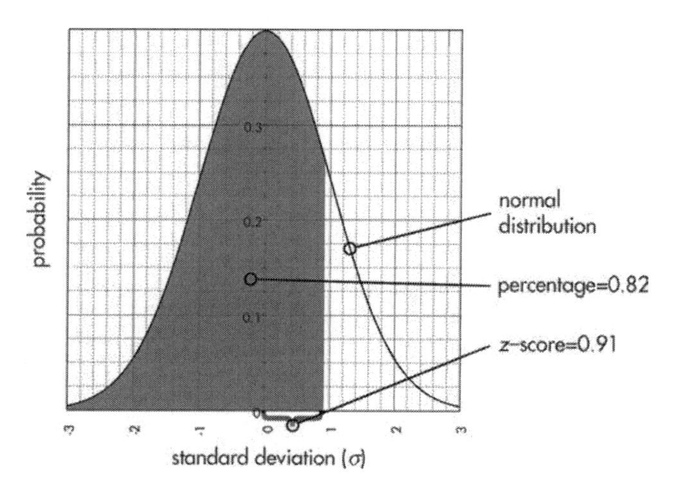

Figure 3.9 The standard normal distribution and the z-score.

Since the result is on a ratio scale, it is independent of scale and σ can, therefore, be set to an arbitrary value (e.g. unity). The ratio-scale scores obtained in this way also have confidence intervals that follow from Case V assumptions, and the 95% confidence interval can be computed as

$$\mathrm{CI}_{95} = z_{\mathrm{BA}}\sigma\sqrt{2} \pm 1.96\frac{\sigma}{\sqrt{N}} \tag{3.3}$$

where N is the number of observations that the z-score is based on.

Applying Case V assumptions to the z-scores shown in Table 3.2c gives the pair-wise stimulus differences shown in Figure 3.10a. Then, to get a single value per stimulus that expresses its degree of exhibiting the property that was judged, the mean of the individual differences can be computed. This expresses the difference of each stimulus from the mean of the set of stimuli. Then, computing confidence intervals according to Equation (3.3) gives the results shown in Figure 3.10b for the 11 observations recorded in Table 3.2. According to these results, stimulus B has most of the judged property, and even though stimulus C has a higher score than A, they are not significantly different from each other at the 95% level (i.e. the score of one is within the confidence interval of the other).

Mosteller, Stevens and Sun later added further cases – Va, VI and VII respectively (Engeldrum, 2000: 96; Sun, 2002) – to cater for other assumptions being made about the response distributions to stimuli in a given experiment or across a series of experiments.

Table 3.2 Examples of: (a) frequency matrix, (b) percentage matrix and (c) z-score matrix.

(a)

	A	B	C
A		9	6
B	2		4
C	5	7	

(b)

	A	B	C
A		82%	55%
B	18%		36%
C	45%	64%	

(c)

	A	B	C
A		0.91	0.11
B	−0.91		−0.35
C	−0.11	0.35	

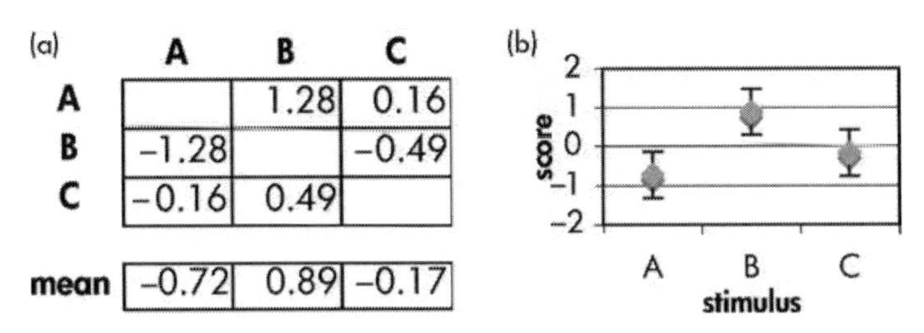

Figure 3.10 Pair comparison results with confidence intervals analyzed using Thurstone's model.

An alternative method of analyzing pair comparison data is based on the *Bradley–Terry* (BT) model (Bradley and Terry, 1952; Bradley, 1984; Handley, 2001). The assumption here is that, when comparing two stimuli (e.g. A and B), the probability that A is chosen over B is $\alpha_A/(\alpha_A + \alpha_B)$, where α_A and α_B are positive-valued parameters representing the degree to which stimuli A and B respectively exhibit the criterion of choice (Hunter, 2004). The probability matrix (e.g. Table 3.2b), therefore, is the result of the parameters (scores) combined in the way shown in Table 3.3 and the aim is to estimate α_A, α_B and α_C given the percentages obtained in the pair comparison experiment.

When evaluating the desired reproduction properties the 'players' are reproductions (stimuli) and 'ability' is the extent to which a reproduction exhibits the chosen property. Given the probabilities (i.e. percentages of A having been chosen over B, like those in Table 3.2b), general maximum-likelihood techniques can be used to estimate α scores of the stimuli evaluated. In other words, estimates of scores can be made computationally to agree with the probabilities obtained from combining the individual observers' choices recoded in a psychovisual experiment.

Fundamentally, the difference between the Thurstone and the BT models is a choice of underlying probability distribution function, which in the Thurstone case is a normal distribution and in the BT case a logistic one. While this difference in framework does not have a significant effect on the scores that are computed for each stimulus, the BT model allows for a more robust estimate of confidence intervals, as they, too, are based on the experimental data. Using Case V of the Thurstone model makes confidence intervals simply a consequence of assumptions made independently of experimental outcomes.

Applying the BT model (Hunter, 2006) to the sample data in Table 3.2b yields the following scores for the three stimuli: $\alpha_A = 0.7$, $\alpha_B = 2.2$ and $\alpha_C = 1.0$. Their confidence intervals can then be obtained by estimating the degrees (likelihoods) to which the estimated scores agree with the probabilities from which they were derived. This is done by computing the negative inverse of the Hessian matrix of the log-likelihood and evaluating it at the estimated scores (i.e. maximum likelihood estimators – MLEs). Taking

Table 3.3 Probability matrix expressed as score ratios.

	A	B	C
A		$\alpha_B/(\alpha_B + \alpha_A)$	$\alpha_C/(\alpha_C + \alpha_A)$
B	$\alpha_A/(\alpha_A + \alpha_B)$		$\alpha_C/(\alpha_C + \alpha_B)$
C	$\alpha_A/(\alpha_A + \alpha_C)$	$\alpha_B/(\alpha_B + \alpha_C)$	

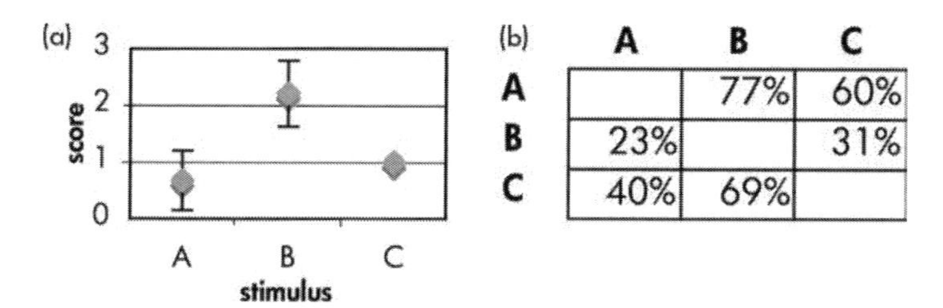

	A	B	C
A		77%	60%
B	23%		31%
C	40%	69%	

Figure 3.11 (a) Scores with confidence intervals obtained using the BT model. (b) Probabilities implied by estimated scores – compare with Table 3.2b.

the square root of the diagonals of the resulting matrix then gives an estimate of each score's confidence interval (Hunter and Lange, 2004); and as the method only considers differences, it will give such estimates for all but one of the scores. The scores and confidence intervals computed for the sample data in Table 3.2b using the BT model are shown in Figure 3.11, where it can be seen that they are indeed similar to results from the Thurstone model (Figure 3.10).

Category judgment The direct result of category judgment experiments is a set of categories assigned to each stimulus by the group of observers, and the aim of the analysis is to provide a single score per stimulus. The analysis can be carried out according to *Torgerson's law* (model) *of categorical judgment* (Torgerson, 1958) or by computing mean category values (Bartleson, 1984; Sun, 2002) from the category judgments made by the individual observers.

The choice of analysis depends on whether it is assumed that observations were made on an equally spaced scale (i.e. mean category computation) or that the scale used by observers is actually not spaced evenly and that even spacing needs to be established by suitably transforming the responses (i.e. Torgerson's categorical judgment model). While the latter is well suited to cases when large numbers of observations are available, it may give erroneous results when applied to small samples, which is often the case when reproductions are evaluated psychovisually. Another way of looking at the difference between these two methods of analysis is that in the mean category case one believes that observers have made their judgments on an evenly spaced scale of categories, whereas the alternative transforms judgments so that their statistics agree with such an assumption.

Using the *mean category* analysis (Bartleson, 1984: 475), the mean of all the category judgments made by the group of observers is computed and confidence intervals can also be obtained by computing the judgments' standard deviation σ. The 95% confidence interval is then computed as $\mathrm{CI}_{95} = \pm 1.96\sigma/N^{0.5}$, where N is the number of judgments from which the confidence interval is computed. Just like with ranking data, it is important here, too, to look at the distributions of individual judgments for each stimulus, as this analysis (or any analysis that computes a single representative from a set of values) is only meaningful for unimodal distributions.

Applying Torgerson's model is somewhat more complex, as it requires transforming the raw category judgments to a scale where they form normal distributions for each sample. Unlike Thurstone's law of comparative judgment (on which Torgerson's law of categorical judgment is based), it is the relative position of stimulus scores with respect to *category boundaries* on a psychological continuum rather than with respect to one another that is assumed here. The aim of the following analysis, therefore, is to compute such a set of category boundaries with respect to which the experimental data exhibit the assumed normal distributions. In other words, making some intervals between neighboring categories narrower and others wider will result in normal distributions for the recorded judgments, whereas keeping the inter-category intervals of equal width (which is what observers were instructed to use) will not.

The following method is used to obtain the scores of n stimuli from category judgments made using m categories (Bartleson, 1984; Lo, 1995):

1. An $n \times m$ *frequency matrix* is calculated from raw experimental data, where each column contains the frequency of each stimulus being judged as belonging to the corresponding category.
2. As the law of categorical judgment is concerned with the percentages of times that a given stimulus (in this case image) is assigned to a position below a given category, an $n \times (m - 1)$ *cumulative percentage matrix* is calculated.
3. The cumulative percentage matrix is transformed into a *z-score matrix.*
4. An $n \times (m - 2)$ *difference matrix* between adjacent columns is then calculated followed by obtaining the mean for each column.
5. Next, *boundary estimates* between the m categories are determined by setting the origin (which is between categories m and $m - 1$) to zero and adding adjacent mean values from the difference matrix.
6. *Scale values* in an $n \times (m - 1)$ matrix are then calculated by subtracting z-scores (step 3) from boundary estimates (step 5). This is done as the z-scores represent the distance from each of the category boundaries. For each reproduction, the mean scale value is then computed. Note that the stimulus exhibiting the judged property most will have a rank of one and the one having least of it will rank nth.

Confidence intervals of scores obtained in this way are calculated in the same way as for the Thurstone analysis of pair comparison (Equation (3.2)) and are a consequence of assumptions rather than based on experimental data. An example of the different results obtained from the same raw experimental data using the two methods of analysis is shown in Figure 3.12. The Torgerson analysis here suggests that in this experiment the observer responses are consistent with intervals that are not evenly spaced, and the implicit interval boundaries are also shown.

Given the techniques discussed here – ranking, pair comparison and category judgment – an important question is the degree of agreement that can be expected between the results obtained using the different experimental methods. For example, how do the ranks from a pair comparison evaluation relate to the ranks from a ranking or category judgment experiment? An example of such a comparison is the correlation values reported previously (Morovič, 1998: 152) between the results of two experiments – a pair comparison and a category judgment one – evaluating the pleasantness of the same set of seven GMAs. To compare the two results, the coefficient of determination r^2 (i.e. the square of the correlation coefficient) can be used, as it expresses the percentage of variation in one set of data that can be accounted for by another set (Wikipedia, 2006d). An $r^2 = 1.0$ means that all of one set of data can be obtained from the other and that they

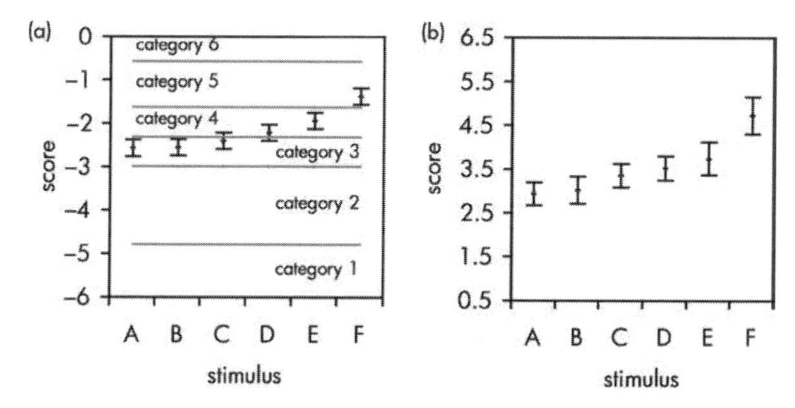

Figure 3.12 Scores and 95% confidence intervals obtained using (a) Torgerson's law of categorical judgment and (b) mean category analyses.

differ only by scaling and a shift; $r^2 = 0.7$ means that 70% could be accounted for but 30% could not. Here, the overall agreement between the two methods was found to be very strong, i.e. $r^2 = 0.90$, except for some original images, where the judgments made about their reproductions came out fairly differently ($r^2 = 0.57$).

Inter- and intra-observer differences Finally, it is also worth considering the questions of how well an observer can repeat their own judgments over time (i.e. *intra-observer differences*) and of how well an observer's judgments agree with that of the whole set of observers who took part in an experiment (i.e. *inter-observer differences*).

Two alternative approaches can be taken in both cases. First, it is possible to compute the difference of a particular judgment from either the central tendency of all of an observer's judgments or from those of the whole set of observers. Applied to category judgment, this would involve the computation of differences between individual judgments and a mean judgment and then taking the mean of the resulting set of differences. However, when dealing with ranking or pair comparison data, such an approach is inappropriate, as distance is not meaningful for this type of data.

Second, the statistical significance of the difference between one set of responses and another (e.g. the representative of either all of an observer's responses or those of the whole observer group) can also be determined. For ranking experiments, the only applicable type of statistical test is a nonparametric one (i.e. one that does not assume that samples are from populations with normal distributions); as the comparisons are made between sets of values where there is a correspondence between their members (i.e. the first value of one set corresponds to the first value of the other set), the *Wilcoxon rank-sum* test (Wild and Seber, 1999) is recommended.

This test 'consists of combining the two samples into one sample . . ., sorting the result, assigning ranks to the sorted values (giving the average rank to any "tied" observations), and then letting T be the sum of the ranks for the observations in the first sample' (TAMU, 2006). If the two populations have the same distribution, then the sum of the ranks of the two samples should be close to being the same. The magnitude of this sum then expresses how different the two samples are. While the Wilcoxon rank-sum test is also applicable to analyzing the results of category judgment experiments, inter- or intra-observer differences are more difficult to assess for pair comparison data.

For further detail on the analysis of psychovisual experiments, see Bartleson (1984) and Engeldrum (2000); and for a general discussion of relevant statistics, consult Wild and Seber (1999).

3.4 CASE STUDY: EVALUATING PRINTED REPRODUCTIONS OF A DISPLAYED IMAGE

To get a more practical understanding of desired reproduction properties and their evaluation, this section will apply the theory presented in this chapter to the specific case of wanting to quantify the degree to which alternative printed reproductions match an original on a computer display in a photographic studio.

As the two media, i.e. display and print, have different color gamuts and are viewed under nonstandard viewing conditions, the most appropriate 'match' reproduction property is subjective accuracy, since none of the measurable properties would take the gamut difference into account and might also struggle with being representative with respect to the actual viewing conditions in the photographic studio even for in-gamut colors.

With the desired reproduction property chosen, the next stage is to set up a procedure for its evaluation. As the viewing conditions are predetermined, the key to doing the evaluation reliably is to keep them constant throughout the evaluation process; while telespectroradiometric measurement would be ideal, it is rarely available in this type of setting. Alternatively, it is advisable to at least recalibrate the display on which the original is shown (if calibration is normally used) and to measure the level of ambient illumination using a lux meter. In the known viewing environment, the next step is to give the observers (e.g. colleagues, clients, etc.) instructions on what the evaluation will involve. Doing this in the environment where the evaluation will take place also gives time to the observers' visual systems to adapt to the viewing conditions.

As category judgment is a relatively quick technique that can also provide information about how accurate the reproductions are in an absolute sense, it will be used here; the following instructions are an example of what can be read out to the observers: 'Please, look at the displayed original and its alternative reproductions and for each reproduction judge how accurately it represents the original. Give your opinion on a scale of numbers from one to nine where one represent the most accurate image and nine represent the least accurate image you can think of. Use numbers between one and nine to represent equal intervals of accuracy so that the difference between any neighboring categories be the same.' The observers, who can either individually or as a group view the original and its reproductions (presented in the way in which they are typically viewed in the studio), can then be given a questionnaire in which to record their category judgments for each reproduction.

The results of the observers' judgments can then be analyzed first by viewing their distributions for each reproduction to detect any multimodality and second by computing mean category values (and confidence intervals) again for each reproduction. The resulting subjective accuracy scores will give information about how the given set of reproductions is considered by the group of observers who took part in their evaluation.

Finally, it is worth pointing out that such a structured psychovisual evaluation is not always the most appropriate approach to follow, and while it is clearly suited for academic research, its application to professional practice is often unnecessary. More specifically, if what matters most is how the desired reproduction property is judged by a single person (e.g. a professional photographer or artist or their client) or if some reproductions are clearly and far inferior or superior to their alternatives, then evaluation by a group of observers will not give new or more precise information and becomes redundant.

3.5 SUMMARY

What we want reproductions to be like when making them is a key decision that needs to be made explicitly. This will allow for a reliable determination of how good our reproductions are. Given a choice of desired reproduction property, the degree to which alternative reproductions exhibit it can be evaluated either using measurement-based or psychovisual methods, depending on the chosen property, and an overview of such methods was given here.

Finally, note that the discussion of both types of evaluation method was applied here to assessing reproductions of a single original rather than of GMAs. The additional challenges that algorithm (rather than single image) evaluation brings will be discussed in Chapter 12.

ACKNOWLEDGEMENT

A portion of text in paragraph 3.2.5 has been reproduced by permission of the International Color Consortium (ICC).

4
Color Reproduction Data Flows

Achieving a chosen desired reproduction property (Chapter 3) is the role of color reproduction or color management systems, and their internal data flows will be looked at here with the aim of locating and delimiting the gamut mapping stage in them. A view will also be provided of what else (besides gamut mapping) is involved in color reproduction. To arrive at where gamut mapping fits in, we will first look at the conceptual stages of color reproduction and then review alternative approaches to color management that implement them in practice.

Note that the discussion here will also be restricted to digital methods and to those aspects of color management architectures that are relevant to gamut mapping. The reader interested in a more extensive coverage of color management can consult Giorgianni and Madden (1998), Sharma A (2003), Fraser (2004) or Morovič and Lammens (2006). Detail on analogue approaches can be found in Yule (1967) or Hunt (1995b), with greater focus on printing provided by Mortimer (1991) or Kipphan (2001).

4.1 DEVICE COLOR SPACES

Before looking at the stages of color reproduction, the question of how the color output of imaging devices can be controlled or of how such devices report captured color information will be addressed first.

With output devices (e.g. printer, displays, projectors) the final stage of control is in the form of data that determine the intensity of effect from a given colorant (or other method of generating a visual stimulus, such as the filters applied to the modulated output of the backlight in LCDs). For example, a display with phosphors emitting red, green and blue lights will, at the last digital stage of the process, contain a set of three values that will result in corresponding intensities of output from the three colorants. RGB values of [0, 128, 255] (on a 0–255 scale) will then result in the least possible intensity of red

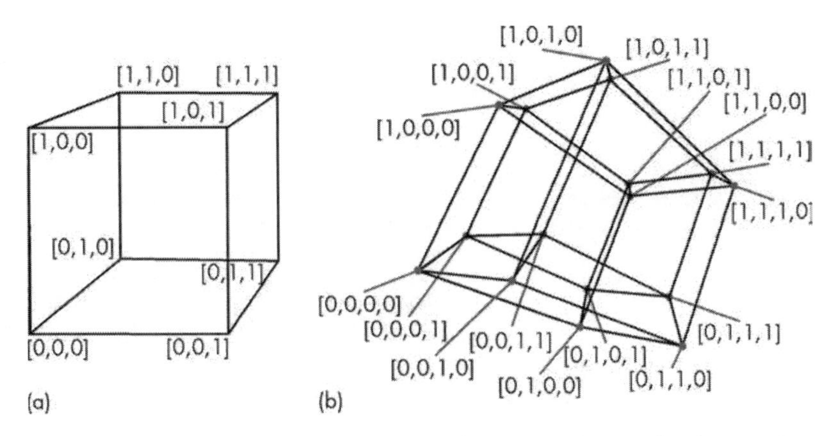

Figure 4.1 Two-dimensional projections of valid color space values in the device color spaces of (a) three- and (b) four-channel devices (values are normalized to the [0, 1] range in each dimension; note that all edges at a vertex are orthogonal to each other).

output, the greatest possible intensity of blue output and an intensity of output from the green channel that is somewhere between that channel's least and most intense outputs. The reader is advised to treat such values with caution and not to rush to a conclusion that in most cases would be incorrect, i.e. that the green channel outputs at half its maximum intensity. Values like these are referred to as *device color values*, which can also be interpreted as coordinates in a space that has as many dimensions as the device has colorants. Furthermore, as the values in these spaces are valid only in specific intervals in all of their dimensions, typically device spaces are cubes or hypercubes (Figure 4.1). Such spaces are referred to as *device color spaces* and will be prefixed by a lower case 'd' in this book. dRGB, dCMYK (*c*yan, *m*agenta, *y*ellow and blac*k* used in printing), dRGBW (the RGB plus white channels of a projector) and dCMYKcm (CMYK plus diluted versions of the C and M inks) are examples of such spaces. The question of how to relate their values to output intensity will be addressed later in this chapter.

In the case of input imaging devices (cameras, scanners, etc.), device color spaces express the intensity of signal recorded in response to light passing through the device and incident on its sensors. A dRGB of [0, 128, 255] here expresses that no more than the least amount of detectable light was incident on the sensor through a red filter, that no less than the greatest amount of detectable light was received through the blue filter and that something intermediate happened for green.

Coming back to color reproduction, it will be obvious that device color values form the terminals of any such data flow. The starting point of the color reproduction data flow will be the source device's device color values and the end point will be values generated for the destination device's device color space.

Note, however, that while device color values are indeed the endpoints of the *digital* workflow and proper to color imaging devices, color reproduction is judged by viewing objects either generated by imaging devices or captured by them. Therefore, the extreme points of the color reproduction process are not sets of digital data in device color spaces but are a viewer's *experiences in response to analogue, visual stimuli.* These stimuli, generated or captured by imaging devices, are present in corresponding *color reproduction media* (Section 2.8.3) that are viewed in a particular way and under particular conditions and result in the visual experiences that are judged when their owners are asked to evaluate color reproduction quality.

For example, dRGB values sent to a projector will result in a particular stimulus being created by projecting a mixture of red, green and blue lights onto a given screen, determined by those dRGBs. The color experiences had by a viewer will then depend on the resulting stimulus, but also on a variety of other factors (e.g. ambient light, viewing geometry – see Section 2.10). Note, then, that changes to dRGB values will only affect a small part of the factors that together result in color experience and that the best dRGB values will depend on the entire color reproduction setup that includes how the result is viewed, rather than only on the projector device to which they are applied. Different dRGB will be needed if, for example, an attempt is made to maintain the visual result across different states of the other factors (e.g. different screen, different ambient illumination).

A key property of device color spaces is also that their coordinates do not, by themselves, have any colorimetric or color appearance meaning, but only attain it as a consequence of being applied to a specific device, with specific settings and components and viewed in a specific way. What this means is that dRGB = [10, 20, 30] does not encode any particular color experience, but instead only specifies values in three channels of a device that has (nominally) red, green and blue channels. Applying these same RGBs to different displays or printers will result in different color output and they would be the result of different colors being captured when output by different scanners or digital cameras. Device color space values, therefore, are akin to quantities in recipes, where the ingredients are not specified, but only labeled by category. A recipe stating that a drink is made by mixing 20 units of a thin liquid and five units of a thick one is about as much use as having a color specified by device color values.

For output imaging devices, device color spaces can be advantageous, though, in that they address all of the range of stimuli that a device can generate and nothing but that range (i.e. they are bijective with respect to the range). This, in turn, makes them very well suited for encoding colors generated using a given device, since the chosen bit-depth is used for encoding only the device's gamut and nothing but that gamut. In contrast, encoding an oblong in a color appearance space like CIECAM02 Jab or CIELAB is very wasteful, since many of the encoding's values do not even refer to possible colors, never mind about those addressable by a given device. To illustrate the degree of encoding inefficiency of color appearance spaces, Braun and Spaulding (2002) have shown that only 41% of encoding a CIELAB oblong delimited by *LAB*s of [0, −128, −128] and [100, 127, 127] (i.e. the ICC *profile connection space*) is used for representing possible surface colors and an even smaller proportion (18%) encodes colors of typical surfaces.

Of increasing relevance is also the potential to provide a layer of abstraction above a device's color channels by setting up a *virtual device color space* on top of the one that is *native* to the device itself. For example, dRGB can be used to address all of the color stimuli that can be generated using: a display with red, green and blue phosphors (where dRGB is also the native device space); a projector that, in addition to RGB channels, also has a white channel (where dRGB is not native); a printer that uses cyan, magenta, yellow and black inks or also other inks like dilutions of some of CMYK. Each of these devices deals with the assignment of native device color values to values in the virtual (typically lower dimensional) device color space, to make the dRGB space addresses all of its color range regardless of the device's set of colorants. This allows for providing the same device color interface regardless of a device's actual set of colorants. The flipside, though, is that some of the control that is possible when addressing colorants directly is given up.

Analogously, dCMYK can also be used as a device color space, e.g. printers that use CMYKcm colorants often do not accept dCMYKcm inputs. Instead, they insist on dCMYK, which they internally transform to dCMYKcm. Note also that device color spaces can be encoded with different bit-depths, and while 8 bits (giving $2^8 = 256$ levels) is most common, some devices have internal encodings that use up to 16 bits (i.e. $2^{16} = 65536$ levels) per channel.

In summary, device color spaces are a means of controlling or encoding device output for output and input devices respectively and are by themselves not suitable for color specification or communication. What they provide, though, is an efficient addressing of a device's color abilities and, as such, are used as end and starting points of color reproduction data flows.

4.2 CONCEPTUAL STAGES OF COLOR REPRODUCTION

Given the above discussion of the end points of color reproduction, the coverage of desired reproduction properties in Chapter 3 and the overview of factors involved in viewing and comparing colors in Section 2.10, let us turn now to the conceptual stages involved in translating color data from a source device into color data for a destination device (Berns, 1992; Morovič, 1998: 142). Here, we can identify three larger conceptual blocks:

1. predicting what the source (i.e. original) looks like;
2. making changes to the original appearance that are either wanted in themselves as a consequence of the chosen desired reproduction property or necessitated by the gamut mismatch between source and destination;
3. predicting destination (i.e. reproduction) device color values that are to match the appearance desired after the first two blocks (Figure 4.2).

Blocks 1 and 3 here are inverses of each other, since one predicts appearance from device color and the other predicts device color from appearance. Note also that appearance prediction has two stages, the first relating to a device (i.e. linking device color to corresponding stimulus) and the other relating to the viewing of a stimulus (i.e. linking stimulus to color experience). Finally, block 2 also has two distinct conceptual components, the first aiming to better achieve desired reproduction objectives and the second making changes necessitated by gamut differences.

Note also that the process described here is a conceptual one: it shows the functionality that an actual color reproduction process needs to provide, but that it may deliver in (at least superficially) other ways. In other words, having a color appearance model component in the above process does not mean that every color reproduction system needs to have an explicit color appearance model stage, but that it needs effectively to deliver the taking into account of color appearance factors. Also, the placing of color enhancement after the color appearance model stage does not mean color enhancement is to be applied

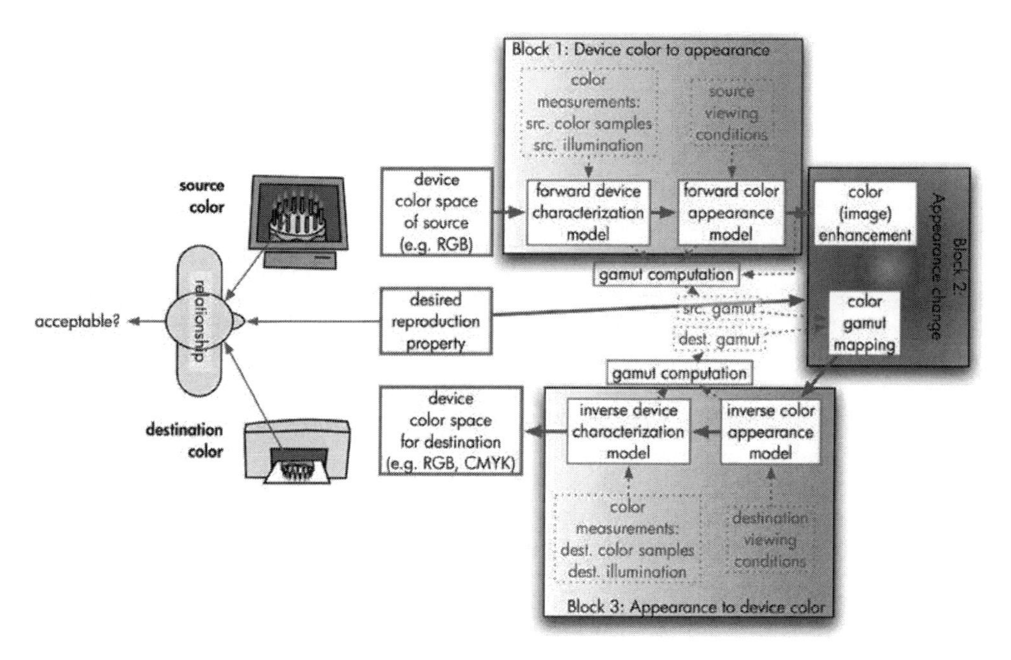

Figure 4.2 Conceptual stages of color reproduction.

only to color appearance attributes, but that it conceptually follows a taking into account of color appearance.

4.2.1 Device characterization and calibration

Given a set of device color space coordinates (e.g. dRGB = [10, 20, 30]) from a source device, the first step of the color reproduction process is to determine what stimulus the device will generate when receiving it as an input (for output devices), or what stimulus the device was presented with when recording it and providing it as an output (for input devices).

This translation of device color space inputs to stimulus outputs is the role of *device characterization* and a wide variety of solutions are reviewed in Bala (2003) with great clarity. In all cases, characterization models have their parameters determined from pairs of (a) device color inputs and (b) color measurements of corresponding stimuli. Given these parameters, a characterization model can perform two functions:

- For a device color input, predict the color stimulus that would be measured if it were sent to the device (e.g. given dRGB = [10, 20, 30], predict CIE *XYZ* tristimulus values that would be measured were it displayed on an LCD) or the color stimulus that the device would have to be presented with to record it (e.g. what CIE *XYZ* or spectral reflectance measurements would a surface have to give if a given scanner recorded dRGB = [10, 20, 30] for it?). This direction of prediction is referred to as *forward*.
- For a stimulus that is to be generated using a device, predict what device color inputs to send to it (e.g. given CIE *XYZ* = [30, 20, 10], predict what dCMYK values to send to a

printer to match it). Note that this direction (the *inverse*) is less useful for input devices, although it, too, can be computed.

A simple example is the gain, offset, gamma (GOG) model developed by Berns *et al.* (1993a,b) for characterizing CRT displays (CIE, 1996). *XYZ*s measured from a CRT are predicted from device RGBs sent to it by performing two transformations: the first takes dRGBs and transforms them to RGBs that are luminance-linear (i.e. whose values have a linear relationship with resulting luminance measured on the specific display) and the second is a 3×3 matrix multiplication that transforms the luminance-linear RGBs to *XYZ* (compensating for the difference in primaries between the chosen display and the *XYZ* system). The parameters of these two stages of the dRGB to *XYZ* transformation can be computed from measuring the display's primaries and a sequence of samples from the $dR = dG = dB$ ramp. Having computed parameters, the model can also be inverted, i.e. *XYZ*s are first multiplied by the inverse of the 3×3 matrix from the forward direction and then the inverse of the luminance linearization is applied to the result to get to dRGB. For this model to perform well the following assumptions made by it need to hold for the device it characterizes (Brill, 1992):

1. spatial independence – the output at a particular location depends only on that location's input (i.e. neighboring pixels do not affect each other);
2. channel independence – the output of a channel (e.g. red) is independent from the input to the channels at that same spatial location;
3. approximate homogeneity – spatial variation is negligible (i.e. the whole CRT can be calibrated from averages of a few pixels);
4. channel constancy – the relative SPD is independent of digital-to-analog converter (DAC) values applied to the electron guns (this implies constant chromaticity of phosphors);
5. additivity – the output of two channels activated together is the sum of their individual outputs.

Given that these assumptions are likely to hold only for some CRT displays, the model is not applicable to other devices. Models of other types of device, like printers, projectors, scanners and digital cameras, are analogous though, in that it is always measurements that are used for determining parameters and that the model attempts to have behavior like that of the device being modeled.

The accuracy of characterization models can be determined just like the accuracy of color reproduction discussed in Section 3.3.1. Instead of samples being taken from a specific original, though, the starting point is a set of samples from the entire gamut of a device. These samples, e.g. in the device color space, are then taken and the model is used to predict CIE *XYZ*s for them. The result can be compared with actual measurements made for the same device color samples and color difference can be computed to express accuracy.

In addition to the level of accuracy achievable using a given characterization model, a strong condition to its predictive powers is the state in which the device is to which it applies. If the device has not changed since measurements were made from which model parameters were determined, then all is well and the only limiting factor is the degree to which model assumptions match the device's actual behavior. However, if the device has changed – whether because of changes to its components (e.g. ageing), to its environment (e.g. change of temperature, relative humidity), or to its settings (e.g. having a display's brightness setting increased) – then the relationships that the model attempts to represent may no longer hold, and while its predictions might have worked in the original state, they now cease to apply.

To address the potential mismatch between the state in which a device has been characterized and the state in which it is at a later point in time, the process of *device calibration* is used. Simply put, the role of device calibration is to take a device in whatever state it is and restore it to a predefined state. Here, the predefined state can be a standard one (e.g. sRGB (IEC, 1999) for displays) or the state of the given device at an earlier time. For a more detailed discussion of device calibration, see Bala (2003).

4.2.2 Color appearance model

From a description of the source stimulus, a *color appearance model* (CAM) is used next to predict its appearance (Fairchild, 2005). This is done on the basis of measurements of a stimulus plus measurements of and information about its viewing conditions. Examples of information that CAMs use are:

- the luminance of the surround (i.e. the part of the field of view around a stimulus), to predict dynamic response;
- reflectance of the background, to predict simultaneous contrast effects;
- the tristimulus values of the adapted white and information about whether adaptation is partial or complete, to predict chromatic adaptation and information about the type of viewing environment (e.g. average – stimulus and surround are similar in luminance, dark – stimulus has significantly higher luminance than surround).

While there are several available models for predicting color appearance, currently the most advanced one, able to predict the appearance of color stimuli and suitable for use in color reproduction, is CIECAM02 (CIE, 2004b). With CAMs, the forward direction refers to predicting appearance from stimuli (plus viewing condition information) and the inverse direction is that of predicting stimuli from appearance and viewing condition information. Work is also under way on extending CIECAM02 to take into account some spatial phenomena and make it perform better for complex images (e.g. photographs). Here, Fairchild and Johnson's (2004) *iCAM* model is the most prominent example to date. Furthermore, the CIE has a technical committee (TC8-08) on *Spatial Appearance Models* where work in this area is promoted and coordinated.

4.2.3 Color and image enhancement

Given the appearance of the source color under its viewing conditions, it is in some cases desirable to alter it before considering its reproduction using a destination device. Note that, by definition, this stage is not used if the reproduction objective is some form of accuracy (e.g. CIE subjective accuracy – Section 3.2.3), as there it is the appearance of the source color that is the reproduction process's aim.

However, when preferred reproduction is wanted, there can be reasons for altering source color appearance. One such change that can lead to more preferred reproductions is a *hue shift*, where the source hues are moved based on the primaries and secondaries of the source and destination media that makes them more similar (Johnson, 1992). Note that a 'primary' of an imaging system is a color obtained by fully applying one of its colorants (e.g. a print's 100% yellow, a display's 100% red) and a 'secondary' is a color obtained by fully applying two of a medium's colorants (e.g. printing 100% of both yellow and magenta, displaying 100% of both red and green). The importance of these colors is that they play key roles in determining the shape of a medium's gamut. For example, if a display's yellow secondary is greener than the yellow primary of a printed destination medium, then keeping hue unchanged can result in a reproduction of a bright, pure source yellow as a darker, less chromatic greenish yellow. If, instead, the source hues are changed to move the source towards the destination, then the reproduction can preserve more of the source's brightness and purity at the expense of some hue change.

Another type of color enhancement is to make colors look more like their *ideal prototypes*, which applies to memory colors in particular (e.g. the colors of skin, grass, sky, etc.). Here, the preference is often for a representation by specific color appearances rather than for matching a particular original (Yendrikhovskij, 1998). The color of grass is typically preferred to have a certain appearance, and if what was in the original differs then a match is not desirable. Another aspect of making images more preferable is to adjust their colorfulnesses to an optimal level, regardless of what there were in the original (Yendrikhovskij *et al.*, 1998; Boust *et al.*, 2005); and the same can also be done for contrast (Wen *et al.*, 2001).

Color adjustments can also depart from photorealism altogether and attempt a mimicking of *artistic styles* such as watercolor, oil painting, pop art, pointillism, etc. and there are software applications that provide the automatic generation of such effects. Changes are made both to color and spatial properties of an original to make it imitate a chosen style (e.g., Adobe® Photoshop®, Corel® Paint Shop Pro®, Virtual Painter®, etc.). Finally, enhancements can also be of the form of *defect removal* – e.g., compensation for incorrect exposure for photographic images, removal of noise, sharpening, red–eye removal, etc. and there is a large body of literature on such techniques.

Note that enhancements to source content can also be performed in color spaces other than those of color appearance models (Holm, 2005). However, in any case their final output can be expressed in color appearance terms and serve as the input to the next stage of the process.

4.2.4 Color gamut mapping

Given the color appearance that is desired in the destination (i.e. either the appearance of the source or a modified, enhanced version of it), the next step is to ensure that it can be matched. Therefore, it is first necessary to know at least the destination gamut (and in some cases also the source gamut), and a means of determining it is required. Based on the gamut(s), a transformation (*color gamut mapping*) is applied to all source colors to make them end up inside the destination gamut.

4.2.5 Completing the process

Starting from source device color values, the process has taken us to a color appearance that is desired in the destination medium and it is next necessary to determine what stimulus, under the destination's viewing conditions, has that appearance. A color appearance model is used in the inverse direction and results in the stimulus that is to be produced using the destination device. The characterization model of the device that is to be used for generating it is used again in the inverse direction to predict the appropriate device color inputs, which are finally sent to the destination device. The two colors (or color images) – i.e. the source, which is the starting point of the reproduction process, and the destination, which is its end – are then viewed and their relationship is judged with respect to the chosen desired reproduction property.

Given these conceptual stages of color reproduction, let us next look at alternative color management architectures that embody them with different degrees of modularity and control.

4.3 CLOSED-LOOP COLOR MANAGEMENT

The most direct way of getting from a source to a destination device color space is to set up a direct transformation, which in its simplest form would consist of a look-up table (LUT) where destination color values are stored for a regular sampling of the source color space.

Setting up such an LUT between a pair of imaging systems can be done by generating a uniform sampling of the device color spaces of both and then manually picking a device color combination from the destination's sampling for each of the source samples. For example, an LUT between the dRGBs of a display and the dCMYKs of a printing system could be built by printing out a uniform 9^4 sampling of print dCMYKs, then viewing each of a 9^3 sampling of display dRGBs and picking one of the dCMYK samples to represent it in print. The result would be a list of source dRGBs from a regular grid in dRGB and a corresponding destination dCMYK for each grid point. Having such an LUT then allows for an assignment of destination device color values to source device color values by taking each source color (e.g. each pixel of a source image), looking it up in the LUT and either finding that color directly or using neighboring LUT colors and interpolation to generate a corresponding destination color (Figure 4.3).

While this is tremendously laborious and may sound far-fetched given current color management approaches, such methods were in fact successfully used in some proprietary, closed-loop color systems made by companies like Crossfield and Scitex in the 1980s and 1990s.

Examples of interpolation methods are tetrahedral and trilinear in three dimensions and their equivalents in higher dimensional spaces (e.g. quadrilinear interpolation in four dimensions). Tetrahedral interpolation, for example, takes a given source device color and

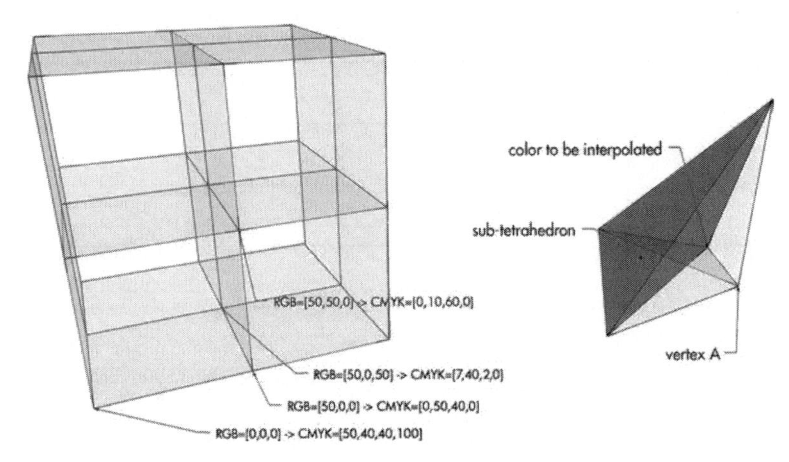

Figure 4.3 Left: example of a 33 LUT from dRGB to dCMYK; right: the tetrahedron from the LUT on the left that has values shown for its vertices and the sub-tetrahedron in it obtained by substituting vertex A by a color to be interpolated.

finds four entries in the LUT that form a tetrahedron around it (e.g. the smallest one). Given such four source-device colors (e.g. those with their values shown in Figure 4.3), the resulting tetrahedron would be subdivided into four sub-tetrahedra – each consisting of one of the original tetrahedron's faces and the device color to be interpolated (shown shaded in Figure 4.3 (right)). The volume of each of these four sub-tetrahedra can then be used as the basis of computing a weighted sum of the destination device color values stored for each of the interpolating tetrahedron's vertices. Specifically, the volume of the sub-tetrahedron formed by substituting vertex A of the interpolating tetrahedron by the color to be interpolated is divided by the interpolating tetrahedron's volume and forms a weight for the destination color corresponding to vertex A (Figure 4.3 (right)). This method is just one of many interpolation techniques used, and the reader interested in their choice and/or implementation is advised to consult Kasson *et al.* (1995) or Bala (2003).

While the color management approach described here can give very good results if set up by skilled professionals, it is practically limited to closed systems, i.e. where source and destination are known and controlled by the same color reproduction system and where only a small number of systems are involved. The first constraint means that it is not suitable for cases where only one of source or destination is known and color communication is necessary with unknown systems. The second constraint, i.e. being limited to a small number of systems, derives from the combinatorial explosion due to the use of direct source–destination mappings. Given n color reproduction systems, there are $n(n-1)/2$ pair combinations, which means that it is also that number of transformations that are needed if each system is either a source or destination, or twice as many if each can be both source and destination. Add to that the fact that a different transformation is needed for each state of a reproduction system (e.g. a printer using different substrates requires as many transformation as substrates and these need to be redone from time to time, since imaging systems and their components change with use) and even for simple setups the number of transformations quickly runs into the hundreds.

As can be seen, all of the conceptual stages of the process are rolled into a single transformation here and no explicit and direct control over its components is provided. Device characterization, compensating for viewing condition and gamut differences and any desired enhancements are all attempted via the setting up of source and destination device color pairs. What this means is that it is very difficult for an amateur to attempt such a task and the high quality results that can be obtained are the fruit of professional experience earned over many trial and error cycles.

4.4 sRGB COLOR MANAGEMENT

Instead of building mappings between device color spaces, the next level of color reproduction modularity is to establish a colorimetrically defined intermediate color space in the process. The transformation from system A (e.g. a scanner) to system B (e.g. a printer) then proceeds from A's device color space to the intermediate color space and from there to B's device color space. The intermediate color space, therefore, can also act as the space in which color information is encoded for communication. In other words, system A contains a transform from its device color space to the intermediate space and then outputs the intermediate space's values for use by other devices. System B is then set up to receive the intermediate color space's data and internally contains a transformation from that space to its own device color space.

The immediate benefits of this strategy are that color communication only requires either as many mappings to or from the intermediate space as there are systems involved or twice as many – depending on whether systems are either source or destination or whether they can serve as both. Adding a new system can be done without knowledge of the other systems it needs to communicate with, and the mapping(s) with the intermediate space can be set up by hand or involve manual adjustments of some automatically determined starting point.

With this approach, the most popular intermediate space is undoubtedly the *sRGB* color space (IEC, 1999). Unlike dRGB spaces (notice the plural!), the sRGB space (singular) is defined in colorimetric terms and, by also having viewing conditions as part of the specification, each combination of sRGB values represents a specific color appearance. Furthermore, the colorimetric definition of sRGB was chosen to correspond to the colorimetry that would be obtained from a typical CRT display. Given sRGB values, CIE *XYZ* tristimulus values are obtained by first transforming each of the three dimensions of sRGB via one-dimensional transformations and then applying a 3×3 matrix multiplication to the result (for details see IEC (1999) or Stokes *et al.* (1996)). In other words, the dRGBs of a typical CRT display, set to a specific white point, with specific brightness and contrast settings and viewed in a specific way would result in colorimetry close to the sRGB specification. The advantage is that if sRGB values are displayed on a typical display as device RGBs then the result would not be far off their correct reproduction.

While the benefits of sRGB are obvious when displays are involved, the fact of having an exact colorimetric interpretation also makes it suitable for color communication between other devices (e.g. digital camera to print). One limitation that sRGB approaches have, though, is that they do not provide the user with choices about color reproduction alternatives, such as the choice of desired reproduction property, and this makes them more suitable for use in consumer imaging products as opposed to those used by some imaging professionals. Note that this is not to say that sRGB workflows cannot give professional-quality results (in fact, they often do), but only that they restrict control.

Not providing alternative choices of desired reproduction properties (e.g. accurate versus preferred) also makes the development of gamut mapping for such workflows a much greater challenge than if several reproduction alternatives are given. This is because a single mapping would need to provide good results when applied to different types of content (e.g. photographs, business graphics, corporate identity colors etc.), each of which would naturally require either some form of accuracy (e.g. the corporate identity colors) or preference (e.g. the photographs). As the different types of content occupy overlapping parts of color space (i.e. part of an original photograph can be of the same color as an original corporate identity color), designing a single mapping that provides good results for each of them is quite a challenging and delicate task.

In summary, sRGB color communication workflows make two choices: first, that RGB is used to communicate color information between devices; and second, that RGB is given a unique colorimetric interpretation. Each device then does the best it can internally either to encode its native color information in sRGB so that its output is pleasing (e.g. scanners, digital cameras) or to provide pleasing color output given sRGB input (e.g. printers, displays, projectors). Only RGB content gets passed between devices, which means that it integrates very simply and transparently with other elements in color reproduction workflows, such as operating systems and software applications (Figure 4.4). A consequence is also that it appears to a user as if no color management was done; and for the users at whom sRGB solutions are aimed, this is actually a benefit. The flip side, though, is that each device can follow only a single desired reproduction property. This is significantly more challenging than if a choice of desired reproduction property were communicated alongside the color data, and it means that a single choice needs to be made to please all tastes. Finally, it is also worth bearing in mind that other colorimetrically defined encodings can be used instead of sRGB to set up analogous workflows. Adobe RGB (Adobe, 2006), which has a larger color gamut than sRGB, is a popular alternative.

Turning to the relationship between conceptual color management stages and sRGB color management, we can see that the mapping from device color to sRGB involves

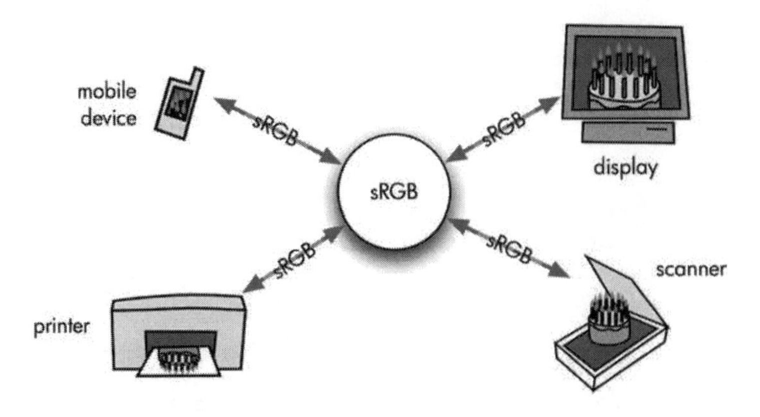

Figure 4.4 An sRGB workflow example.

blocks 1 and 2 of the conceptual process (i.e. everything up to gamut mapping into the sRGB gamut). Going from sRGB to device color, in turn, involves blocks 2 and 3 (i.e. enhancement can be applied again by the destination device and gamut mapping to its gamut is then followed by the inverse CAM and device characterization stages). In terms of implementation, the mapping to and from sRGB can either be in the form of a single LUT each (i.e. between device color and sRGB), in which case it is sometimes referred to as a *color map*, or it can be done by having the conceptual stages implemented explicitly. In any case, the transformations are done with little or no control provided to the end user over their effect; this is due to a key market for this architecture being home and office users, who do not want to interact with color management and who want it to 'just work.'

4.5 ICC COLOR MANAGEMENT

An important alternative to sRGB color management is the framework set up by the ICC, which is currently the *de facto* standard for explicit color management for the professional reproduction of still images.

The ICC was established in 1993 by eight imaging companies, 'for the purpose of creating, promoting and encouraging the standardization and evolution of an open, vendor-neutral, cross-platform color management architecture and components' (Stokes, 1997). To this end, the solution proposed by the ICC is one where the color reproduction process is divided into two transformations: First, a forward transformation that takes device color data and transforms it into a colorimetric description for specific viewing conditions (called the *PCS*). Second, an inverse transformation that takes such a colorimetric description and transforms it back into device color space data. So far this is not too different from the sRGB case, as there, too, the interchange space (sRGB) is colorimetric.

Color communication between devices is then achieved by being able to perform (up to) both parts of the transformation for each of the color reproduction devices among which color is to be managed (Figure 4.5). The parameters based on which the forward and inverse transformations are performed for a given color reproduction medium are stored in a data file referred to as the *ICC device profile*. Its detailed specification and the specification of the entire architecture can be found in ICC (2004). Note that a specific profile is needed for each color reproduction medium rather than just for each imaging device. For example, a printer printing on plain paper will need a different profile to that same printer printing on glossy paper.

The PCS, through which all color communication takes place, is defined by the ICC (2004: 8) as 'the reference color space in which colors are encoded in order to provide an interface for connecting source and destination transforms.' The color spaces that can be used for the PCS are CIE *XYZ* and CIE *LAB* for a reference viewing environment, defined by the graphic arts ISO 3664 (ISO, 2000) viewing condition P2 standard (D50 light source; 500 lx illuminance; 20% surround reflectance). In both *XYZ* and *LAB* the PCS is an oblong in that space, which in the *LAB* case has *LAB*s of [0, −128, −128] and [100, 127, 127] delimiting it. No *LAB*s outside this region can be encoded as they are and need to be clipped. This, however, is not an issue for most color reproduction media, but can be a challenge when attempting to encode specular highlights (i.e. reflections of a light source

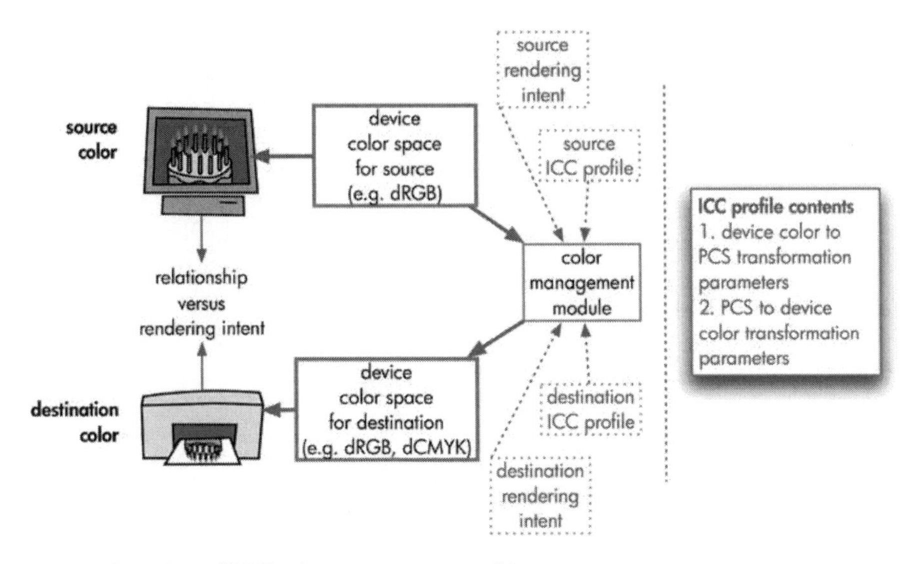

Figure 4.5 Overview of ICC color management architecture.

from a glossy surface, which are brighter than the perfect diffuser under the same viewing conditions and would, therefore, have a lightness exceeding 100).

Unlike the sRGB approach, the ICC framework allows for choosing from among alternative desired reproduction properties and refers to them as *rendering intents*. Note here that rendering intents (just like desired reproduction properties) are either a descriptive or an ideal property for the reproduction (see Chapter 3 for detailed discussion) – in other words they are a chosen aim. Nonetheless, there is an occasional misidentification of rendering intents with GMAs, as the latter are frequently the main tools for attempting to achieve the former. The difference between the two is important though, since applying a given algorithm does not guarantee achievement of the chosen reproduction property (rendering intent). Instead, whether the choice of rendering intent was achieved by the chosen gamut mapping is something that needs to be evaluated. Such confusing of the means with the end then easily hampers successful color reproduction.

The four rendering intents of the ICC framework (see also Section 3.2.4) are as follows:

1. The *media-relative colorimetric* intent prescribes that the CIE *XYZ* values of a device color to PCS or PCS to device color transformation be scaled to map the corresponding medium's white point to the PCS white point. This represents full adaptation to the medium white point (i.e. the medium white point, e.g. a print's blank paper, will look white, without any color cast) and, therefore, is useful for reproductions between media to which observers are fully adapted. A popular use of this rendering intent is also in conjunction with black point compensation (BPC) (Adobe, 2006), where the source luminance range is linearly scaled to the destination luminance range. This rendering intent (with BPC) is particularly popular in professional photography, as it has the greatest potential to result in as similar a print to an original viewed on a display. Note, though, that the quality of the result depends a lot on choices made by the ICC profile-building software that controls the GMA used.

2. The *ICC-absolute colorimetric* intent leaves tristimulus values unchanged (i.e. they are scaled to make the perfect diffuser or transmitter – i.e. a material that reflects or transmits all light at all wavelengths – have a Y value of 100) and is useful for reproducing artwork (where the intention is to also reproduce the effect of the original's illumination and substrate), spot colors and proofing.
3. The *perceptual* intent is useful for a preferred or pleasing reproduction of images, particularly pictorial or photographic ones – especially where source and destination media are substantially different. To allow for more control in providing preferred color reproduction, the ICC specifies a reference medium for this rendering intent. This medium is an ideal reflection print with a specific dynamic range (i.e. black and white points) and gamut (Section 11.3) and its purpose is to provide a more suitable lightness range for the connection space than the [0, 100] range of the encoding oblong. Note, however, that the definitions of reference medium dynamic range and gamut are absent from previous versions of the ICC system and have only been introduced in its current, fourth version (v4).

   The gamut defined for the reference medium – the reference medium gamut (RMG) – provides greater control over color reproduction, which occurs in two stages via the PCS (Koh *et al.*, 2003). With the RMG, the ICC approach has become more similar to the sRGB one, with the difference that its reference gamut is print gamut shaped, whereas the sRGB one is a display gamut. The key benefit of the connection or interchange space having a realistic gamut as opposed to the ICC PCS's oblong is that it allows for the forward direction to map to a known, realistic gamut and, even more important, for the inverse direction to map from a known, realistic gamut to the destination. This means that only the reference gamut needs to be mapped to the destination, rather than the entire encoding oblong, large parts of which do not even represent possible colors (Braun and Spaulding, 2002).
4. The *saturation* intent is also vendor specific, involves compromises such as trading off preservation of hue in order to preserve the vividness of pure colors and is useful for images containing objects like charts or diagrams.

In summary, color transformations in the ICC framework are performed between devices on the basis of device profiles via the PCS and rendering intent choices, and when color data are to be communicated it is necessary to provide them alongside the color data itself (Figure 4.6). For further detail, see the ICC website (http://www.color.org), which also includes useful white papers on a number of color management topics.

The relationship between the ICC framework and the conceptual stages of color reproduction is very much like that of the sRGB approach, where the forward direction of mapping in an ICC profile contains conceptual blocks 1 and 2 and the inverse contains blocks 2 and 3. Gamut mapping, therefore, is combined with device characterization and color appearance modeling when data structures of ICC profiles are populated on the basis of color measurements, information about viewing conditions and a choice of desired reproduction property (i.e. rendering intent). Note that key benefits of the ICC approach over the sRGB approach are that it allows for a choice to be made from among four rendering intents and that it is easier for a user to build an ICC profile for the individual color imaging medium they have (e.g. using ICC profile building software like

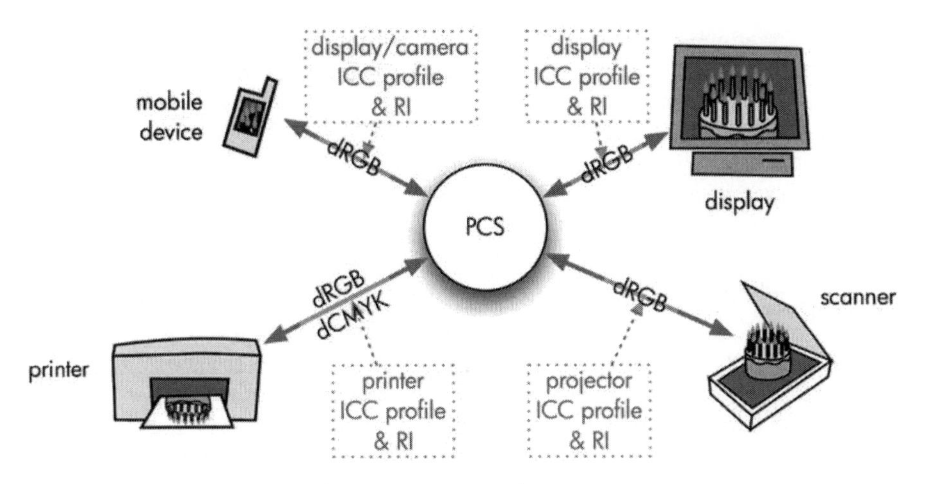

Figure 4.6 ICC workflow sketch (RI: rendering intent).

GretagMacbeth ProfileMaker, Monaco EZcolor, etc.) than to update their sRGB workflow's color maps in a given device (which is often impossible).

4.6 WINDOWS COLOR SYSTEM COLOR MANAGEMENT

An emerging alternative to ICC color management is the *Windows Color System* (WCS) introduced by Microsoft as part of their *Vista* operating system (Microsoft, 2006). The main differentiator of this approach is that it implements most of the conceptual stages of the color reproduction process as separate modules that can also be provided by third parties. The stages of the WCS architecture (Figure 4.7) can be seen to be virtually

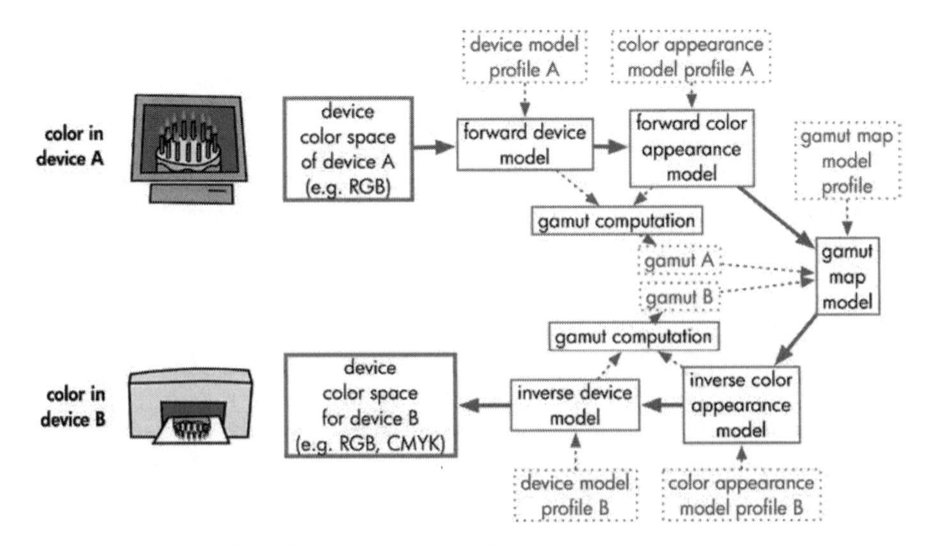

Figure 4.7 Stages of WCS color management architecture.

identical to the conceptual stages shown in Figure 4.2, with the difference that color enhancement is not provided in WCS and would instead have to be part of the gamut mapping module.

The key point to note about WCS is that it provides individual access to conceptual stages and that it allows for the substitution of the device and gamut map models in addition to setting the parameters of the default ones. It is worth commenting on the questionable use of the term 'model' in relation to gamut mapping here, as gamut mapping in the vast majority of cases is not descriptive (when model would be appropriate) but prescriptive. That is also why the gamut mapping literature talks about gamut mapping *algorithms* rather than *models*, and the former terminology will also be used here.

More important, it is essential not to be blinded by the various color management architectures' components, e.g. the single LUT of closed-loop approaches versus the modularity of WCS, and to think that they employ different conceptual approaches underneath. In fact, all of the approaches discussed so far can (and in most cases do) use device characterization models, color appearance models and GMAs – it is only that they are explicit and individually controllable in WCS that is the difference. As WCS is a very recent entry into the color management arena, it remains to be seen how this conceptually appealing approach will fare in practice.

4.7 IMPORTANCE OF GAMUT MAPPING IN COLOR REPRODUCTION

Finally, let us considered the importance of gamut mapping in different applications of color reproduction. The reason for doing this is to make explicit the degree to which gamut mapping choices matter in different contexts, since it might otherwise appear as if gamut mapping was always of equal importance. It might also seem that it was either, at one extreme, only an optional, marginal component or, at the other extreme, a core, essential one all the time.

First, let us consider those color reproduction scenarios where gamut mapping is redundant and an identity mapping provides the best result. The most obvious such case is one where some form of accurate reproduction is desired and where the destination gamut encloses the entire source gamut. The proofing of offset-lithographic presses on some digital ink-jet printers or the emulation of LCD displays on multi-primary projectors are examples here. Yet others are accurate reproductions of originals even from source media that have larger gamuts than the chosen destination medium, but where a given source image's colors are enclosed in the gamut of the destination medium. Examples of such source image content can be watercolors or many images showing natural scenes containing skin tones, as these do not tend to have large color gamuts that would exceed those of some destination media.

Second, there are cases where source gamuts do exceed those of the chosen destination, but they do so only to a limited extent; as gamut differences are limited, the choice of one gamut mapping approach versus another can lead to very small and often invisible differences among corresponding reproductions. For example, originals printed on glossy paper and their reproductions on semi-glossy paper or also the reproductions of some displayed images in ink-jet prints with large color gamut belong to this category. While gamut mapping choices can have an effect on how well the reproduction achieves the chosen desired reproduction property in such cases, it is typically the other components of the reproduction process that are more important, and gamut mapping just needs to be chosen so as not to cause problems.

Finally, there are cases where source and destination gamuts are significantly different from each other, such as film transparencies or displays versus print, newsprint versus ink-jet print on glossy paper, outdoor scenes or oil paintings versus displays, etc. Here, the choices made by gamut mapping need to overcome significant differences, and the impact of choosing one algorithm over another can be dramatic. Part of the challenge of developing gamut mapping solutions is that it would be ideal for them to work across all of the scenarios described here and also for all types of content present in them (e.g. business graphics, photographs, etc.). This, alas, has not been possible to date.

4.8 SUMMARY

Gamut mapping takes place in the context of a color reproduction process that can be implemented by means of a color management system. A variety of such systems were reviewed here and their relationship to the conceptual stages of color reproduction was discussed with the aim of showing where gamut mapping takes place. Finally, the different degrees of importance that gamut mapping can have in color reproduction and that range from marginal to fundamental were addressed.

The key point to take away from here is that there are a variety of color management architectures that implement the conceptual stages of color reproduction in different ways, but that all of them need to provide gamut mapping functionality in at least one place.

5
Overview of Gamut Mapping

Having reviewed the basics of color science, discussed the nature of desired color reproduction properties and looked at color reproduction workflows and their implementations in color management systems, we have finally arrived at the heart of this book, i.e. gamut mapping. The aim of this chapter is to give an overview of gamut mapping by looking at its definitions, examining its aims, surveying its components and the factors that can affect it and finally giving some thought to its future. In the following chapters we will take a closer look at some of the key prerequisites of gamut mapping (i.e. computational geometry, gamut computation, color spaces) and then delve into the details of alternative GMAs, their use in color management and their evaluation.

5.1 DEFINITIONS

The definition of gamut mapping that will be used in this book is that provided in the CIE's (2004d: 2) *Guidelines for the Evaluation of Gamut Mapping Algorithms*:

> **Colour gamut mapping:** a method for assigning colours from the reproduction medium to colours from the original medium or image (i.e., a mapping in colour space).

where the terms *color gamut, color reproduction medium* and (original) *image* are defined as follows:

> **Image:** two-dimensional stimulus containing pictorial or graphical information whereby the original image is the image to which its reproductions are compared in terms of some characteristic (e.g., accuracy).
>
> **Colour reproduction medium:** a medium for displaying or capturing colour information, e.g., a CRT monitor, a digital camera or a scanner. Note, that in the case of printing, the colour reproduction medium is not the printer but the combination of printer, colorants and substrate.

> **Colour gamut:** a range of colours achievable on a given colour reproduction medium (or present in an image on that medium) under a given set of viewing conditions – it is a volume in colour space.

This clearly places gamut mapping at the heart of the conceptual color reproduction process, where it provides the link between the appearance of an original (source) and the appearances possible in the reproduction (destination). A key aspect of this definition is that it refers to color and not to colorimetry, spectral properties, or some other domain. By referring to color as the inputs and outputs of gamut mapping, what is implicit is that viewing conditions are taken into account outside it and also that the specific ways of addressing different imaging devices, which give rise to the media between which gamut mapping is applied, are dealt with elsewhere. Some ambiguity remains in this definition, though; and whether image enhancement is part of gamut mapping or not remains open. The term *accurate gamut mapping* is often used to specify that only the overcoming of gamut differences is being referred to.

An alternative way of drawing lines in the color reproduction process is that provided by ISO 22028-1 (ISO, 2004a), which puts forward the following definitions:

> **colour re-rendering**
>
> mapping of picture-referred image data appropriate for one specified real or virtual imaging medium and viewing conditions to picture-referred image data appropriate for a different real or virtual imaging medium and/or viewing conditions
>
> NOTE Colour re-rendering generally consists of one or more of the following: compensating for differences in the viewing conditions, compensating for differences in the dynamic range and/or colour gamut of the imaging media, and applying preference adjustments.
>
> [. . .]
>
> **gamut mapping**
>
> mapping of the colour-space coordinates of the elements of a source image to colour-space coordinates of the elements of a reproduction to compensate for differences in the source and output medium colour gamut capability
>
> NOTE The term 'gamut mapping' is somewhat more restrictive than the term 'colour-rendering' because gamut mapping is performed on colorimetry that has already been adjusted to compensate for viewing condition differences and viewer preferences, although these processing operations are frequently combined in reproduction and preferred reproduction models.[1]

[1]The terms and definitions taken from ISO 22028-1:2004 (ISO, 2004a) *Photography and graphic technology – Extended colour encodings for digital image storage, manipulation and interchange, Part 1: Architecture and requirements*, clauses 3.12 and 3.22, are reproduced with permission of the International Organization for Standardization, ISO. This standard can be obtained from any ISO member and from the website of the ISO Central Secretariat at www.iso.org. Copyright remains with the ISO.

This view of gamut mapping, which places it inside the broader concept of re-rendering, assigns a more constrained meaning to gamut mapping than will be used in this book and more constrained than has been used in the published literature on gamut mapping. In fact, of these two ISO-defined terms, color re-rendering is much more like what will be meant by gamut mapping here and ISO gamut mapping is more similar to our accurate gamut mapping. In any case, it is useful to be aware of such differences in terminology (Holm, 2005) and of the fact that some lines that delimit stages in color reproduction are not unique choices but only some of several alternatives.

Another set of approaches that ought to be compared with gamut mapping are reproduction techniques – often referred to as *tone mapping* – forming part of *high dynamic range* (HDR) imaging. These methods (Reinhard *et al.*, 2005) aim to take an original that contains stimuli covering a very large dynamic range, like that of natural, outdoor scenes with luminances ranging from 6×10^3 to 6×10^{-5} cd m^{-2} (Hunt, 1995b: 787), and map them into the much more limited dynamic ranges of typical imaging media (with luminance ranges in the region of 300:0.001 cd m^{-2}). While this does sound a lot like gamut mapping (i.e. the mapping between media of difference gamuts), an important difference is introduced by the fact that HDR images are not well described by the color appearance models used in color reproduction. Hence, the GMAs developed for the mapping of color appearance attributes are inadequate when provided with absolute tristimulus values (i.e. denoted as $X_L Y_L Z_L$ by Hunt (1995a: 58)) and no means of predicting the appearance they represent.

The reason for this gap is twofold. First, HDR images are not ones that the visual system experiences using a fixed state of adaptation (e.g. imagine looking at a lawn with a tree in its middle on a sunny day: looking at the grass in the shade will show detail there and make the directly lit grass appear very bright and showing less detail; that same part of the scene will show good detail when viewed directly and the grass in the shadow will now look dark and will lack detail – see also Figure 2.15) and a given scene can look different depending on where a viewer focuses in it. The reason for this is that the human visual system has a relatively limited *simultaneous dynamic range* of around 150:1 (Vos, 1984; Seetzen *et al.*, 2004). In other words, if the difference between two sides of an edge exceeds this luminance ratio, its perceived contrast will not continue increasing. Contrast across greater dynamic ranges of up to 10^5:1 (Johnson and Fairchild, 2006; Rizzi *et al.*, 2007) can, nonetheless, be seen thanks to adaptation (Section 2.7). Second, color appearance models are derived from data collected under viewing conditions where stimuli would only come from the much more limited dynamic range of typical imaging media; therefore, their predictions of HDR stimuli are of unknown accuracy. As a consequence, there is currently separate development of gamut mapping and HDR imaging solutions that could converge if the above differences were resolved.

For completeness' sake it is also worth mentioning that the broad concept of gamut mapping is also used in the context of *color constancy algorithms* (e.g. Forsyth, 1990; Finlayson and Hordley, 2000), i.e. algorithms that take an image captured under unknown illumination and attempt to estimate that illuminant or transform the image to an equivalent one under a fixed, known illuminant. While gamuts are mapped here too, the sole purpose of the mapping is to compensate for an unknown illuminant rather than to overcome differences between an original's and a reproduction's gamut. As the focus in

Table 5.1 What gamut mapping is versus what it is not.

Gamut mapping is	Gamut mapping is not
• Part of color reproduction • Effected by color management system transformations • Designed with the aim of assigning reproducible colors to desired ones • A mapping/transformation in a color space • Partly art • Partly science • Prescriptive • An algorithm	• A rendering intent (e.g. Lavendel, 2003: 224) • Pure science • Pure art • Descriptive (e.g. like a device characterization or color appearance model) • A model • Transformation from one color space into another (e.g. Heidelberger Druckmaschinen AG, 2003)

this book is on gamut mapping as used in cross-device or cross-media color reproduction, its use in color constancy algorithms will not be addressed further.

Finally, Table 5.1 contrasts a few correct statements about gamut mapping with some common but incorrect ones that occasionally appear in the literature.

5.2 AIMS OF GAMUT MAPPING

The high-level aim of gamut mapping follows from the discussions of color reproduction in Chapter 4 and desired reproduction properties in Chapter 3:

To assign reproduction colors to original colors to result in a reproduction that has a maximum of the chosen desired reproduction property.

This translates into the aim of gamut mapping being to make such an assignment that maximizes the chosen desired reproduction property (e.g. preference or subjective accuracy). A consequence of this being the aim of gamut mapping is also that it can in some cases be predominantly *science*, in others much more *art* and in most a balanced mixture of the two – depending on the choice of desired reproduction property. For accuracy of some form, gamut mapping becomes science, when it is pleasantness or preference it is a mixture of art and science, and when it is Field's creative reproduction it is primarily art.

If we then look at the aims set for gamut mapping in the literature, we can see some commonality in the reproduction features that are considered as important contributors to its success.

An early paper by Stone *et al.* (1988) realizes that 'the relationship between ... colors [... is] more important than their precise value,' and that it is this relationship that GMAs ought to preserve. This is a key observation, and its significance lies in moving away from simply considering the effect of gamut mapping on individual colors to taking into account how it affects relationships (e.g. How do the gamut mapped lightnesses of this pair of colors compare to the original ones? Is the darker one still darker and by a similar amount?). A number of heuristically determined objectives were also identified, including the preservation of the original's gray axis, aiming for maximum luminance contrast, reducing the number of out-of-gamut (OOG) colors, minimizing hue shifts and preferring an increase in saturation rather than a decrease. Note that this list represents *a priori* choices (referred to by Stone *et al.* (1988) as 'widely accepted graphic arts and psychophysical principles'), where the reasons for making them were in most cases based on experience from traditional (analogue) color reproduction. Even though such experience can be of great value, its maxims need to be looked at carefully when used in an environment that enables far greater control over color attributes than was previously possible. Such a change of framework from traditional to digital does indeed put a question mark on the results of some previous research. Therefore, care is advised when considering research from the pre-digital era; for example, for a difference in the understanding of lightness compression, see Morovič (1998: 99–102).

For subjective accuracy, the aim of gamut mapping can also be stated as desiring 'a good correspondence of overall colour appearance between the original and the reproduction by compensating for the mismatch in the size, shape and location between the original and reproduction gamuts' (Morovič, 1998).

Another way of looking at the aim of a gamut mapping is for it to preserve some function of the original while adjusting it to a different destination gamut. For example, if the original expresses an idea that an artist is trying to communicate, then a high-level aim of the gamut mapping applied to that artwork is for the reproduction to communicate the same as the original did. When a reproduction needs to sell, whether someone will be prepared to pay for it in the end is crucial in terms of whether the gamut mapping worked. In practice, this communication of an idea and salability get translated into some desired reproduction property that can either be measured or psychovisually evaluated, but in the end the true aim is to allow for an idea's communication and/or the reproduction's sale. On the other hand, the aim of gamut mapping when applied to the creation of a contract proof is quite different and needs to ensure that the signed proof represents the production copies as accurately as possible and that it can be reliably used to settle any potential disputes between buyer and print service provider.

5.3 GAMUT MAPPING ALGORITHM CONTEXT

While there are numerous GMAs (Chapter 10) that differ from each other in many ways, they all share the same context (Figure 5.1). Gamut mapping always takes place in some color space, whose dimensions (e.g. predictors of lightness, chroma and hue) are used to express source color information and where destination color needs to be provided in the end. This space is also where the destination and in some cases also the source gamuts are described, and it can be a composite of other component spaces (e.g. Zeng, 2001). Further detail on color spaces for gamut mapping, e.g. CIELAB or CIECAM02, will follow in Chapter 6.

As discussed previously, the choice of desired reproduction property (Chapter 3) plays a controlling role in gamut mapping. Gamut boundaries, whose computation will be introduced in Chapter 8, provide key parameters for all gamut mapping approaches, since they delimit where colors need to end up and where they start out. These inputs to a GMA form the basis for determining parameters of the source color information's transformation in addition to any predetermined parameters or parameters supplied manually by a user (e.g. the user might want to force a predefined mapping for a specific,

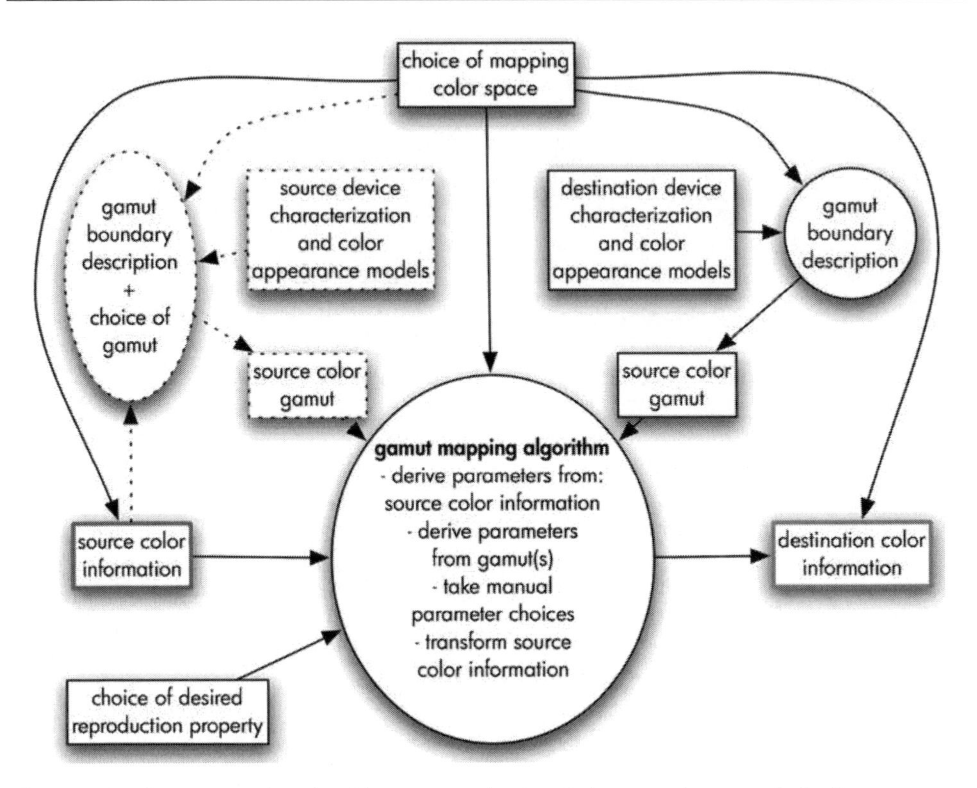

Figure 5.1 Gamut mapping algorithm context (optional elements shown as dashed).

single color). The end result is a transformation applied to the source color information, the aim of which is to maximize the desired reproduction property while resulting in colors that are inside the destination color gamut.

5.4 TYPES OF GAMUT MAPPING

GMAs can be categorized depending on the type of source–destination gamut relationship they address (Sara, 1984: 88) and depending on the type of color information they apply to (Figure 5.2).

Of the various possible source and destination gamut relationships and color information types to which mapping is applied, the most common ones are those where the destination has a smaller gamut than the source and where the mapping is done either on a color-by-color basis or where it is applied to an image and spatial relationships are taken into account. Gamut mapping has also been applied to spectral data in the same gamut context or to individual colors in other gamut contexts. Note that the case where there is a gamut overlap, and where both reduction and expansion are possible, is often treated as if the destination were smaller than the source and those parts of the destination gamut that are outside the source are simply not used. Where a GMA makes such a choice, it will be

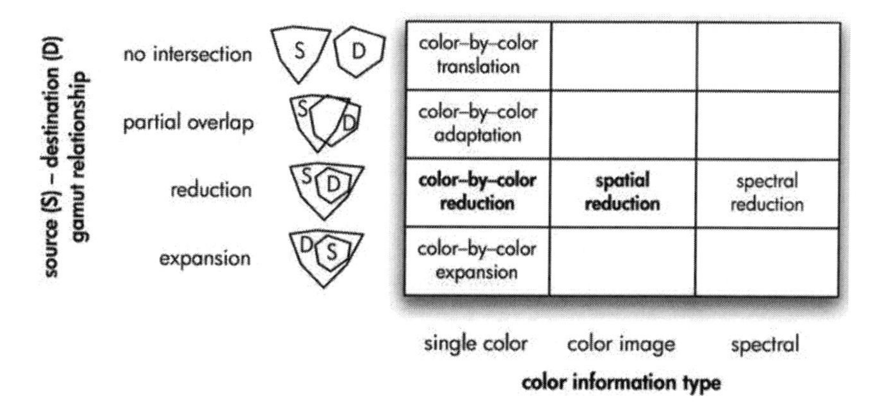

Figure 5.2 Combinations of color information and gamut relationship types (bold labels show most common cases and thin ones show infrequent ones).

considered here to be of the reduction type. Alternatively, some algorithms apply the same transformation as used for reduction to expansion also and simply let the parameter values determine whether it is the former or the latter (e.g. if along a line of mapping the source gamut is inside then the destination the result will be expansion and if it is outside then it will be compression).

Finally, gamut reduction can be divided into two subtypes: *compression* and *clipping*, where compression can even change source colors that are already inside the destination gamut and clipping only changes colors outside it (i.e. *OOG* colors).

5.5 BUILDING BLOCKS OF GAMUT MAPPING ALGORITHMS

Given a pair of source and destination gamuts in a color space and a source color to be gamut mapped, there are a number of approaches that can be followed (Chapter 10). This multitude of GMAs is made up of a more limited set of building blocks, though, that result in the various ways of mapping a source color into a destination gamut when combined. These can be grouped as follows: mappings onto a point in the mapping space (e.g. a color or a spectrum) with predetermined properties; mappings along a path; interpolation or morphing techniques; spatial operations; and choice of gamuts to map between.

5.5.1 Mapping onto point with predetermined properties

Conceptually, the simplest and most obvious approach to gamut mapping is one where all source colors that are inside the destination gamut are left unchanged and colors that are OOG are mapped onto those destination gamut surface colors that are closest to them. This approach, called *minimum* ΔE (MINDE), has been used many times since Sara's 1984 work and will be looked at in detail in Chapter 9.

In addition to being a GMA in its own right, mapping to the nearest color on the destination boundary is also used as a component in more complex algorithms (often as the last step to take care of small gamut differences not dealt with by previous stages). Note, here, that the idea of mapping to the closest color can be implemented in many color spaces and using many color difference metrics (e.g. Katoh *et al.*, 1999). It is also possible to apply different weights to differences in the dimensions of a color space when computing a distance (e.g. Katoh and Ito, 1996) or even to allow for no difference in one of the dimensions (typically hue; Murch and Taylor, 1989).

Another type of mapping that belongs to this category is to express a source color's location relative to the source gamut and to map it to the destination color with the same relative location in the destination gamut (e.g. the TRIA algorithm; Morovič, 1998: 110). Motomura's (2000) color-categorical mapping, which requires for the reproduction color to belong to the same color name category as the source color, also has aspects of mapping towards a predetermined property, though only in a looser sense.

Finally, the predetermined property can also be a manual choice. For example, Sara (1984) defined a gamut mapping that consisted of manually choosing destination colors for 26 source colors and then interpolating for the remainder of the source color gamut on the basis of the 26 manual mappings (Section 9.5.4).

5.5.2 Mapping along a path

The most popular building blocks of GMAs are mappings along a path, and there is a lot of variety in both the path (Figure 5.3) and in how mapping takes place along it

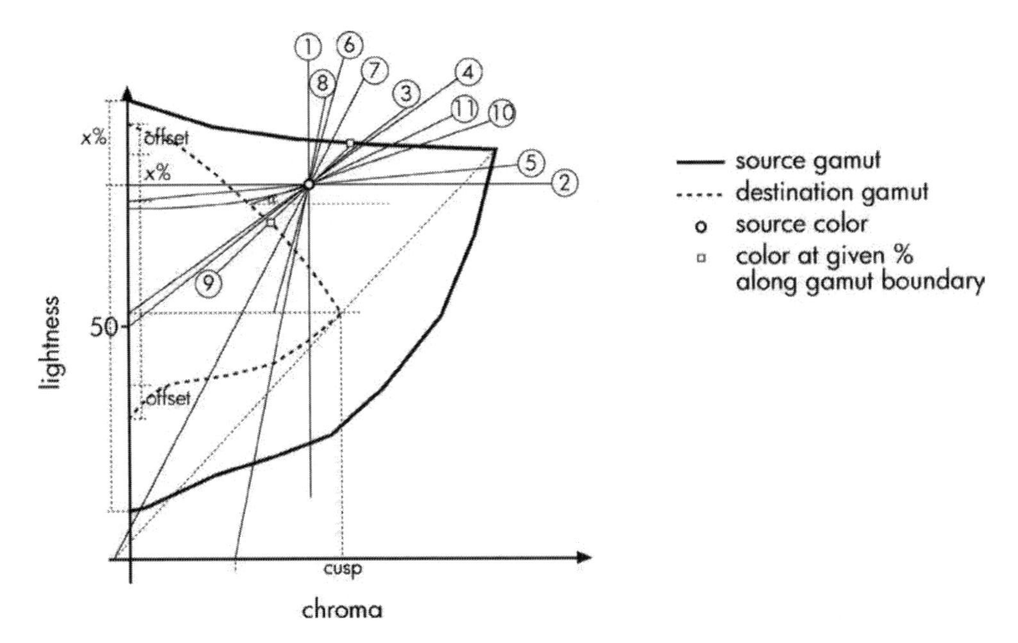

Figure 5.3 Gamut mapping paths.

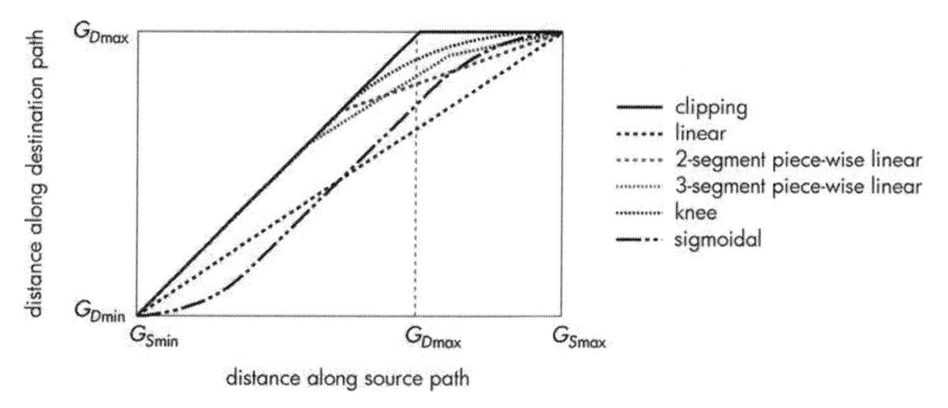

Figure 5.4 Mapping types along a path.

(see Figure 5.4). Note that these paths are typically in a color appearance space, but they can also be applied in a space representing spectral reflectance or power distributions.

Path types

The simplest case of a path here is a *line*; among mapping lines, the simplest ones are those that vary only in one dimension of a color space, typically:

1. Lightness.
2. Chroma.

Other lines used in gamut mapping are ones that, in addition to the source color, also contain one of the following (numbers correspond to those in Figure 5.3):

3. The centre of the lightness axis (e.g. in CIECAM02 $Jab = [50, 0, 0]$) or the center of the destination gamut's lightness range (Sara, 1984).
4. The point on the lightness axis that has the same lightness as the destination gamut's most chromatic color at the source color's hue (Johnson, 1992). Note that the color at a given hue that has the largest chroma is called the *cusp.*
5. A hue-dependent point on the lightness axis determined by the relationship of the source color's lightness and the source gamut's lightness range (Marcu and Abe, 1996).
6. A point having the same lightness as the cusp at a given hue and a percentage of its chroma (Ito and Katoh, 1995).
7. The point on the lightness axis obtained by intersecting it with the line joining the source and destination cusps (Johnson, 1992).
8. The point on the chroma axis having half the destination cusp's chroma (Johnson, 1992).

The mapping line can also be defined to:

9. Have a fixed slope, typically relative to the chroma axis (Ruetz, 1994; Lee *et al.*, 2000).

10. To connect specific points on the source and destination gamuts, e.g. points that have the same relative distance along the source and destination boundary at a give hue (MacDonald *et al.*, 2001).

Alternative paths along which mapping can take place are:

11. The curves proposed by Herzog and Büring (1999, 2000) or a circle in planes of constant lightness with its center at the origin (i.e. a mapping that changes hue while preserving lightness and chroma).

Note that all but the last one of these paths are in planes of constant hue, as they are either components of hue-preserving GMAs or are used in such planes after hue changes have been made.

Mapping types

Given a mapping path, the source color is mapped on the basis of its location along it and on the basis of the locations of either only the destination gamut or also the source gamut along that path. The basic challenge here is to make a choice for representing a path in color space of a certain length (source) by another path of a different length (destination).

A visual example of this challenge is shown in Figure 5.5, where the leftmost column of color patches has a certain lightness range and other columns represent different ways of

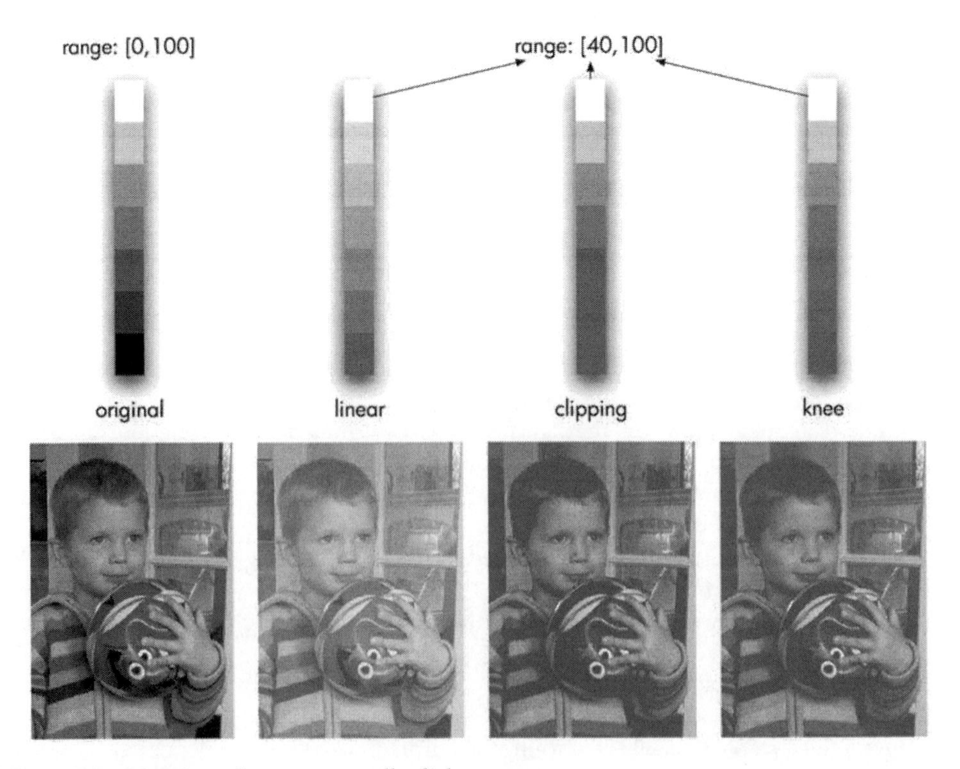

Figure 5.5 Mapping a larger to a smaller lightness range.

mapping that range into a much more limited one. The first mapping from the left attempts to preserve the differences in the original, the second attempts to match each patch as closely as is possible and the third tries to balance these two requirements. The effect on complex images is then shown at the bottom of the figure. Note how the three mapped images have very different appearances while occupying the same lightness range. This clearly illustrates that it is both the available gamut and the method used for populating it that determine a reproduction's appearance and that they both play significant roles. In other words, the size of the gamut alone does not determine how much of an original's appearance can be preserved, but it is also the choice of GMA that is decisive.

The mapping from source to destination color along a path can be performed in different ways, too, where O will indicate the origin of the path, S the distance of the source color from the origin along the path, G_S the distance of a source gamut boundary intersection with the path, G_D the distance of an intersection of the destination gamut boundary, D the distance computed for the destination color and min and max indicate whether the intersection has the smallest or largest distances respectively among the boundary's intersections along a path. The following, then, is a list of typical mappings:

1. *Clipping.* Here, distances in the destination gamut's range are maintained and those outside it are clipped to its end points:

$$D = \begin{cases} G_{\mathrm{D\,min}};\ S \leq G_{\mathrm{D\,min}} \\ S;\ G_{\mathrm{D\,min}} < S < G_{\mathrm{D\,max}} \\ G_{\mathrm{D\,max}};\ S \geq G_{\mathrm{D\,max}} \end{cases} \tag{5.1}$$

   A common application of clipping, for example, is towards the lightness axis along paths of constant lightness and hue in a lightness, chroma and hue space (e.g. the *JCh* space of CIECAM02). In this case $G_{\mathrm{D\,min}}$ becomes the origin and all distances are chroma values (e.g. Ito and Katoh, 1995). Clipping towards some other point on the lightness axis is also commonly used, where $G_{\mathrm{D\,min}}$ is again zero and distances are from that lightness axis point.
2. *Linear.* Here, the source distances are linearly scaled to fit into the destination range as follows:

$$D = G_{\mathrm{D\,min}} + (S - G_{\mathrm{S\,min}})\frac{G_{\mathrm{D\,max}} - G_{\mathrm{D\,min}}}{G_{\mathrm{S\,max}} - G_{\mathrm{S\,min}}} \tag{5.2}$$

   A similar mapping (Equation 10.1) is often applied to lightness (Buckley, 1978) when the G_{min} points are the black points of the two gamuts and the G_{max} points are the whites, and it has also been commonly used along other paths.
3. *Piece-wise linear.* The idea here is to subdivide the source and destination ranges into typically two (Gentile *et al.*, 1990) or three (Sara, 1984: 115; Montag and Fairchild, 1997) intervals and to use linear mapping within them. For two segments there are two pairs of min and max points – one for each interval, whereby the max points of the first interval are the min points of the second one. In fact, clipping and linear mapping are also types of piece-wise linear mapping – the former having two intervals (the first is an identity and the second is constant) and the latter having a single interval.

4. *Knee function.* This is a nonlinear function whose tangent for lower values is the identity function and which for higher values becomes the clipping line (Stone and Wallace, 1991). The effect of this function is to provide a smoother transition than clipping would provide while maintaining values at the beginning of the range. This mapping is typically applied along paths towards some point on the lightness axis. Note that two-part, piece-wise linear mappings are sometimes also referred to as knee functions.
5. *Sigmoidal.* Instead of preserving values at one end of the scale, a sigmoidal mapping attempts to preserve the middle of the range and to squeeze values closer together at both the high and low ends of the range. This type of mapping has been used in traditional analogue photographic color reproduction (Yule, 1967) and has also been applied to lightness mapping in digital techniques (Buckley, 1978; Braun, 1999).

5.5.3 Interpolation or morphing techniques

GMAs can also include *interpolation* or *morphing* techniques, which are typically used in addition to the previous groups of building blocks. A common way to employ these techniques is to have some of a source gamut's colors mapped to the destination gamut using other building blocks and then to use interpolation to provide a mapping for the remaining colors. Examples of this approach are the already-mentioned method by Sara (1984) and Spaulding *et al.*'s (1995) *UltraColor* algorithm.

5.5.4 Spatial operations

When gamut mapping is applied to an image, some algorithms perform mappings that are determined only by the color of each pixel on its own, while others take the pixel's neighborhood in the image into account (Figure 5.2). These latter algorithms use spatial operations in addition to the building blocks described above. This often involves a decomposition of an image into a combination of sub-images, each of which contains spatial information at different spatial frequencies (Figure 2.14). Different mappings are then applied to these sub-images before they are combined to give the complete destination image. In general, such decomposition can be performed either in the *frequency domain* or the *spatial domain.*

The first of these is based on the fact that images can be represented as a weighted combination of periodic functions of different frequencies and orientations; such a representation can be obtained via a *Fourier transform* (Gonzalez and Woods, 1993: 81–128). For example, Figure 5.6 shows an image and the Fourier spectra of two of its parts: one with predominantly higher spatial frequencies (in the left part of the image) and one with lower frequencies (on the right). Here, pixels in a Fourier spectrum's visualization that are closer to its center represent lower spatial frequencies, the direction in which a pixel is from the center shows the direction of spatial change it corresponds to, and the lighter the pixel the more dominant the corresponding frequency component in the image is. An example of the use of the Fourier transform in gamut mapping is the work of Meyer and Barth (1989).

Figure 5.6 Fourier transformations (left) and spatial frequency band decomposition (right) of an image (center).

On the right side of Figure 5.6 is an example of spatially decomposing an image into several frequency bands: the top left shows the lowest frequencies and the remaining three images show bands of increasing frequency (where a mid-gray represents zero). When all of these four images are added up, the original image is obtained again. Such decomposition can be obtained both using Fourier methods in the frequency domain and spatial operations directly in the spatial domain. For an example of the use of the latter in gamut mapping, see Balasubramanian *et al.* (2000).

In all of these cases the basic idea is to allow for different treatments of different spatial frequencies and/or for the forcing of the preservation of fine spatial detail following a gamut mapping operation.

5.5.5 Choice of gamuts to map between

Given a source image (or other set of source colors) and a destination medium, there are several gamut combinations that the above building blocks can be applied to.

First, there is a choice of three gamut combinations that can be used to determine parameters for mapping components: the destination gamut alone (this applies only to clipping algorithms), the gamuts of source and destination media or the gamuts of the source image (Stone *et al.*, 1988) and destination medium. What is apparent here is that the destination gamut always needs to be known (since it sets limits to where the mapping needs to end up), but that there are alternatives as far as the source is concerned.

Using a source image's gamut as the starting point of gamut mapping will result in compression only being performed to the extent needed by that image, and it has been shown in several studies that it results in more accurate reproductions (Gentile *et al.*, 1990; Pariser, 1991; Montag and Fairchild, 1997; Wei *et al.*, 1997). Using medium gamuts,

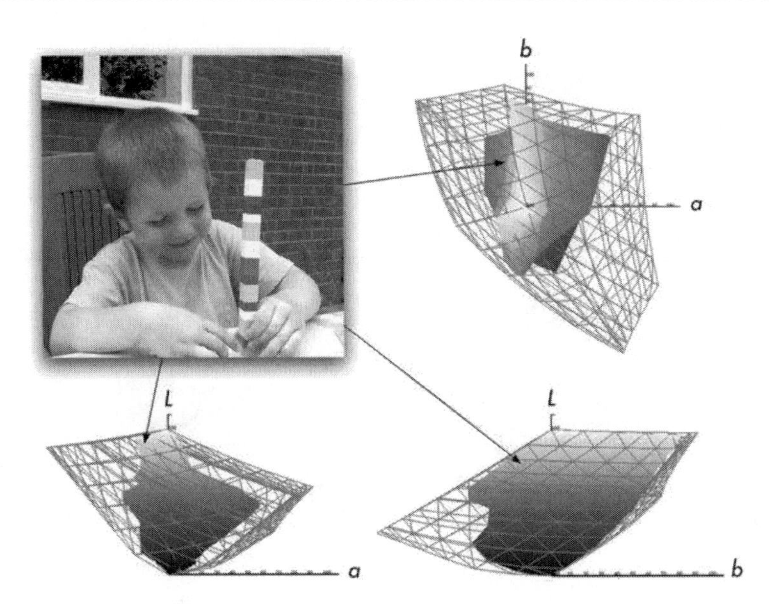

Figure 5.7 The gamuts of a CRT display (mesh) and of an image (solid) shown on that display. The gamuts are shown in CIELAB for D50 and the 2° observer – each (half)axis is 110 units in length.

on the other hand, allows for the transformation between a pair of media to be computed only once and then applied to a large number of images. Furthermore, a given source color will always be reproduced by the same destination color. This can be of importance when an object is present in a number of images – using the source medium gamut will make all of its reproductions the same, whereas using the individual images' gamuts could result in unwanted differences.

An example of a source medium gamut and that of an image in it is shown in Figure 5.7. It can be seen here, for example, that if the image gamut were used as opposed to the medium gamut, then less compression would be necessary virtually for all hues, as the image has a much smaller gamut than the medium in which it is present across the board. Using the source medium's gamut could result here in unnecessary compression for some GMA–destination gamut combinations.

Second, given a choice of gamuts, either all of the gamuts' volumes can be used or part of them can be excluded from gamut mapping and kept unchanged. An example of the latter case is the definition of a *core* region (Sara, 1984) where no change is made to source colors and compression is applied only between the complement of the core in the source and the complement of the core in the destination.

5.5.6 Summary of GMA building blocks

Figure 5.8 gives an overview of the four types of building block discussed here and the hierarchical structures of their sub-types. The GMAs that will be described in Chapter 10 are simply different combinations of blocks from the range shown here, and future work on gamut mapping will consist both in the development of new blocks and of new combinations of existing ones.

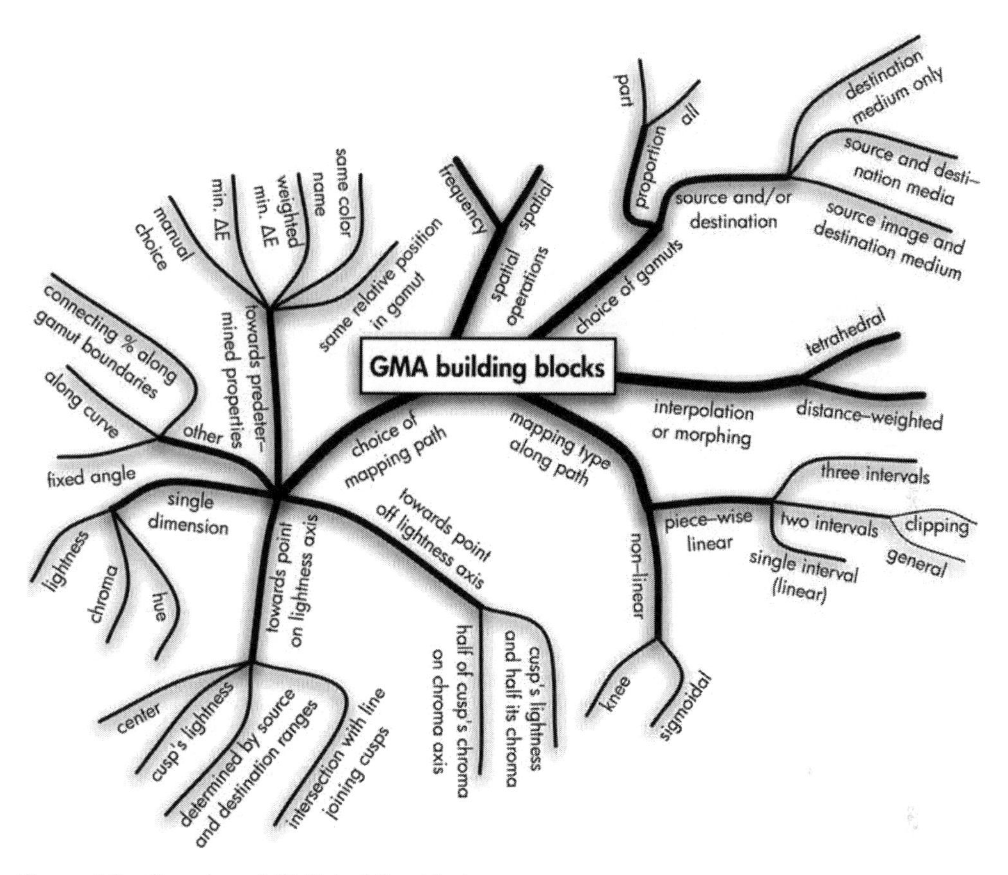

Figure 5.8 Overview of GMA building blocks.

5.6 FACTORS AFFECTING GAMUT MAPPING

The success or failure of a particular GMA depends on a range of factors, and an algorithm that works extremely well for one constellation can fail miserably for another. The range of factors that have already been shown to impact GMA performance will be sketched next. Note, though, that the following list is likely to be incomplete and that the continuing emergence of new aspects that affect gamut mapping is a key obstacle to its systematic study and evaluation.

5.6.1 Desired color reproduction property

First and foremost, the choice of desired color reproduction property has a deciding impact on how the performance of a GMA is judged, and algorithms that under given circumstances result in the most pleasing reproductions can at the same time fail if judged from the point of accuracy.

Figure 5.9 Linearly compressing an original (left) in CIELAB's lightness predictor (center) and CIEXYZ luminance (right).

5.6.2 Mapping color space

The factor closest to a GMA is the color space to whose dimensions it is applied. Even though most GMAs are described in terms of changes to appearance attributes (e.g. chroma should be reduced without changing lightness or hue), they need to be implemented as changes to appearance attribute predictors of a color space (e.g. CIELAB C^* is reduced while L^* and h^*_{ab} are kept unchanged). The implications of this gap and the range of color space options for gamut mapping will be looked at in detail in Chapter 6.

For now, Figure 5.9 just shows the effect of applying the same linear mapping to the single dimension of two different color spaces. Even though the mapping (linear scaling) is the same, the effect is very different, which shows the fundamental interrelationship between a mapping and the space in which it is performed.

5.6.3 Gamut computation

Given a mapping in a color space, the next factor that influences its effect is the way in which the gamuts that determine its parameters are computed. Due to the ambiguity of what the gamut of a set of colors is (as will be discussed in Chapter 8), there are several valid answers that can be given and each of them will place the gamut boundary at somewhat different locations in color space. As a consequence, mapping will end up in different places too and the extent of compression will also be affected. The end result is that differently computed gamut boundaries can affect the appearance of reproductions generated on their basis.

5.6.4 Gamut differences

Given a certain choice of mapping, color space and gamut computation, the magnitude and nature of source–destination gamut differences will have a significant impact on the

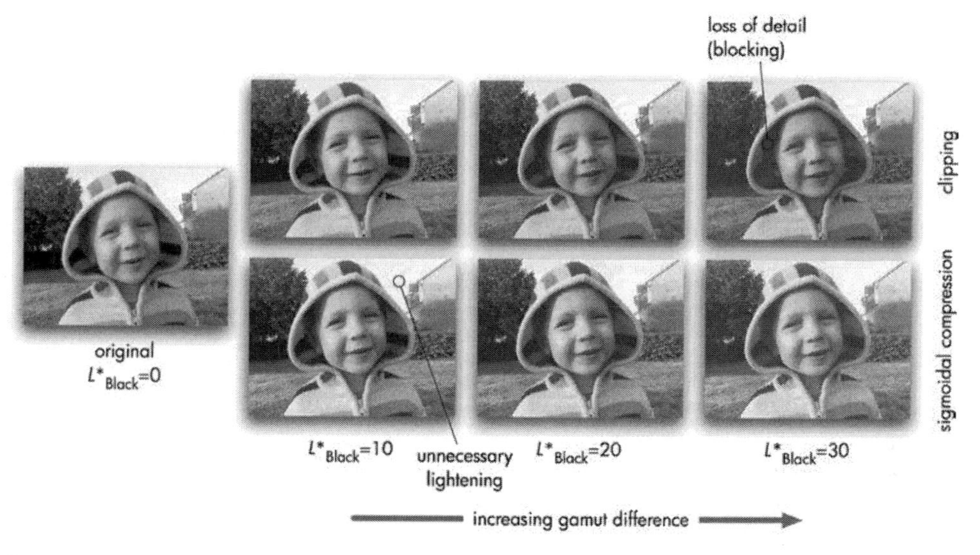

Figure 5.10 Clipping versus sigmoidal compression in CIELAB L^* for various degrees of gamut difference.

success or failure of a GMA. Methods that work well for small differences often fail for large ones, and methods that do a good job of overcoming large differences can introduce unnecessary changes when gamuts are similar. Figure 5.10 gives an example where, when differences are small, clipping is the better strategy than sigmoidal compression, which here introduces unnecessary changes. When differences are large, however, clipping results in loss of detail that sigmoidal compression can preserve. Note, however, that this is simply a single example, rather than an attempt at generalization.

5.6.5 Source image content

In addition to the gamut differences between media, the source image's content also plays a role in how successful a GMA is. Figure 5.11 shows the effect of clipping on a pair of images: one that has predominantly light colors (i.e. a *highlight* image) and one in which colors are dark (i.e. a *shadow* image). More on the effect of image content on gamut mapping will follow in Chapter 12.

5.7 WILL GAMUT MAPPING BECOME REDUNDANT?

There is clearly a lot of complexity in gamut mapping: it involves making a plethora of choices and there is currently no robust solution for it (Chapters 10–12). Whether gamut mapping is successful or not depends on an ever-expanding list of factors. In the face of such a thorny situation, there are at least two obvious approaches: invest resources to understand the problem better and try to develop robust solutions, or see whether the problem can be made to go away. Clearly, if all media could be made to have the same

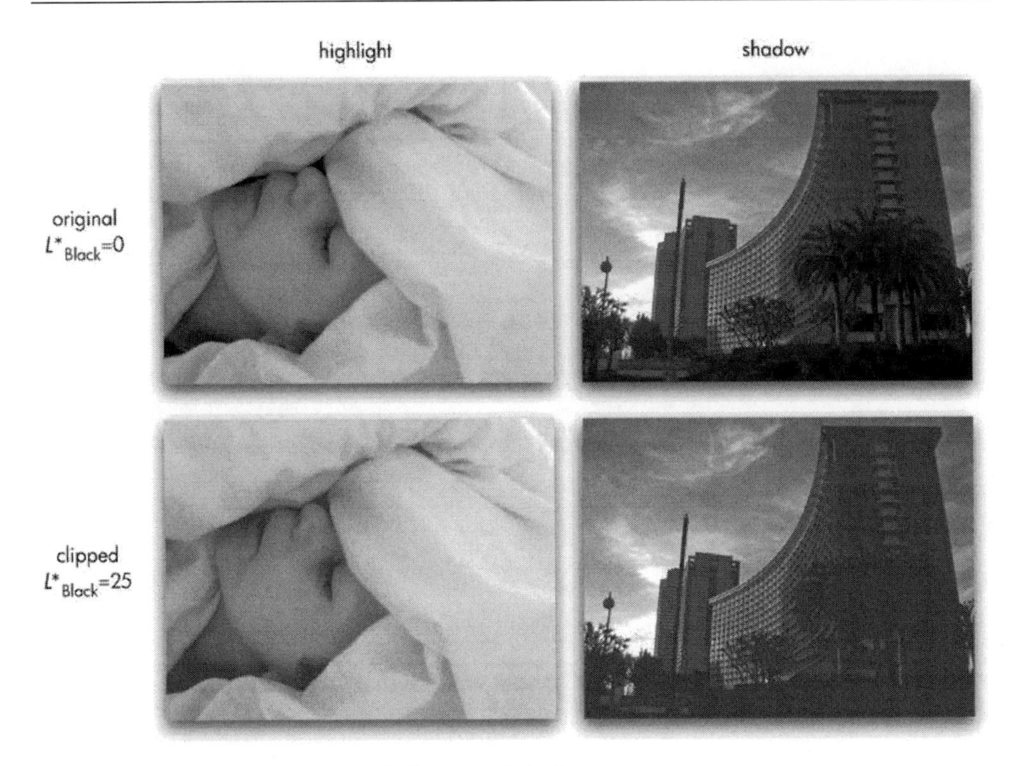

Figure 5.11 Effect of clipping on highlight and shadow images.

color gamut, then the difficulties posed by gamut mapping would become nonissues and gamut mapping would become a redundant oddity of the past. All that would remain would be to have accurate characterizations of each medium, and color reproduction would simply be a transcoding rather than the reinterpretation that it is today.

Before delving any further into the intricacies of gamut mapping, let us, therefore, turn to the question of whether gamut differences between media are likely to disappear in the face of ever-advancing color imaging technologies. To do this, it is useful to make the sources of color gamut differences explicit and to look at the likelihood of each one of them disappearing in turn.

The first source is the multitude of *colorant* (primary) sets used to bring about different imaging media. Here, developments in the materials science of colorants (and substrates – in the case of printing) could indeed result in increases in the gamuts of all media to the point of coming close to their theoretical limits. This does have two caveats though. The first is that imaging systems would have to use *many* more colorants than they use today. The reason for this comes from the fact that there are currently two dominant types of color reproduction approach, i.e. additive and subtractive, and that these differ from each other in how their primaries and secondaries relate. In the additive case, secondaries are the sum of primaries and, therefore, will result in appearances of similar or higher brightness than the primaries. In the subtractive case though, secondaries are the result of two primaries both subtracting some part of the illuminating light and, therefore, they are likely to be darker than the primaries that result in them. Hence, even if an additive and a subtractive medium both had the same primaries, they would have different secondaries and, therefore, gamut shapes. Using many primaries (i.e. significantly more than the typical three chromatic ones) would reduce the impact of such a difference. The second caveat is that, in the case of print, only the *substrate* leading to the largest gamut would result in a possible match to the gamuts of other media. Since there are good reasons for choosing substrates that would not have this property (e.g. matte papers for fine-art reproductions), gamut differences are likely to remain even if colorants reach theoretical limits.

The second source of gamut differences is the variety of *viewing conditions.* As long as originals and reproductions need to have specific desired reproduction properties when viewed under different viewing conditions, there will be gamut differences between them. The reason for this is that the theoretical limits of what color appearances are possible under different viewing conditions differ, and even if an imaging medium can give access to all that is possible under a set of viewing conditions it may not be able to match what is possible under others. For example, there are deep reds that can be seen at sunset that are not possible under the viewing conditions at midday, or blues seen under a fluorescent light source cannot be experienced under a tungsten one.

It is evident, therefore, that gamut differences will persist in spite of all possible efforts to improve the capabilities of imaging devices, and robust ways of dealing with them will remain a necessary and permanent part of color reproduction.

ACKNOWLEDGEMENT

A portion of text in paragraph 5.1 has been reproduced by permission of the International Commission on Illumination (CIE).

6
Color Spaces for Gamut Mapping

Many descriptions of GMAs start out with stating that they intend to achieve their aim by establishing specific relationships between the *appearance attributes* of a source and a destination. For example, in the majority of cases the intention is to keep a source's hues unchanged. Some methods then intend to scale the source's lightnesses linearly to fit them into the destination's range. Other methods want to achieve a mapping that changes source colors' chromas while maintaining their lightnesses and hues. Yet others want to map source colors onto those colors in a destination gamut that look most similar to them one by one.

None of these intentions, however, can be acted upon directly, as it is not appearance attributes themselves that can be mapped. Even though such attributes are effortlessly available to any viewer, they are not immediately quantifiable – i.e. there is no way of directly measuring the brightness, colorfulness, hue, or other appearance attribute of a stimulus. Instead, what can be measured are amounts of energy present in different parts of the visible spectrum (Section 2.5), and the computation of color appearance attributes from such measurements is a process of prediction rather than conversion. Given physical measurements, a multitude of approaches has been developed over the years to predict appearance attributes from them, and the continued research in this area is further evidence of their predictive status. As a consequence of such variety, this chapter will examine the issues involved in choosing a color space for gamut mapping and give an overview of some of the most frequently used ones.

Note also that the term *color space* does not necessarily refer to a representation that is derived from measurement (e.g. CIELAB), but can also be the space that results from making psychophysical magnitude estimations. For example, when a person looks at a colored surface and reports that it has a lightness of 30, a chroma of 20 and a hue of 10% red and 90% yellow, they are providing coordinates for that surface's color in their own color space. In fact, these psychophysical color spaces are the ones that truly merit the label 'color space' and those derived from measurement are merely attempts at approximating them.

6.1 IMPLICATIONS OF MAPPING APPEARANCE PREDICTORS

Before looking at color appearance spaces used for gamut mapping, it is worth making the implications of their predictive nature explicit, where there are three key aspects that warrant closer attention: cross-contamination, linearity and uniformity. In their description, the prefixes 'e' and 'p' will refer to 'experienced' and 'predicted' respectively; e.g., eL is experienced lightness and pL is predicted lightness. Note also that any space that has dimensions predicting color appearance attributes will here be considered to be a *color appearance space* – not only the spaces that form part of color appearance models (CAMs). Therefore, both CIELAB and CIECAM02 will be referred to as color appearance spaces, even though the range of conditions for which these two can make predictions is very different (Moroney, 1998) and the latter is much more broadly applicable and has uses well beyond providing a space for gamut mapping.

Cross-contamination refers to cases where changes to the predictor of one appearance attribute also result in changes experienced for another appearance attribute. For example, if a given algorithm intends to maintain hue, then this is implemented by maintaining the hue *predictor* value in a given color space. If, however, this hue predictor does not accurately predict hue, then the reproduction will have unwanted properties. Imagine a highly chromatic blue color in an original image and an algorithm that intends only to reduce its chroma and leave lightness and hue unchanged. In spaces where there is cross-contamination, changing the chroma predictor's value will also change the other appearance attributes of the color (Figure 6.1). If such a transformation were made in the CIELAB color space, then the resulting color's hue would change from blue to purple in addition to having reduced chroma, even though the CIELAB hue angle would remain unchanged. In other words, in CIELAB, the hue and chroma predictors are cross-contaminated in some parts of color space (notably around blue).

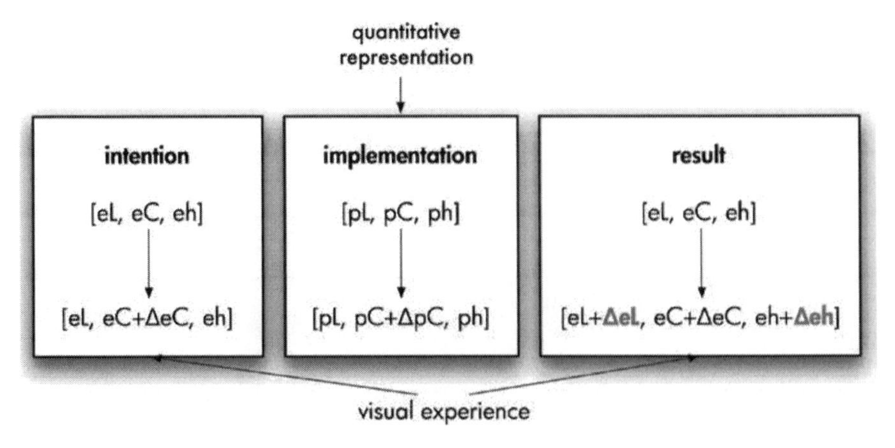

Figure 6.1 Effect of cross-contamination on mapping.

Linearity refers to linearity between changes in an appearance attribute predictor's value and changes in the perceived magnitude of that attribute (or linearity in differences rather than absolute values, as it is only this that applies to hue). Using the above notation, for two lightness differences ΔL_1 and ΔL_2 it must be the case that if $\Delta eL_1 = k\Delta eL_2$ then $\Delta pL_1 = k\Delta pL_2$, too, for there to be linearity between the visual experiences and their predictions. In other words, if the visual difference between one pair is k times that of another pair, then the predicted differences also need to differ k times. The consequence of this is that if a GMA intends to scale chroma linearly and it applies linear scaling in a space that does not linearly predict chroma, then the end result will be an effective scaling that is not the intended one.

The third aspect, i.e. *uniformity*, requires that, in addition to linearity of each attribute's predictor, the units of the predictors be the same and that a unit distance in any direction, anywhere in the space, represents the same perceived difference. For lightness and chroma, if $\Delta eL = k\Delta eC$, then it must be the case that $\Delta pL = k\Delta pC$. Since hue is angular, the uniformity requirement translates to a requirement for metric hue difference, where metric hue difference ΔH (a distance in color space) is that part of the Euclidean distance ΔE between two points (colors) in a color space that is not accounted for by lightness and chroma differences: $\Delta H = (\Delta E^2 - \Delta L^2 - \Delta C^2)^{0.5}$ (Figure 6.2).

Figure 6.2 also shows a pair of orthogonal dimensions, redness–greenness and yellowness–blueness, that are in most appearance spaces denoted as a and b respectively and that can be computed from chroma C and hue h (which form polar coordinates) as follows:

$$a = C^* \cos(h) \tag{6.1}$$

$$b = C^* \sin(h) \tag{6.2}$$

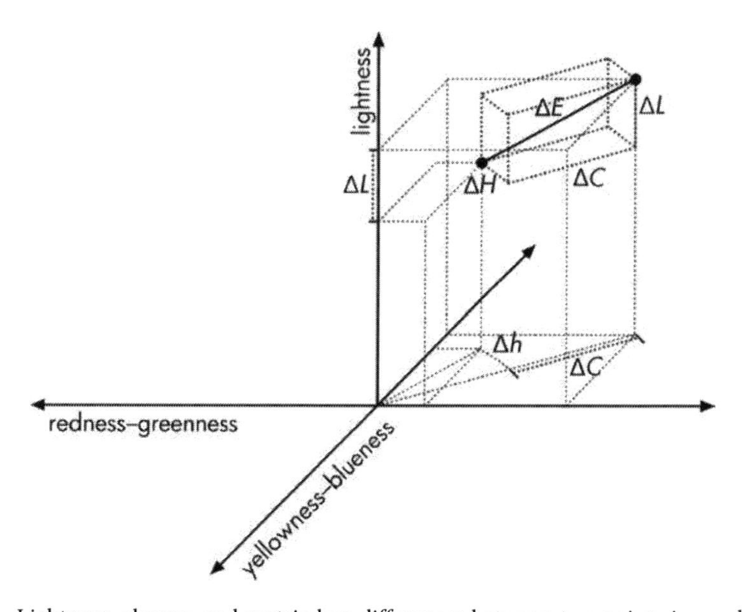

Figure 6.2 Lightness, chroma and metric hue differences between two points in a color space.

Care needs to be taken when doing this conversion in practice to ensure that the *ab* coordinates obtained from *Ch* are located in the correct quadrant. Even though it is the chroma and hue attributes and their predictors that are chosen when deciding which space to perform a mapping in, their orthogonal equivalents are frequently used in at least part of a mapping's computation.

For algorithms that are defined in terms of an intended effect in color appearance, the best color space for their implementation is one that most accurately predicts color appearance attributes, that is linear and uniform and that does not suffer from cross-contamination. One point to note, though, is the use of mapping parameters in this context. For example, let us take an algorithm that uses a constant chroma expansion of 20%, where this expansion factor was determined in a psychovisual experiment and in terms of some predicting color space (e.g. CIELAB). If the algorithm is later applied in a more accurate color space (e.g. CIECAM02), then the previously optimized parameter may no longer be optimal and new psychovisual determination may be needed. This is because parameter values are derived not in the space of experienced appearance attributes, but are predicted ones; therefore, they also depend on the inaccuracies of the space for which they are determined.

When applying a GMA defined in one space in another one, it is important, therefore, to distinguish between those aspects of the algorithm that are defined in terms of experience (e.g. hue–preservation) and those that are expressed in terms of the color space used (e.g. setting a 30% limit to CIELAB C^* expansion).

6.2 WHICH APPEARANCE ATTRIBUTES' PREDICTORS TO MAP

Even given a color appearance space that perfectly predicts all color appearance attributes, the choice of which ones to apply a mapping to remains, and the principal choice is between two sets: an absolute one (brightness, colorfulness and hue) and a white point relative one (lightness, chroma and hue). Which of these sets to use will depend on the chosen desired reproduction property and on the extent to which both the chromaticity and absolute intensity of the light source under which the original is seen need to be reproduced.

If the original's appearance is to be reproduced in an absolute way, then brightness and colorfulness are used and the resulting reproduction is an attempt at showing, under the reproduction's viewing conditions, what the original looked like under its viewing conditions. For example, when reproducing art, it is the appearance of a painting as seen in its original surroundings that is desired in the reproduction. Instead of wanting the reproduction to look like the original artwork would in one's living room, it is also the effect of the original viewing conditions that is desired, since it is under those conditions that the original has its canonical appearance.

If, on the other hand, it is the original's appearance relative to its viewing conditions that is of importance, then lightness and chroma are the right choice and the result is an attempt at showing what the original would have looked like if it was viewed under the reproduction's conditions (less color inconstancy – Section 2.7.2). In most cases it is this approach that is taken, since original viewing conditions do not in themselves have value. The result is also a maximization of the gamut available for a reproduction and can even

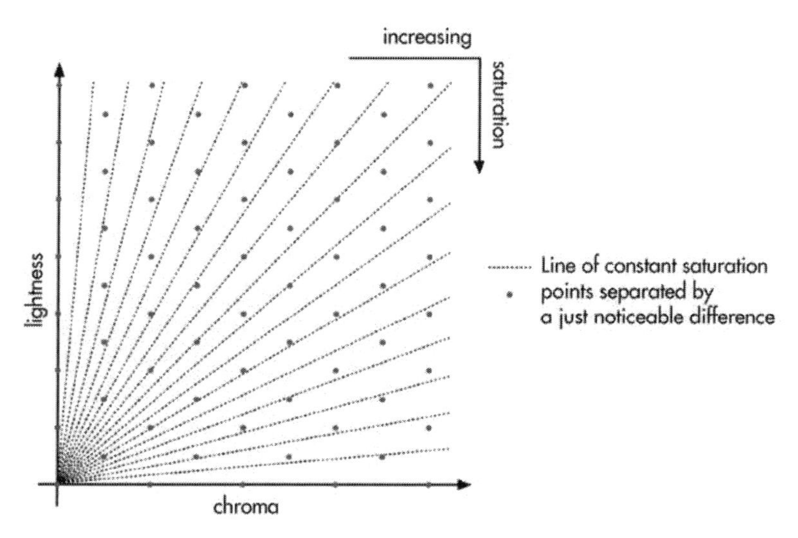

Figure 6.3 Saturation versus JNDs.

lead to enhancements of the original's appearance. These would otherwise be inhibited by the use of absolute appearance attributes – reproducing brightnesses of a dimly lit original in a brightly lit reproduction would lead to lower contrast than when using lightness (revisit Figure 2.2 for an example).

In both cases of attribute choice there is also the option of using saturation instead of chroma, since it has the advantages of being an object's illumination-invariant property (e.g. all surfaces of a directionally lit cube have the same saturation but different chromas and lightnesses) and of being used in shadow series of colors in art. In particular, the former property of saturation is an appealing one in the context of gamut mapping as it would allow for the easy mapping of an original, OOG color to another one that the original would have had under different illumination conditions. Nonetheless, saturation has not been used in published gamut mapping work as it does have the challenge of not being perceptually uniform. Being the ratio of chroma to lightness (and colorfulness to brightness), even only JNDs would correspond to large saturation differences at low lightness levels, whereas even perceptually very different colors could have very small saturation differences at large lightnesses (Figure 6.3). There is also the problem of saturation being undefined at zero lightness.

6.3 OVERVIEW OF COLOR APPEARANCE SPACES

With the above considerations in mind, and given the evolving nature of color science, it is not surprising to see that existing gamut mapping work has been performed in a multitude of color spaces.

Of these, the most common one by far is CIELAB (CIE, 2004a), which predicts lightness, chroma and hue using its L^*, C^* and h^*_{ab} predictors. This is partly due to its having been available since 1976 and partly due to the majority of gamut mapping work being centered on subtractive media, for which CIELAB was originally developed. Its additive counterpart, CIELUV (CIE, 2004a), was also used in a small number of gamut mapping studies. Arguably the most serious shortcoming of CIELAB as a gamut mapping space is the cross-contamination between its appearance attribute predictors. As mentioned previously, taking a highly-chromatic blue color's L^*, C^* and h^*_{ab} values and only reducing its C^* value will result in a color of purple hue. This problem of hue change as a result of chroma predictor change is referred to as a problem of *hue linearity, hue uniformity* or *hue constancy* in the literature and has been extensively studied and documented (Hung and Berns, 1995; Ebner and Fairchild, 1998a; Moroney, 2000a).

In addition to a direct evaluation of hue linearity by psychovisual means, it is also possible to understand it by plotting previously collected hue-linear data in the various color spaces. If the hue-linear data fall on straight lines in a space, then that space is hue linear; if not, then the extent of nonlinearity can be seen and quantified. A popular set of data where there are samples of the same perceived hue but differing levels of perceived chroma and lightness is the *Munsell color order system* (Munsell, 1941), whose basis is a large set of visual judgments about reflective surface color samples. The result is a set of physical samples that are visually evenly spaced in lightness (Munsell *value*), chroma and hue and where the coordinates of each sample have been determined from visual judgment. Figure 6.4 shows the Munsell chroma and hue representations of a set of samples (MCSL, 2006) with a Munsell value of 3 (i.e. CIELAB L^* of 30) and the chroma and hue predictions made for the same samples using a number of color spaces. Note that all of these are representations of a single set of 331 colors and the difference is only in the color space used.

In addition to the CIELAB and CIELUV spaces and the Munsell notation discussed above, Figure 6.4 also shows the result of using the chroma and hue predictors of two CAMs: CIECAM97s and its successor CIECAM02 (CIE, 2004b). Here, there are clear differences in *hue uniformity* and the often-noted weaknesses of CIELAB are less noticeable in the other three color spaces. Of note in the context of gamut mapping is also the IPT color space (Ebner and Fairchild, 1998b), which was specifically developed to provide good hue uniformity, and the experimental data from which it was derived was also used to build a hue linearization of CIELAB in which gamut mapping was successfully performed (Braun *et al.*, 1998).

The obvious question that arises from the above comparison is why gamut mapping is not performed in the Munsell space. The reason is that the Munsell color order system is defined by a set of principles for visual judgment and embodied in a set of physical samples: the *Munsell Book of Color*, made and sold by Gretag-Macbeth (www.gretagmacbeth.com). As such, it is not directly possible to take measurement of an arbitrary stimulus and compute its Munsell notation. Having said this, there has been work on achieving such a computation by interpolating between measurements of the Munsell system's samples, and this has also been used for gamut mapping (Marcu, 1998; McCann, 1999).

The chief limitation of such approaches is that the original Munsell set has a gamut that does not enclose the gamuts of all imaging media and that extrapolation needs to be used to go beyond the physical sample set's gamut. It is important to note here that, while a limited gamut is apparent when using interpolation to compute color space coordinates, it underlies all current color spaces. In the end, every color space (even though it may be

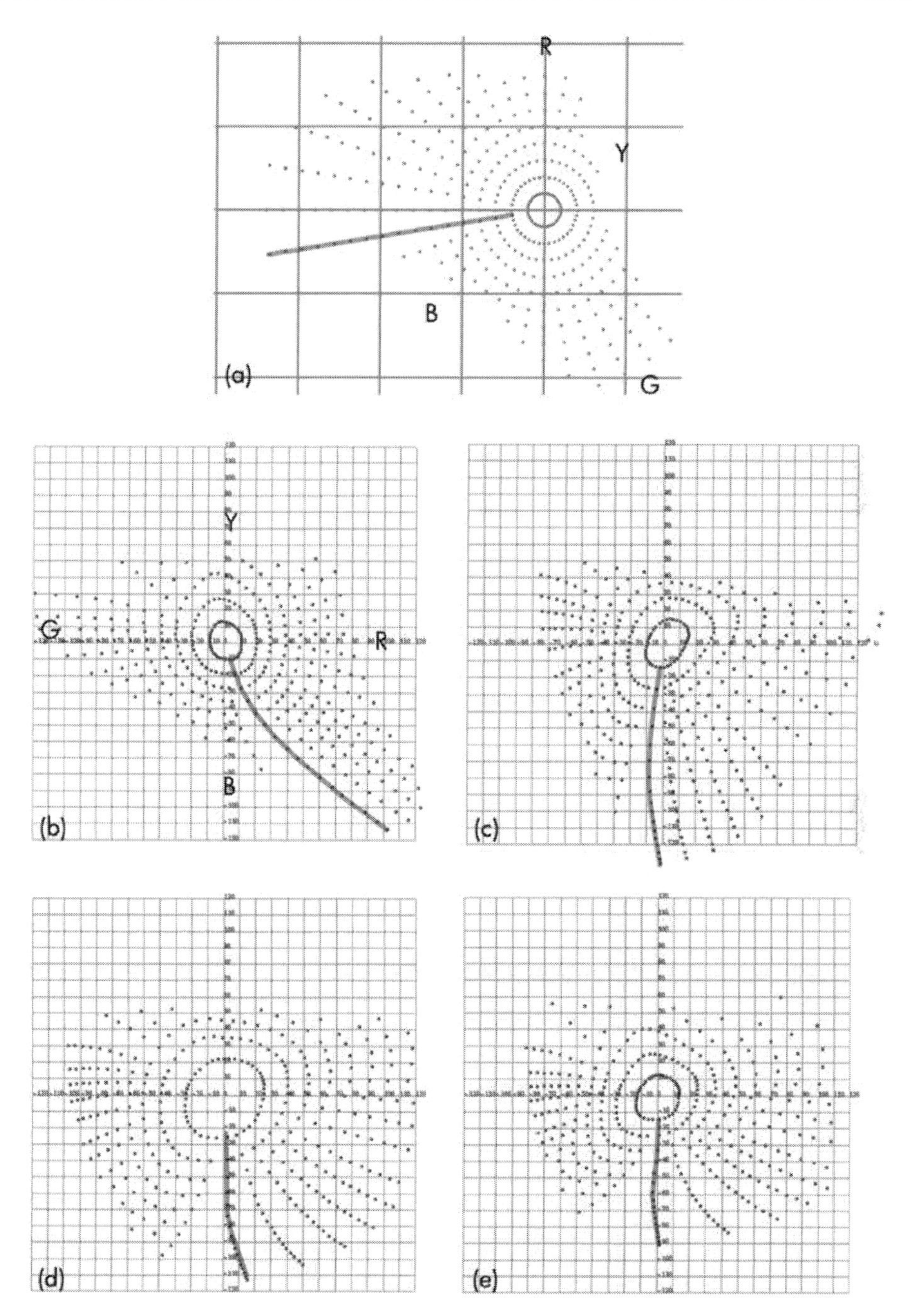

Figure 6.4 Set of (a) Munsell colors and their representations in (b) CIELAB, (c) CIELUV, (d) CIECAM97s and (e) CIECAM02. Transparent lines show a set of blue samples of the same perceived hue and this is the same for (a), (b), (c) and (d). Note that the Munsell representation also indicates the location of red (R), green (G), yellow (Y) and blue (B) regions.

defined by equations whose ranges often exceed even the gamut of theoretically possible stimuli) is derived from or justified by psychophysical data of judgments made about stimuli. These stimuli in turn have a specific *gamut*, and the predictive power of a color space outside that gamut is like that of any extrapolation. Nonetheless, color spaces defined as sets of equations rather than by interpolation in a database of stimulus–coordinate pairs are more widely used, not least due to their greater ease of application.

For completeness it is also worth mentioning other color spaces in which gamut mapping has been done: the color space of the Hunt (1995a) CAM used for gamut mapping by Hoshino and Berns (1993); the Coloroid color space (Nemcsics, 1987) used by Neumann and Neumann (2004); the LABHNU space (a predecessor of CIELAB) (Richter, 1980) used by Laihanen (1987); WUV (Hunt, 1995a: 74) used by Buckley (1978); ATD (Guth, 1989) used by Granger (1995).

In addition to the uniformity, linearity and noncross-contamination properties of a color space, it is also important to realize how differently various color spaces represent gamut differences. To this end, the gamuts of an sRGB display (IEC, 1999) and a *SWOP* (*Specifications for Web Offset Publications*) print (www.swop.org) are shown in Figure 6.5 in the CIELAB, CIELUV, CIECAM97s and CIECAM02 color spaces. It is plain from this visualization that taking the same mapping and applying it to *LCh* predictors in these color spaces is likely to give visually different results for each space. For example, applying the same linear chroma scaling in CIELAB versus CIECAM02 will result in very different

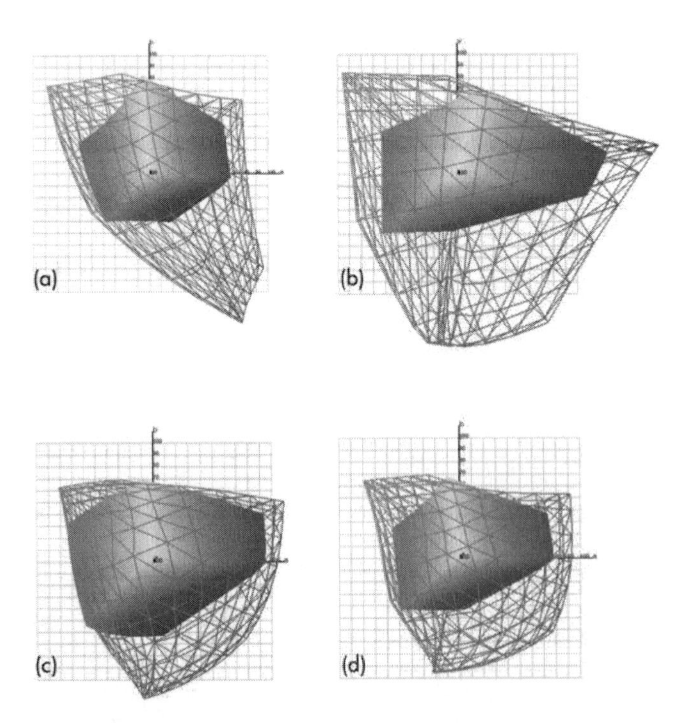

Figure 6.5 Comparing the sRGB (mesh) and SWOP (solid) gamuts in the *ab* planes of (a) CIELAB, (b) CIELUV, (c) CIECAM97s and (d) CIECAM02. Grid squares are 10 × 10 units in size.

compression ratios being used: whereas the blues in CIELAB have C^* values about four times as large in the source as is possible in the reproduction, in CIECAM02 the same blues are predicted to have only about twice as much chroma in the source as is possible in the destination. Remember that both of these predictions are about the same pair of gamuts that have a single, specific experienced relationship.

Considering this fact of a single experience being represented differently in different color spaces, Heckaman and Fairchild (2006a) have asked how well a color space's representation of gamut size relates to that gamut's experience – the *perceptual gamut.* In their psychovisual study, CIECAM02 was found to make good predictions of observer judgments and it will be interesting to see how other spaces fare versus their experimental data.

Using CIECAM02 for gamut mapping requires a careful treatment of cases where the model is used to make predictions for data that do not represent possible stimuli. As is the case with all CAMs, its parameters are set to make predictions of color appearance for actual stimuli (as those are the only ones that can be presented to observers in psychophysical experiments). CIECAM02, in particular, is not well behaved when used to make predictions for tristimulus data that are not physically realizable. What the equations that make up CIECAM02 then result in are abrupt changes in appearance attribute predictions or in impossible predictions (e.g. negative lightness); while this is not an issue for measured stimuli, it can cause problems when implementing gamut mapping. This is because gamut mapping often needs to be applied to stages of a transformation process, some of which can result in appearance values that are physically not possible. Another case where such data need to be processed is the population of ICC profile LUTs (Section 4.5), where those that map from the *PCS* to a device color space do so for a grid that samples an oblong in either CIEXYZ or CIELAB and where many of the samples have values that no physically realizable object would have as measurements.

In practice, the solution to this property of CIECAM02 is to pre-clip inputs to the CIECAM02 model to values that will give valid outputs (Tastl *et al.*, 2005; Kuo *et al.*, 2006). For example, Kuo *et al.* (2006) propose to clip to a convex polyhedron in XYZ defined by the following four inequalities in the xy chromaticity space:

$$\begin{aligned} 0.468391x + 0.767661y &\geq 0.078684 \\ -0.27662x + 1.13699y &\geq -0.04641 \\ x + y &\leq 1 \\ x &\geq 0 \\ y &\geq 0 \end{aligned} \tag{6.3}$$

6.4 MAPPING IN NONAPPEARANCE SPACES

In addition to wanting to achieve specific relationships between source and destination appearances, some gamut mapping methods also set out to perform at least part of their work in other domains.

The first of these other types of space is *CIEXYZ.* A linear scaling in this space is the basis of, for example, *black point compensation* (BPC; Adobe, 2006), popularly made available in Adobe's imaging products. What this step, which precedes other gamut mapping operations, does is to scale the source's XYZ range to that of the destination and, as it is done using media-relative colorimetry (i.e. the adopted or reference white is the white point of each given medium), the white points are already represented by the same XYZs and it is only the difference in the black points that is overcome as follows:

$$\begin{bmatrix} X_{\mathrm{BPC}} \\ Y_{\mathrm{BPC}} \\ Z_{\mathrm{BPC}} \end{bmatrix} = \begin{bmatrix} X_{\mathrm{W}} \\ Y_{\mathrm{W}} \\ Z_{\mathrm{W}} \end{bmatrix} - \begin{bmatrix} \frac{Y_{\mathrm{W}} - Y_{\mathrm{BPD}}}{Y_{\mathrm{W}} - Y_{\mathrm{BPS}}} & 0 & 0 \\ 0 & \frac{Y_{\mathrm{W}} - Y_{\mathrm{BPD}}}{Y_{\mathrm{W}} - Y_{\mathrm{BPS}}} & 0 \\ 0 & 0 & \frac{Y_{\mathrm{W}} - Y_{\mathrm{BPD}}}{Y_{\mathrm{W}} - Y_{\mathrm{BPS}}} \end{bmatrix} \begin{bmatrix} X_{\mathrm{W}} - X_{\mathrm{S}} \\ Y_{\mathrm{W}} - Y_{\mathrm{S}} \\ Z_{\mathrm{W}} - Z_{\mathrm{S}} \end{bmatrix} \tag{6.4}$$

where $X_{\mathrm{W}}Y_{\mathrm{W}}Z_{\mathrm{W}}$ is the white point (e.g. D50 in the ICC PCS), BP stands for black point, S for the source and D for destination.

By applying a linear scaling in XYZ to make the source black point's luminance Y map onto the destination black point's luminance (both normalized), the result is a change in source colorimetry that is akin to a change in illumination intensity for reflective or transmissive objects (Holm, 2005). The change, therefore, is like some natural color changes we encounter in our environment, which is in itself a desirable property for a transformation to have. One of the key challenges in gamut mapping is to avoid making artificial-looking reproductions, and doing something that mimics a natural event can certainly help. Note also that some early gamut mapping work was implemented in its entirety as a mapping in CIEXYZ (e.g. Stone *et al.*, 1988).

The second type of nonappearance spaces that have been used in gamut mapping to perform at least part of the transformation are variants of density. The reason for this is that a lot of the research into color and image reproduction using analogue means was conducted with reference to densitometric data, since such data were successfully used for process control in photography and printing and since it also bears a closer relationship to appearance than luminance-linear spaces. Linear and sigmoidal density mapping have been found to work well especially for black-and-white images or the intensity-related aspect of color images.

Density in general is defined as $D = \log(1/R)$, where R is reflectance on a [0, 1] scale and the various types of reflectance differ only in what filter they use to integrate light across the visible spectrum. There are various filters designed for measuring photographic materials (e.g. Status T, Status A; Hunt, 1995b: 294–304), and when the color matching functions are used as filters the result is *colorimetric density*, which can also be computed as

$$D_A = \log(A_{\mathrm{W}}/A) \tag{6.5}$$

where $A \in \{X, Y, Z\}$ and W denotes the white point.

In digital gamut mapping, examples of using density as the mapping space are some of the algorithms proposed by Buckley (1978).

6.5 CHOOSING A SPACE FOR GAMUT MAPPING

The choice of mapping color space is in some respects as important as the choice of the mapping itself. If the mapping is designed for an ideal appearance space and the space in which it is implemented has serious shortcomings, then the result will not be successful even though the mapping would have worked had the color space been ideal. As a consequence, there is always a degree of ambiguity about whether a given (especially unsatisfactory) result is due to the mapping or the mapping space used.

With this in mind, it is clearly advisable to use the most accurate color appearance space available at any moment if the aim is to apply specific changes (or nonchanges) to appearance attributes and the challenge of using algorithms defined for older spaces lies in how to transport their optimized parameters. At present, the most accurate color appearance space is CIECAM02, and advice for setting its parameters and using it in the context of color management can be found in the work of Tastl *et al.* (2005).

At the same time, it is also worthwhile bearing in mind that not all of a GMA needs to be performed in a single color space, as is the case with BPC (which uses *XYZ* before applying a mapping typically defined in a different space) or, as was proposed by Zeng (2001), where mapping takes place in different spaces for different hues.

This also leads to the very important distinction between an *encoding* and a *mapping* space, whereby the two do not need to be the same. While an encoding space is used only to store color information, a mapping space is one in which changes are made to it. For example, while ICC profiles can contain LUTs that store CIELAB or CIEXYZ values, those spaces are simply encoding (and interpolation) spaces and the data for such LUTs is typically computed by performing mappings in other color spaces. Hence, even though a color reproduction system uses a specific CAM (e.g. CIECAM02), there is nothing to stop the gamut mapping being performed in IPT and all that needs to be added to the process is a transformation from CIECAM02 *JCh* to IPT coordinates and then after gamut mapping has been performed back to *JCh*.

7 Basic Computational Geometry for Gamut Mapping

Colors can be represented as points in color spaces (Chapter 6), and since gamut mapping is implemented as a transformation applied to such points, it is important to have a grasp of the basic geometric operations that underlie GMAs. Rather than being an in-depth introduction to computational geometry, the present chapter intends merely to provide a refresher of the basics that should then make it easier for the interested reader, unfamiliar with this topic, to consult text books on this subject (e.g. Preparata and Shamos, 1985; de Berg *et al.*, 2000) for more in-depth coverage and advanced techniques. This chapter, therefore, is not meant for those familiar with computational geometry, who are advised to proceed to skip it.

At the same time, the basics covered here should be sufficient for the reader to follow Chapters 8–10, which will cover gamut boundary computation and the details of various GMAs. Therefore, the remainder of this chapter will cover: describing points, lines and planes; intersecting lines and a line or a plane; computing the normals of planes and planes orthogonal to lines; and the triangulation of point sets.

7.1 SPACES, POINTS, LINES AND PLANES

The most basic entity that gamut mapping deals with is a single color, which in a color space is represented as a *point*. Each point is in turn defined by its *coordinates*, which express the point's position in terms of each of the space's *dimensions*. In gamut mapping, the three most common types of space are orthogonal, cylindrical and spherical.

Color Gamut Mapping Ján Morovič

7.1.1 Spaces and points

An n-dimensional *orthogonal* space is one whose n dimensions are such that each of them is orthogonal to all the others. For example, the CIECAM02 *Jab* space has three orthogonal dimensions (i.e. $n = 3$): J (lightness predictor), a and b (where a and b are the orthogonal equivalents of C (chroma predictor) and h (hue predictor)) and each color is represented as a point whose coordinates express the point's distance from the origin (i.e. the point having zero for each of the n coordinates, e.g. [0, 0, 0] in three dimensions) along straight lines in each dimension in turn (Figure 7.1a).

A *cylindrical* space is a three-dimensional space that has two orthogonal dimensions and whose third dimension is angular. Here, the first two dimensions still express straight-

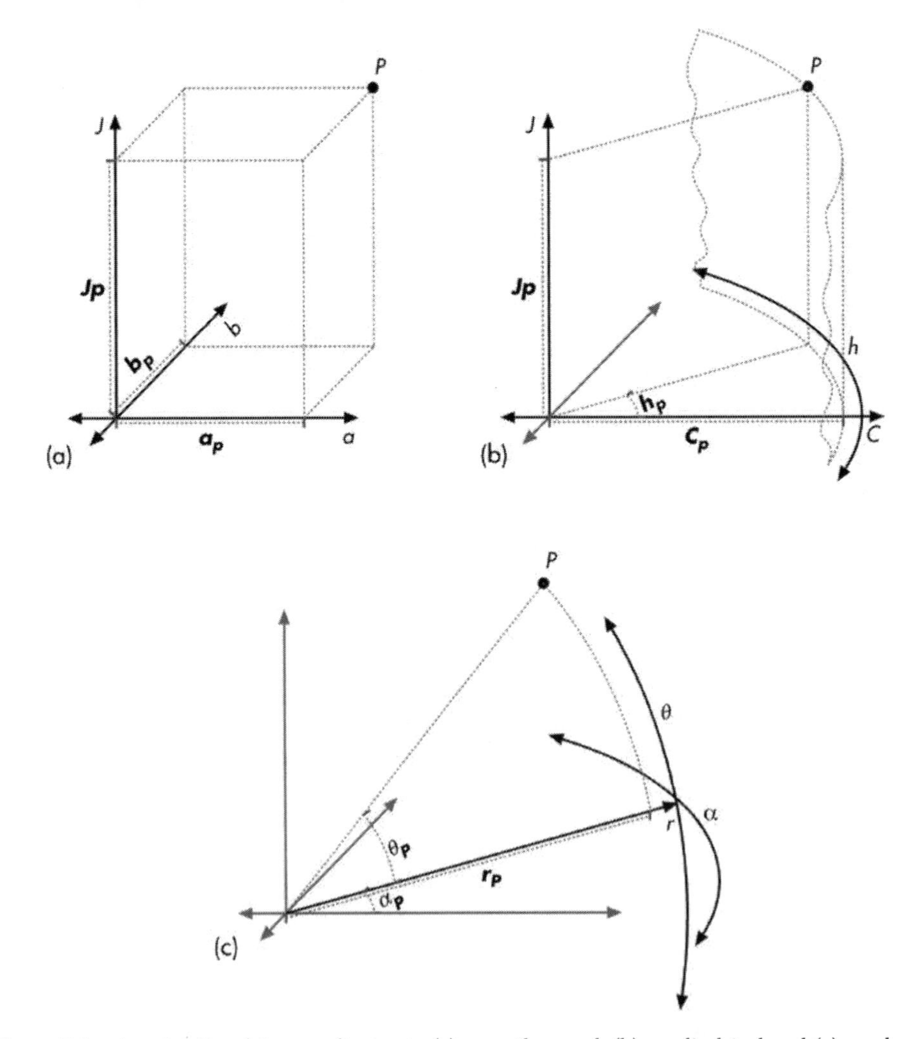

Figure 7.1 A point P and its coordinates in (a) an orthogonal, (b) a cylindrical and (c) a spherical space. Note that it is the same point P relative to the origin in all three cases.

line distance from the origin, but the third dimensions uses angular difference. A practical point that needs to be borne in mind when performing computations in cylindrical spaces is that the angular dimension wraps around (i.e. 361° is the same as 1°) and care needs to be taken when computing angular distances. An example of a cylindrical space is CIECAM02 *JCh* (Figure 7.1b), whose C and h values can be computed from the a and b dimensions of the equivalent orthogonal space as follows:

$$C = (a^2 + b^2)^{0.5} \tag{7.1}$$

$$h = \tan^{-1}(b/a) \tag{7.2}$$

Note that $\tan^{-1}(\ )$ is the inverse of the tangent trigonometric function and that C is Euclidean distance in the two a and b dimensions from the origin. To compute a and b from C and h, revisit Equations (6.1) and (6.2).

A *spherical* space (Figure 7.1c) is then a space that has two angular dimensions and one dimension that expresses straight-line distance from the origin. While there are no color spaces used in gamut mapping that are spherical, it can be useful for some algorithms to take either an orthogonal or a cylindrical space and convert it into a spherical one. For an orthogonal *Jab* space, a spherical equivalent – with angular dimensions α (an angle in the *ab* plane) and θ (an angle in planes orthogonal to *ab*) and distance from origin dimension r – can be computed as follows:

$$r = [(J - J_E)^2 + (a - a_E)^2 + (b - b_E)^2]^{0.5} \tag{7.3}$$

$$\alpha = \tan^{-1}[(b - b_E)/(a - a_E)] \tag{7.4}$$

$$\theta = \tan^{-1}\left\{\frac{J - J_E}{[(a - a_E)^2 + (b - b_E)^2]^{0.5}}\right\} \tag{7.5}$$

where $J_E a_E b_E$ are the orthogonal coordinates of the spherical space's origin. For example, a mapping in *Jab* towards the centre of the *Jab* space can be achieved by placing the center of the spherical space at $J_E a_E b_E = [50, 0, 0]$ (i.e. half-way up the lightness axis) and then simply changing the spherical r value as needed.

7.1.2 Lines

For each point P of a given line, the following equation describes the coordinates of its points in n dimensions:

$$P = A + t\mathbf{u} \tag{7.6}$$

where each point on the line has a different value of t and fixed values of A and $\mathbf{u}$. If a line is determined by two points $J = [j_1, j_2, \ldots, j_n]$ and $K = [k_1, k_2, \ldots, k_n]$, then $A = J$ and the vector $\mathbf{u} = [k_1 - j_1, k_2 - j_2, \ldots, k_n - j_n]$. Furthermore, $t = 0$ for J and $t = 1$ for K, points in the JK line segment have values from the [0, 1] interval, and points further from J than K have values of $t > 1$. For example, if K is a point for which we want to know whether it is

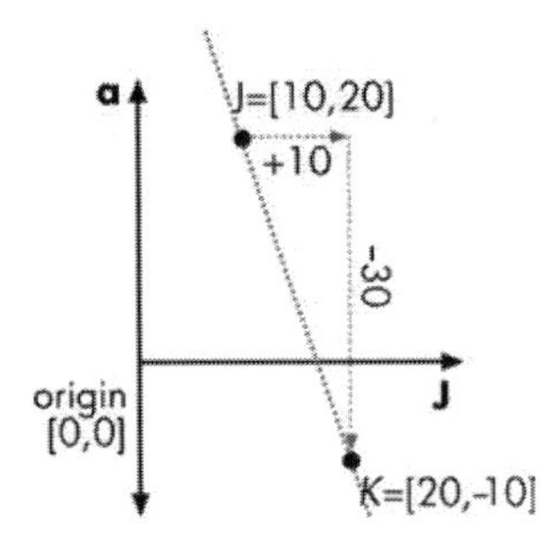

Figure 7.2 Two points and the line they define in two dimensions.

IG or OOG, we can compute the line going from the centre of the gamut and passing through K. If the intersection between that line and the gamut has $t \geq 1$ then K is IG, otherwise it is OOG.

For example, for two colors J and K with Jab values of $J = [10, 20, 30]$ and $K = [20, -10, 40]$, $A = [10, 20, 30]$ and $\mathbf{u} = [20 - 10, -10 - 20, 40 - 30]$, i.e. $\mathbf{u} = [10, -30, 10]$ (Figure 7.2). From Equation (7.6) we can then compute points on the line with desired properties. For example, the point L that is twice as close to K as it is to J can be obtained by setting $t = 2/3$, which then gives $L = [10 + 10 \times 2/3, 20 + (-30) \times 2/3,\ 30 + 10 \times 2/3]$, i.e. $L = [16.67, 0, 36.67]$.

Finally, note that the above *parametric* way of defining a line is not unique; it is only one of several alternatives, each of which has its own attractions.

7.1.3 Planes

For each point Q of a given plane, the following equation describes the coordinates of its points in n dimensions:

$$Q = B + r\mathbf{v} + s\mathbf{w} \tag{7.7}$$

whereby each point in the plane has a different set of r and s values and fixed B, $\mathbf{v}$ and $\mathbf{w}$ values. For a plane determined by three points $J = [j_1, j_2, \ldots, j_n]$, $K = [k_1, k_2, \ldots, k_n]$ and $L = [l_1, l_2, \ldots, l_n]$ (which must not be collinear), $B = J$ and the vectors $\mathbf{v} = [k_1 - j_1, k_2 - j_2, \ldots, k_n - j_n]$ and $\mathbf{w} = [l_1 - j_1, l_2 - j_2, \ldots, l_n - j_n]$.

By adding another point $L = [-10, 30, -20]$ to the example given above we obtain a plane defined by $B = A = J$, $\mathbf{v} = \mathbf{u}$ and $\mathbf{w} = [-10 - 10, 30 - 20, -20 - 30]$, i.e. $\mathbf{w} = [-20, 10, -50]$.

7.2 INTERSECTIONS

Computing the intersection of a pair of entities (e.g. lines, planes) involves the simultaneous solution of the equations that define each of them, since an intersection is described by the equations that define both. While there are many intersection cases in general, two in particular are popular in gamut mapping computations and will be dealt with in more detail next.

7.2.1 Line-line intersections in two dimensions

Since many GMAs operate in constant hue-predictor planes and as they perform changes along lines, they need to know where a line intersects a gamut boundary, which in two dimensions is often represented as a polygon. Each polygon is in turn the combination of line segments, and finding the intersection of a line and the gamut boundary will involve intersecting the mapping line and the lines of the gamut boundary polygon's segments.

Given a mapping line M and a gamut boundary line G in two dimensions (designated as X and Y here), the two lines are defined as

$$A_M + t\mathbf{u}_M \quad \text{and} \quad A_G + u\mathbf{u}_G \tag{7.8}$$

Intersecting the lines then corresponds to finding the point where both sets of equations hold; in other words, the values of t and u that represent the same point – the intersection:

$$A_M + t\mathbf{u}_M = A_G + s\mathbf{u}_G \tag{7.9}$$

This is equivalent to

$$A_M - A_G = s\mathbf{u}_G - t\mathbf{u}_M \tag{7.10}$$

which, when the points and vectors defining the lines are expanded, gives

$$\begin{pmatrix} A_{M_X} - A_{G_X} \\ A_{M_Y} - A_{G_Y} \end{pmatrix} = \begin{pmatrix} u_{G_X} & -u_{M_X} \\ u_{G_Y} & -u_{M_Y} \end{pmatrix} \begin{pmatrix} s \\ t \end{pmatrix} \tag{7.11}$$

Solving this system of two linear equations using standard linear algebra techniques (Anton, 2005) gives one of three results: a pair of s and t values that point to the intersection (Figure 7.3) in each of the two lines (if they intersect); there not being an intersection (if the

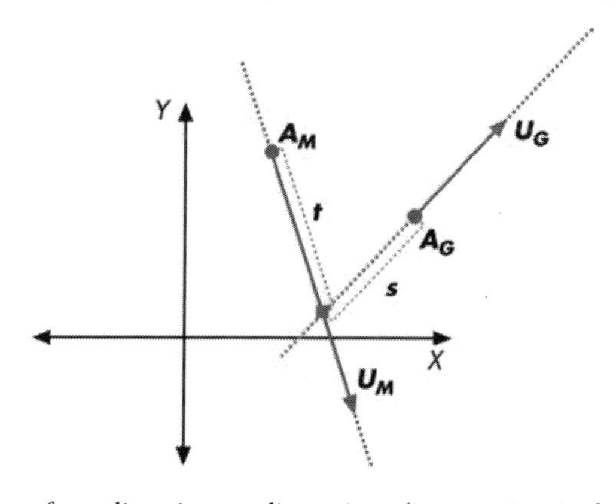

Figure 7.3 Intersection of two lines in two dimensions (intersection is shown by square).

lines are parallel); there being an infinity of intersections (if the lines coincide). When computing solutions to systems of linear equations, care needs to be taken to treat the consequences of finite precision adequately, which here can include the misrepresentation of nearly parallel lines as parallel and, therefore, as not having an intersection.

7.2.2 Line-plane intersections in three dimensions

Analogously, the equations of a line and a plane (Equations (7.6) and (7.7)) in three dimensions can be used for computing the values of the point that results from their intersection:

$$A + t\mathbf{u} = B + r\mathbf{v} + s\mathbf{w} \tag{7.12}$$

The value of t along the line and r and s in the plane obtained by solving the set of three linear equations in Equation (7.12) then gives the position of the intersection point.

The intersection of lines and planes is frequently used in GMAs that do not preserve hue, where the line is often a mapping line and the plane is one of the planes defined by points that delimit a gamut.

7.3 IS A POINT INSIDE OR NOT?

A further step beyond the computation of intersections between entities is the determination of whether the intersection between them lies within a specific region of at least one of them. In gamut mapping it is often important to know not only where two lines intersect, but also whether the intersection occurs between the two points that define one of the lines, i.e. whether it is inside the *line segment* between those points. Another important case is the determination of whether the intersection of a line and a plane is inside the triangle that defined the plane, as such triangles are often used to form gamut surfaces.

7.3.1 Point in line segment

Given the notation used here (Equation (7.6)), it is straightforward to determine whether a point is inside a line segment simply by inspecting its t value. If this value is in the [0, 1] range (where it is zero if the point is at one of the segment's end points and one if it is at the other), then the point is inside the segment; if not, then it is outside. Again, owing to finite precision in computation, it may be desirable to extend this range by a small amount (e.g. 10^{-4}) to avoid false negatives (i.e. in-segment points being deemed to be outside).

7.3.2 Point in triangle

There are a number of ways of determining whether a point is inside a triangle. For example, Scott (2006) describes a method that answers the in-triangle question by realizing that a point is inside a triangle when it is on the same side of each of the lines

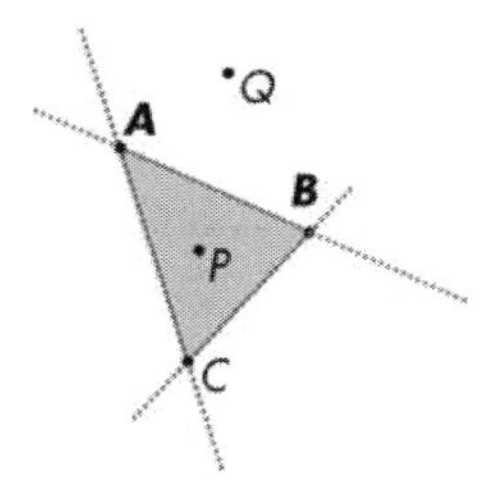

Figure 7.4 A triangle and two points: P inside it, and Q outside it.

that delimit it as the vertex that does not define that line. In other words, in a triangle with vertices A, B and C, if a point P is on the same side of the line defined by AB as C and this is also true for lines AC and BC and points B and A respectively, then P is inside the triangle. For example, in Figure 7.4, P is on the same side of the line passing through A and B, as is C, but point Q is on the other side.

To compute whether two points are on the same side of a line as each other, we can compute the *cross product* of the vectors formed by them and a point on the line and if the *dot product* of the resulting cross products is nonnegative then they are on the same side. For a line defined as $A + t\mathbf{u}$ (Equation (7.6)) and two points O and P, we can first compute two vectors from A to each of O and P, where $\mathbf{o} = [O - A]$ and $\mathbf{p} = [P - A]$. The cross product of vectors $\mathbf{o}$ and $\mathbf{u}$, i.e. $\mathbf{o} \times \mathbf{u}$, in three dimensions ($XYZ$) is then

$$\mathbf{o} \times \mathbf{u} = [u_Y o_Z - u_Z o_Y, u_Z o_X - u_X o_Z, u_X o_Y - u_Y o_X] \quad (7.13)$$

and the cross product of $\mathbf{p}$ and $\mathbf{u}$ is obtained analogously. If we designate the two cross product vectors $\mathbf{o}'$ and $\mathbf{p}'$, then their dot product $o' \cdot p'$ is computed as

$$o' \cdot p' = o'_X p'_X + o'_Y p'_Y + o'_Z p'_Z \quad (7.14)$$

If this scalar is then greater or equal to zero, then both O and P are on the same side of the line defined as $A + t\mathbf{u}$.

Whether a point is inside a triangle or not can also be determined when the plane in which the triangle lies is expressed as in Equation (7.7) whereby B is one of the triangle's vertices and the two vectors ($\mathbf{v}$ and $\mathbf{w}$) are defined by B and each of the two remaining triangle vertices in turn. A point is then inside the triangle if all of r, s and $r + s$ are in the [0, 1] range. The reason for describing these alternative ways of doing the in-triangle test is due to their different relative pros and cons given finite precision and choices made about implementing geometrical computations.

7.4 NORMALS

A *normal* vector of an entity (e.g. line, plane) is a vector that is orthogonal (i.e. is at a 90° angle) to it and in gamut mapping it has particular importance as it can lead to the closest point between one entity and another.

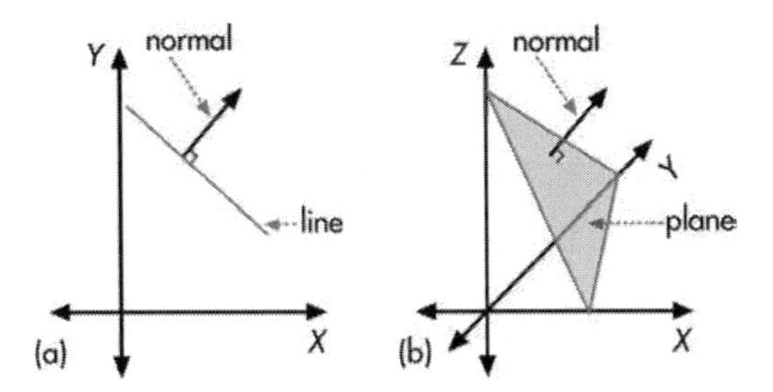

Figure 7.5 Normals to (a) a line in two dimensions and (b) a plane in three dimensions.

7.4.1 Normals of planes

For a point P and a plane α, the shortest distance between P and α is along a line that is orthogonal to α (Figure 7.5b) and that passes through P. Given the parametric definition of a plane shown in Equation (7.7), simply involves computing the cross product of $\mathbf{v}$ and $\mathbf{w}$ (using Equation (7.13)). Defining a line by this cross product vector and P and intersecting it with α then gives that point in α, which is closest to P in terms of Euclidean distance.

7.4.2 Normals of lines

Given a point P and a line l, the shortest distance between l and P is again where a line that is orthogonal to l (Figure 7.5a) and passes through P intersects l. Given a line, its normals can be computed from the vector that is part of its definition (i.e. $\mathbf{u}$ in Equation (7.6)).

Depending on the number of dimensions of the space in which a line is defined, there will be different numbers of normal vectors. In two dimensions (XY) the normal to a vector $\mathbf{u} = [u_X, u_Y]$ is either the vector $\mathbf{u}'_1 = [-u_Y, u_X]$ or $\mathbf{u}'_2 = [u_Y, -u_X]$ (or any nonzero scalar multiple of these) where these two are collinear and differ only in orientation.

In three dimensions, the set of all normals of a line are the vectors that lie in a plane to which that line is orthogonal, which effectively involves doing the inverse of computing the normal of a plane. What we have here is the normal and the two vectors that define the plane to which it is a normal. As such a plane can be defined by an infinity of vector pairs, in practice one strategy is to pick two vectors such that their projections onto two dimensions are made orthogonal by setting suitable constant values for the corresponding members (in Equation 7.15 this was done for X and Y) and then computing values of the third dimension of the two vectors from the line to which a normal plane is desired as follows:

$$\mathbf{v} = \begin{pmatrix} 0 \\ k \\ \frac{-ku_Y}{u_Z} \end{pmatrix} \quad \text{and} \quad \mathbf{w} = \begin{pmatrix} k \\ 0 \\ \frac{-ku_X}{u_Z} \end{pmatrix} \tag{7.15}$$

where $\mathbf{v}$ and $\mathbf{w}$ are vectors defining a normal plane to the line defined by vector $\mathbf{u}$ and k is an arbitrary constant (e.g. 10 in the units of a color appearance space). Care needs to be taken here with regard to the members of **u**, as the above is invalid if $u_Z = 0$ and alternative vectors need be defined in such cases analogously.

7.5 TRIANGULATION

Given a set of points, it is often necessary to construct triangles that fill (*tessellate*) the space occupied by them without overlap, as will be seen in Chapter 8, where the computation of color gamuts is discussed. The reason for this is that, in gamut computation, there is a need both for choosing colors that represent extremes of a color set and then to construct a boundary on their basis. Such boundaries are then intersected with lines along which changes are intended by a gamut mapping algorithm. The step of taking points and defining a surface on their basis typically involves *triangulation*, i.e. the construction of a set of nonoverlapping triangles that span the space occupied by the given points.

As can be seen in Figure 7.6, a given set of points can lead to the choice of alternative sets of nonoverlapping triangles and the question arises as to which of these triangulations to choose. Looking at the alternatives shown here, it can be noted that Figure 7.6b involves some triangles that have very small angles at some vertices and, therefore, look like slivers (Wikipedia, 2006e); Figure 7.6c, on the other hand, has triangles whose angles are more similar to each other at all of a triangle's vertices. The significance of this difference is that 'sliver' triangles are undesirable when it comes to their use in computation due to *finite precision* (i.e. coordinates are expressed on a fine discrete grid rather than a continuum as they are encoded using discrete binary data). The error due to finite precision results in relatively greater differences between an actual triangle and its discrete representation both in terms of its area and where its edges lie; and as these two properties are needed to determine whether a point is inside a triangle or not and where in the triangle a given line intersects it, there is a need to avoid the use of 'sliver' triangles as much as possible. A further disadvantage of 'sliver' triangles is that they involve vertices that are further apart than is necessary and, therefore, define surface based not on local values but ones that span an unnecessarily wide range.

The way to maximize the minimum angle at any vertex is *Delaunay triangulation* (Delaunay, 1934; Preparata and Shamos, 1985), which is also frequently used in practice. The essence of Delaunay triangulation is to construct triangles on the basis of a given point set so that each triangle's *circumcircle* (i.e. the smallest circle enclosing it) contains only three of the point set's points, i.e. the points to whose triangle it is circumscribed (Figure 7.7). For specific point sets this may not always be possible, as sets of (at least) four

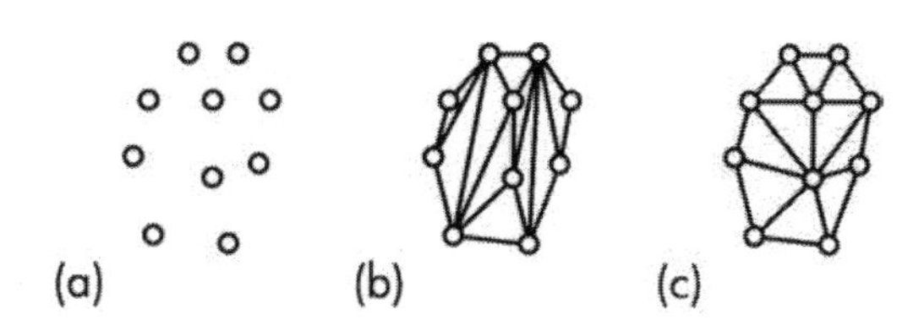

Figure 7.6 Alternative triangulations, (b) and (c), of a point set (a).

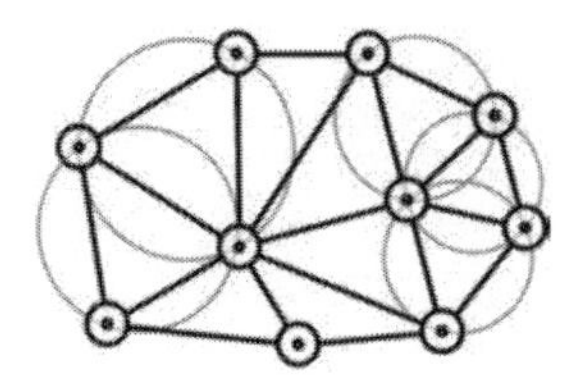

Figure 7.7 Delaunay triangulation of point set showing some of the circumcircles (thin gray lines)

points can lie on the same circle (i.e. be *cocircular*). Such exceptions can then either lead to a choice from among equivalent alternatives (e.g. imagine a square and its circumcircle, which can be triangulated in two equivalent ways, using either one or the other diagonal of the square) or the use of minuscule perturbations that break the cocircularity and allow for the application of the general Delaunay principle (this is typically used in efficient implementations of the algorithm).

Note that this method can also be applied in higher dimensions to tessellate the convex hulls of point sets. In three dimensions, the triangulation becomes a tetrahedralization, i.e. a filling of a point set's convex hull with nonoverlapping tetrahedra whose circumspheres contain no points beyond the tetrahedron's vertices.

7.6 SUMMARY

The above overview of basic geometric entities and the computation of their intersections and normals was aimed at providing a basis to the reader, unfamiliar with the subject, for following the discussion of GMAs in Chapters 9 and 10. Those to whom the content of this chapter was new are advised, though, to obtain deeper understandings of computational geometry and linear algebra before attempting to implement GMAs by consulting, for example, de Berg *et al.* (2000) and Anton (2005) respectively.

8 Color Gamuts and their Computation

The noun *gamut* is defined by the *Oxford American Dictionaries* as 'the complete range or scope of something,' and in the context of color and imaging that *something* is typically *color*. Therefore, *color gamut* will refer to the 'complete range or scope of a set of colors,' which is a more general form of the following definition given by the CIE (2004d: 1):

> a range of colours achievable on a given colour reproduction medium (or present in an image on that medium) under a given set of viewing conditions – it is a volume in colour space.

Note that gamuts can also be considered in other than the color domain. For example, it is possible to talk about the *spectral gamut* of a set of stimuli, i.e. the range they occupy in a space representing spectral power or reflectance rather than color.

In practice, one often comes across statements about the color gamut of a display, a printer, or some other imaging device and it is important to understand the shorthand they imply. As a color gamut is the range of a set of colors, talking about the gamut of a device means referring to a set of colors associated with it in some intrinsic way.

For a display one might think that it is the set of all colors that it can generate by itself, but that is not an entirely accurate interpretation. Instead, it is the set of all colors it can generate under given viewing conditions (e.g. in the dark, or with specific ambient illumination) and for a given viewer (e.g. a particular person or a standard observer), as colors are dependent on more than a device by itself (Section 8.1.1). For projectors and printers the situation is more complicated still, as it is neither printers nor projectors that are viewed themselves. Therefore, all the colors they can generate under given viewing conditions and for a given viewer also depend either on a screen or on paper (or another substrate), using which they generate color stimuli.

Finally, digital cameras and scanners are popularly believed not to have color gamuts; however, applying the same analysis as for output devices, we can see that there are also sets of colors intrinsically associated with input devices. Namely, the subset of all possible colors among which a scanner or digital camera can distinguish and that set (like all sets of colors) has a color gamut. Since that set is intrinsically linked to a camera or scanner (under given conditions), it is its color gamut (Morovič and Morovič, 2003).

The aim of this chapter, then, is to spell out the implications of what a gamut is, point out some differences between how the concept applies to the media brought about by output versus input imaging devices, describe alternative methods of gamut computation, show examples of the gamuts of different sets of colors relevant to color reproduction and, finally, highlight issues specific to the idea of image gamuts.

8.1 CHALLENGES AND IMPLICATIONS OF DEFINITION

The three key questions that need to be asked about color gamuts are: First, what is it that has a gamut? second, is a gamut continuous or discrete? Third, how does sampling relate to gamut convexity/concavity? Misunderstanding any of these is a source of unnecessary difficulties and disappointment when attempting a gamut mapping solution and evaluating its performance.

8.1.1 What is it that has a gamut?

As has been mentioned above, one thing that is commonly ignored when considering color gamuts is that *viewing conditions* are intrinsic to them, since a set of stimuli only has a particular gamut for specific viewing conditions. This follows directly from color being an experience that arises from an interaction of observer, stimulus and viewing conditions (illumination, geometry, surround, etc.) over time (Chapter 2). Being the case for individual colors, it is necessarily also the case for color gamuts, which are simply their ranges. The details of all of the components that contribute to a color gamut need to be specified when talking about it; therefore, general statements about device or medium gamuts are meaningless.

The popular misconception that displays have larger gamuts stems from a misunderstanding of what it is that has a color gamut and statements like '[m]onitors (RGB) can normally render millions of colors while printers (CMYK) can render thousands' (Getty Images, 2004) or '[d]isplays have bigger gamuts than printers' (Suzuki *et al.*, 1999) are good examples of the resulting confusion. While under some conditions displays do have larger gamuts than prints, this is by no means the case in general; in fact, the opposite is often true (see Figure 8.31).

Equally, in general, talking about the color gamut of a printed image is also meaningless, as it can assume a number of forms subject to viewing conditions, observers and viewing sequence, i.e. a printed image has a set of possible color gamuts rather than a single one. Looking at such an image in the dark gives a zero-volume gamut (to use an extreme example), different levels of illumination result in different gamut volumes, illumination chromaticity changes gamut shape as well as volume, and viewing distance and flare in the environment make a difference too. Hence, the question 'What is the gamut of this print?' cannot be answered and should instead be rephrased to 'What is the gamut of this print under viewing conditions X for observer Y when viewed in sequence Z?' This point in particular will be illustrated in Section 8.4 to show that such an understanding of color gamuts is necessary rather than dismissible as being overly pedantic.

8.1.2 Discrete or continuous?

The next challenge is how to express a color gamut, and we will see that this involves an underlying question of whether color gamuts are continuous parts of color space or whether they are simply sets of discrete colors.

The amount of attention given to the question of what range a set of colors has might at first seem odd, since the concept of range is intuitively very simple. This is indeed the case for magnitudes along a single dimension – given a group of people, the idea of the range of their heights is straightforward and can be unambiguously and uniquely expressed by a pair of values (the heights of the shortest and tallest members of the group). Analogously, the range of lightnesses in an image is unproblematic to determine and express as a pair of either lightness judgments made by observers or lightness predictions made by a color appearance model on the basis of measurements.

However, extending the question even to only two dimensions complicates the matter substantially. Take for example the trees standing in a plot of land shown in Figure 8.1 and try to answer the question: 'What is their range?'; or to put it more verbosely: 'Where is the boundary in this plot of land that divides the tree-covered part from the tree-free one?' The most obvious thing about this question is that it is ambiguous and that there is merit in each of at least the indicated answers in the figure. What is important here is that each of the answers spanning from Figure 8.1a (i.e. the smallest rectangle enclosing the trees), via Figure 8.1b (which is like tightly tying a string around the set of trees to delimit them) to Figure 8.1d (which says that there are trees in as many separate parts of the plot as there are trees themselves – each surrounded by a tight boundary) is in some way right and choosing from among them is likely to depend on the intention of the original question.

Returning to color, it should now be apparent that in the three dimensions of color spaces the question of the range of a set of colors is equally ambiguous and methods for computing it will also provide results on a continuum of possibilities. It is worth stating,

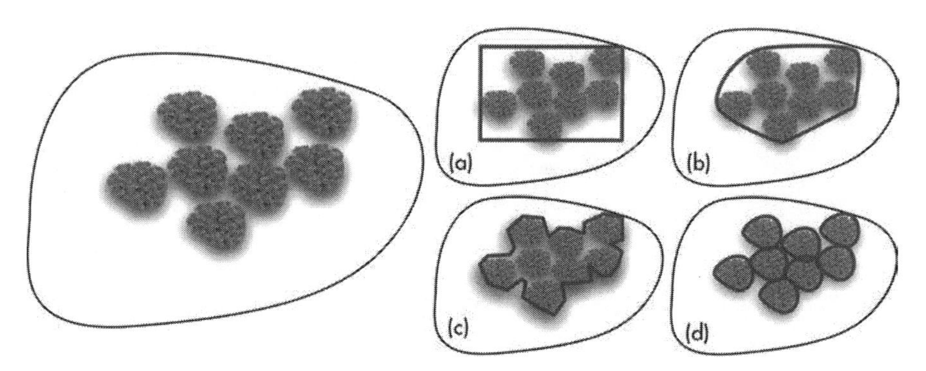

Figure 8.1 A plot of land containing trees and four different ways of drawing a border around them.

though, that the ends of the continuum, as described in the example of trees, are not of equal status when applied to color. While the extreme of drawing boundaries around each individual color in a set does make psychovisual sense, doing the equivalent of drawing a rectangle or tying a string around them is not meaningful. This is because we can never have continuous data about color (either psychophysically or by measurement); therefore, we will only ever have a set of *discrete* colors in a color space. The concept of volume, or *continuity,* is only a shorthand, and its use should be appropriately sparing.

A simple and correct answer to what the range of a set of colors is, is that same set itself, which is equivalent to Rosenblueth and Wiener (1945) saying that 'The best . . . model of a cat is another, or preferably the same, cat'. However, it is often necessary to divide color space into two parts: one that is in the same range as a given set of colors and one that is outside it. Hence, there is a need for a method that takes a set of colors and answers the inside–outside question for any color on its basis. Furthermore, the method needs to work not just for the colors of the set it is based on.

Before moving on and considering the discrete versus continuous question as settled, it is necessary to address a potential doubt. Even if one grants that for sets of colors (that are discrete by definition) the color gamut is discrete too, one might still think that the color gamut of, for example, a display is continuous and that discreteness is only an attribute of the sample on which a computation can be based and not of the entire population. This, however, is not the case either. Taking digital imaging devices, it is apparent that their output is also restricted to a set of stimuli generated on the basis of a finite, discrete (albeit large) set of digital inputs and that these stimuli also form a discrete set. For example, on a digital display that is controlled by 8-bit values having a range of [0, 255] in each of three channels, it is not possible to generate a stimulus that is between the pair obtained by sending [0, 0, 0] and [1, 0, 0] to it. Hence, the display's gamut, too, is discrete rather than continuous and only discretely sampled.

In spite of this underlying discreteness, it is still meaningful to attempt the definition of a continuous boundary in space that separates IG from OOG colors, as there is a distinction between a color being just in-between a pair of the most similar possible discrete IG colors and another color that is further away from any such IG color than its closest possible IG neighbor (Figure 8.2).

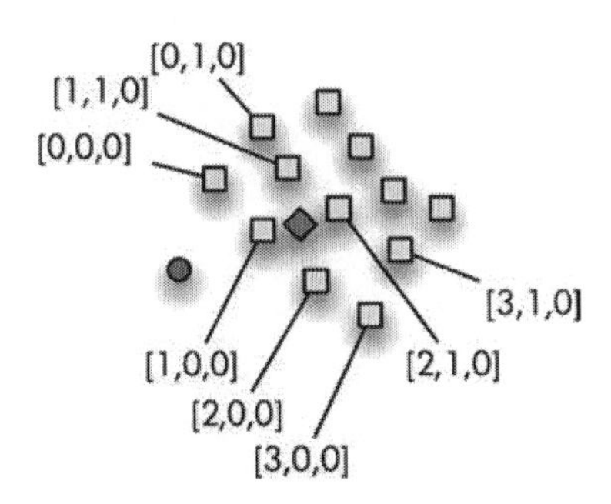

Figure 8.2 IG colors obtained by single digital value differences from an imaging device (square – also showing corresponding digital inputs) and colors from a different source: one that can be considered IG even though it is not one of the discrete population (diamond) and one that cannot (circle).

In summary, color gamuts are fundamentally discrete sets, but they can be – and for practical reasons need to be – represented by continuous volumes. A key example of the need for continuous representations is in fact gamut mapping, where there is a need to know both whether an original color is inside a reproduction gamut and how far it needs to be moved at least when it is outside. As a consequence of fundamental discreteness, though, ambiguity remains about where to place the continuous boundary and how accurate one method is versus another (more on this in Section 8.3.1).

8.1.3 Concave or insufficiently sampled?

A particularly relevant ambiguity regards the relationship of sampling and the placement of a boundary based on it on the one hand and the gamut of the population on the other hand. When the description of the gamut boundary of a given, small set of colors is required (e.g. the 1114 colors of the *Pantone Formula Guide®*, used in graphic design), there is no need for sampling – the gamut can be directly computed from all of the set's colors. However, typically it is the gamut of a multi-megapixel image or of an imaging medium that can receive millions or even billions of different device color inputs and it is impractical to compute the gamut from the entire set of colors that needs to be described. Sampling, therefore, is needed, and the ambiguity we have discussed in the previous section becomes a real issue.

Take Figure 8.3a, for example, which shows the lightnesses and chromas corresponding to a particular, uniform sampling of all possible inputs to a two-channel imaging device (e.g. a printer using two inks, resulting in a duotone print). Given these samples, there

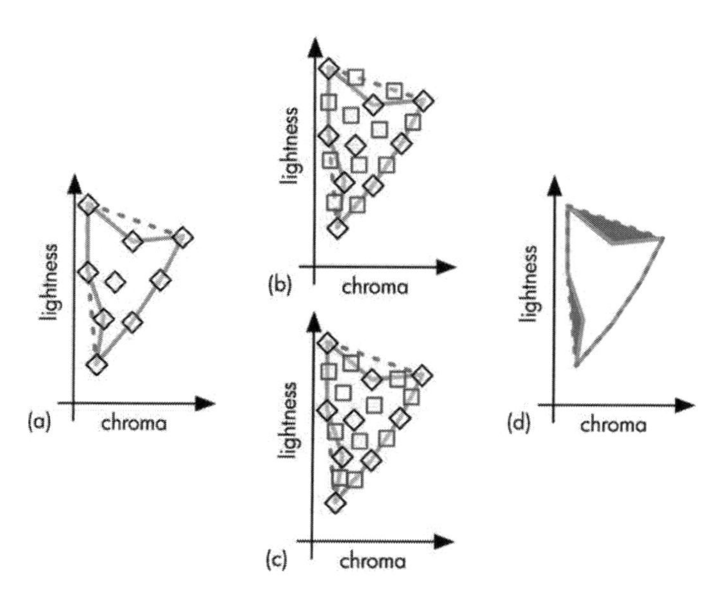

Figure 8.3 Sampling (diamonds) versus describing the population's (diamonds + squares) gamut boundary.

Table 8.1 Implications of how gamut boundary types derived from a sample relate to color population types.

Gamut boundary type	Color population gamut	
	Large (Figure 8.3b)	Small (Figure 8.3c)
Tight (dashed line)	Parts of the population's gamut (shown in gray in Figure 8.3d) are outside the gamut boundary and color reproduction will be to/from the case. This a smaller boundary than is actually will result either in *uncontrolled clipping* (if it is a source gamut) or not using some of the available gamut (if it is a destination)	Good fit
Loose (solid line)	Good fit	Essentially the same problem as having a tight gamut boundary for a large gamut, except that for source gamuts it will result in overcompression and in destination gamuts it will give *uncontrolled clipping*

are many ways of drawing a boundary around them, and two of them are shown in the figure – the dashed line is the 'elastic band' approach mentioned above and the solid one is an attempt to draw the boundary more tightly. Given only this sampling, we have no way of knowing which of the boundaries is closer to the gamut of the entire population of colors we are trying to describe. As a consequence, one of four scenarios can be the case – i.e. the two types of gamut boundary of the sample, let us informally call them *tight* and *loose* here – can relate to the two different types of gamut of the population, i.e. *small* and *large* (Figure 8.3b and c); and Table 8.1 discusses the implications.

A particularly important concept in Table 8.1 is *uncontrolled clipping.* While this phenomenon has different consequences when applied to source versus destination gamuts, in both cases they are undesirable and, especially from the point of view of gamut mapping, disruptive. Uncontrolled clipping is always a consequence of the sample-derived gamut boundary either under- or over-estimating the population's gamut. The consequences of underestimation for a source gamut are that colors outside the estimated boundary will be clipped onto it (or will have to be clipped later in the workflow onto some other gamut) and that this clipping will not be controlled by the GMA. In other words, a source color outside the estimated source gamut is likely to end up outside the destination gamut after gamut mapping as gamut mapping will only be set up to map the source gamut (which, here, is underestimated) onto the destination gamut. So, thinking back to the conceptual workflow (Section 4.2), this means that the inverse CAM will be applied to appearance attributes that are outside the destination gamut, resulting in OOG colorimetry under destination viewing conditions. This will then be the input to the characterization model, where it will be clipped at some point to the gamut of the destination. In the best case this clipping will take place in the colorimetric space (e.g. a mapping onto the point with smallest Euclidean distance on the destination gamut) and in the worst case it will just be range clipped (i.e. each of the device color channel's values will be checked in turn to ensure that they are in the 0 to $(2^n - 1)$ range, where n is encoding bit-depth, and values outside it will be clipped to the range's extremes). In either case, such a source color will not be mapped using the GMA of choice and, therefore, the entire reproduction will not have been obtained as intended. This is likely to cause visual artifacts (e.g. discontinuities, contours) and will certainly interfere with a controlled evaluation of the algorithm used.

The consequences of underestimation for a destination gamut are that not all of it gets used: overestimating a source gamut means that it gets unnecessarily compressed and overestimating a destination gamut would have the same consequences of uncontrolled clipping as the underestimation of a source gamut has. Getting the gamut boundary right, therefore, is of key importance in gamut mapping and controlled color reproduction in general, regardless of the chosen desired reproduction property.

Returning to Table 8.1, we can see that there are two cases where gamut boundaries obtained from a sample match the population well and two where they do not. The problem is only that, when the gamut boundary type is chosen, the type of population is not known and only a sample is available. Therefore, the solution cannot lie solely in the choice of gamut boundary type, but requirements need to be made on the nature of the sampling to which it is applied. As should be apparent from Figure 8.3, if a suitable 'tight' gamut boundary is computed for the population (or the denser sampling), then it results in a good description of both the small and large gamuts, whereas the 'loose' gamut boundary would only work for the large gamut.

Therefore, the best strategy for computing a representative gamut boundary is to use a dense sampling and a 'tight' gamut boundary descriptor (GBD) algorithm. The challenge in practice is then to see how dense is sufficiently dense; for the gamuts of imaging media, for example, a uniform sampling of around 60 (but typically not below 40) samples per dimension tends to work well. For 60 samples per dimension this results in 60^3 (i.e. approximately 2×10^5) samples for a three-channel device and 60^4 (i.e. approximately 10^7) samples for a four-channel device. What this also implies is that, in practice, the sampling needs to be applied to data generated from characterization models (Section 4.2.1) rather than to measured data, unless the measured data are suitably (i.e. nonuniformly) sampled (Green, 2001).

8.1.4 Output versus input devices and their sampling

While the above discussion was conducted in general regarding the color gamut of a set of colors, what is needed in practice are the gamuts of imaging devices (i.e. the gamuts of stimuli that imaging devices can generate or meaningfully capture under given viewing conditions and for a given viewer) and images in them, which in turn requires an understanding of how to sample them.

A digital color imaging device provides a link between digital data and color stimuli and can be of two principal types, depending on whether color stimuli are its inputs or outputs. *Output color imaging devices* (e.g. monitors, projectors, printers) produce color stimuli on the basis of digital data sent to them, whereas *input devices* (e.g. digital cameras, scanners) produce digital data based on sensing color stimuli (Figure 8.4).

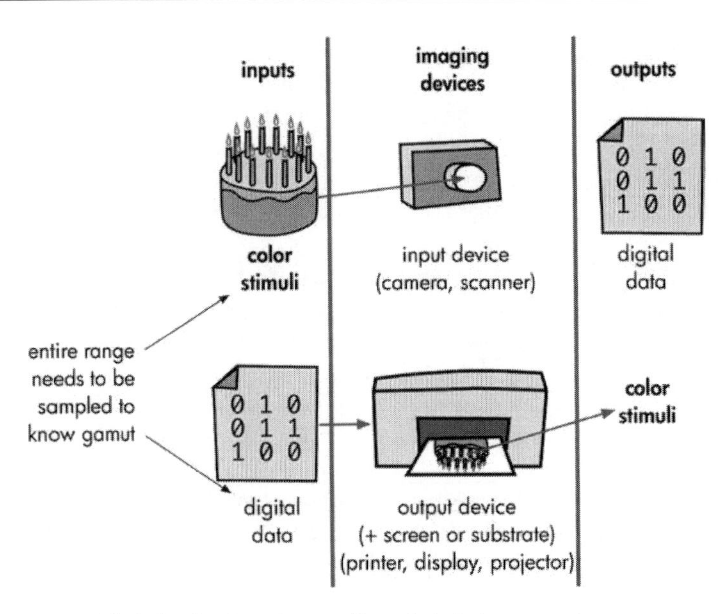

Figure 8.4 Types of digital color imaging media and their inputs and outputs.

Furthermore, the relationship between inputs and outputs of the two types of device depends both on their physical properties (e.g. spectral characteristics of colorants or sensors) and on the parameters of the software used for controlling the corresponding devices. The color gamut of a device, therefore, is always a function of particular states of physical and software parameters plus the other factors already discussed above for color sets in general.

The result then depends on which of the two types a device belongs to (Morovič *et al.*, 2001). For output devices the gamut is the range of color stimuli they can *produce* and for input devices it is the range of color stimuli among which they can *distinguish differences*. In both cases, determining the gamut of a device requires having access to the *entire range of inputs* to it – for output devices this means access to the entire range of digital data that can be input to them and for input devices one needs access to the entire range of color stimuli that can be presented to them for capture. Given access to the entire range of inputs, the gamut is determined on the basis of a device's corresponding outputs.

In the context of input devices, it is also possible to consider the color gamuts of the transformations that are used from captured signal to colorimetry. The colorimetric values that result from such transforms can sometimes be seen to exceed even the spectral locus, and observations like these can help diagnose camera design issues. Such gamuts are called *capture color analysis gamuts* (Holm, 2006), and one could analogously also consider the gamuts of other color transformations (e.g. characterization models) for diagnostic purposes.

Sampling inputs to output devices

Hence, calculating the color gamut of an *output device* consists of sending every possible digital input (or a meaningful sample thereof, e.g. a uniform one) to it, measuring the color of each corresponding output and then calculating a boundary enclosing these colors in a color space (Chapter 6). The generation of inputs to these media is in most cases a trivial matter, as one always has access to their entirety.

Furthermore, if the device is trichromatic (or controlled by three, virtual device color channels – Section 4.1) and the relationship between the medium's inputs and the colors of resulting stimuli (e.g. between RGB and CIELAB for a CRT) is monotonic, then it is sufficient to sample the extremes of the range of digital inputs (e.g. the faces of the RGB cube for a CRT), as these will correspond to the extremes of the resulting stimuli's colors – the gamut boundary. The gamut obtained from sampling only the surface of the device color cube is referred to as the *physical color gamut*, whereas sampling the entire volume yields the gamut in general. Devices whose physical color gamuts match their actual color gamuts are then considered to be *well behaved* (Mahy, 1997, 2002).

When more than three channels are used to control an imaging device, sampling the corresponding n-dimensional hypercube can become more problematic as n increases. Whereas uniform sampling (i.e. taking k^n samples, where k is the number of evenly spaced samples taken per dimension and n is the number of dimensions) for low dimensions is feasible, it becomes prohibitive for higher dimensions. If one were to attempt taking 20 uniformly placed samples per dimension and consider all the combinations in a 10-channel device (e.g. a printer or projector), then there would be 20^{10} samples (i.e. approximately 10^{13}), and this sampling density might not even be sufficient to distinguish concavities in the gamut versus gaps due to sampling. To address this challenge, two sampling techniques have been proposed (Morovič, 2007; Zhao, 2007) that exploit the fact that it is not the sampling of an arbitrary n-dimensional space, but of a device color space.

Sampling inputs to input devices

The reason why complexity arises for the gamuts of *input devices* is that sampling the entire range of possible inputs to them means sampling the entire range of possible color stimuli. This is the case because, to determine the range across which differences in stimuli can be sensed, a set of stimuli with a gamut greater than or equal to the gamut of the given input device needs to be available. As an input gamut to be determined is (by definition) not known, only the entire possible gamut of color stimuli can be *a priori* known to be greater than or equal to the input gamut in question. Once a set of samples from the entire possible gamut of stimuli is available, it is necessary to know the device's responses to each of them and then to determine the device's gamut boundaries.

While the physical generation of a meaningful sampling of the set of all possible colors is impractical at best, it can be achieved by means of the following computational simulation (Morovič and Morovič, 2003).

The set of all possible surface reflectances resulting in visual responses is bounded, as it is determined by spectral reflection or transmission properties that are bounded themselves. While this set, the *object color solid* (OCS – Schrödinger, 1920), does not as such represent all possible stimuli, it does represent all possible stimuli resulting from a particular light source's output being reflected from or transmitted by all possible surfaces. Here, the stimulus that has maximum energy S_{max} is defined by the surface that reflects all light at every wavelength and thereby results in all the energy of the light source being reflected. Note that there is also an indirect way of making the OCS represent all possible stimuli; for details, see Morovič and Morovič (2003).

Computing the gamut of an input device, therefore, will require samples of the OCS, which can be obtained by varying reflectance values independently across the visible spectrum. However, this results in a very uneven distribution in color appearance terms, and to achieve a sufficiently dense sampling would require the calculation of an extreme amount of spectra. Using 31 intervals between 400 and 700 nm (i.e. at 10 nm steps) to represent spectra results in 2^{31} (i.e. approximately 2×10^{9}) samples if only 0% and 100% levels are considered. Wanting to increase the number of reflectance levels sampled even to three increases this number to 3^{31} (i.e. approximately 6×10^{14}), which is unacceptably high in terms of computation cost and still far too crude for sufficiently sampling color appearance space in visual terms.

A much more efficient approach is to sample in terms of a color appearance space and then calculate spectra for each of the sampled color space coordinates. To do this, the gamut boundary of the OCS is first needed in the color appearance space and can be calculated using the following three-stage process:

1. Calculate the gamut boundary for the 2^{31} 0% and 100% samples in color appearance space using a GBD algorithm (Section 8.2). This is a first approximation of the OCS's gamut – GBD'.
2. Generate further stimuli from the OCS by taking the colors in GBD' and scaling their tristimulus values to p levels between 0% and 100%. The resulting XYZ values are all from the OCS, as they correspond to the XYZ values of scaled versions of the spectra from GBD' and as any spectrum from GBD' scaled by p is in the OCS. Note that these scaled XYZ values correspond to stimuli of constant CIE xy chromaticity, differing in luminance only.
3. The OCS's GBD is then calculated from color appearance coordinates computed in step 2 (Figure 8.5).

For details of how to sample the resulting OCS color gamut and from its sampling (and the simulation of a scanner's or camera's response) estimate an input device's gamut, see

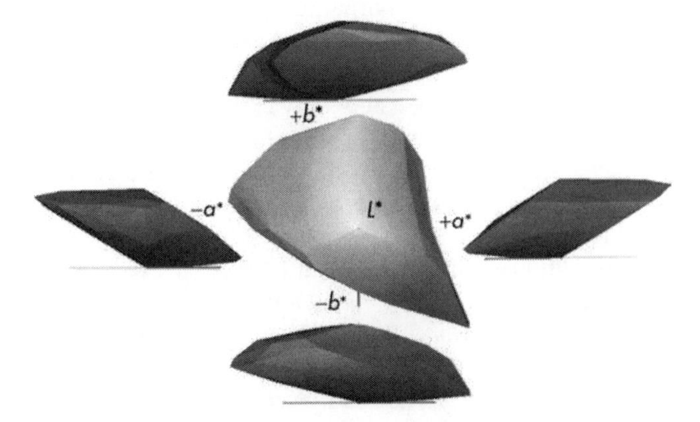

Figure 8.5 The gamut of the OCS in CIELAB for D65 and the 2° observer (labels indicate (half) axes orthogonal to given projection and the size of these (half) axes is 100 units).

Morovič and Morovič (2003). Note that this approach has also been extended to the computation and representation of a camera or scanner gamut in spectral reflectance space (Morovič and Morovič, 2006). Finally, an alternative approach for the colorimetric case has also been proposed by Pujol *et al.* (2003).

8.2 GAMUT BOUNDARY DESCRIPTOR ALGORITHMS

Given the need for delimiting the gamut of a set of colors and the fundamental ambiguity that comes with it, a number of methods of gamut computation have been proposed to date. Before looking at their details, let us first define three key terms (CIE, 2004d, pp. 1):

Colour gamut boundary: a surface determined by a colour gamut's extremes.
Gamut boundary descriptor (GBD): an overall way of approximately describing a gamut boundary.
Line gamut boundary (LGB): the points of intersections between a gamut boundary (as characterized by a GBD) and a given line along which mapping is to be carried out.

Given these definitions, we can see that the methods for computing GBDs come in one of two categories: *medium-specific* and *generic* (Morovič, 2003).

As their name implies, *medium-specific* methods are suitable only for the computation of the GBDs of imaging media. This is the case either because they are tied to particular characterization models (Section 4.2.1), or because they rely on assumptions about the regularities of medium gamuts that are not satisfied by gamuts in general. Methods tied to characterization models include Engeldrum's (1986) approach for calculating GBDs from the Kubelka–Munk equations (Kubelka and Munk, 1931; Sinclair, 1997) that describe how inks mix, and Mahy's (1997) approach for determining the gamut of a printed medium characterized by the Neugebauer equations (Neugebauer, 1937). Other medium-specific methods include Inui's (1993) algorithm using partial differential equations to find the surface in color appearance space beyond which changes to the inputs of a characterization model no longer result in three-dimensional changes in the color appearance space outputs. Another example is also Herzog's (1998) *gamulyt* method, which exploits the fact that the gamuts of most media have the shape of a distorted cube – their vertices being black, white, red, green, blue, cyan, magenta and yellow (Figure 8.6). By representing a medium gamut as a cube distortion, this

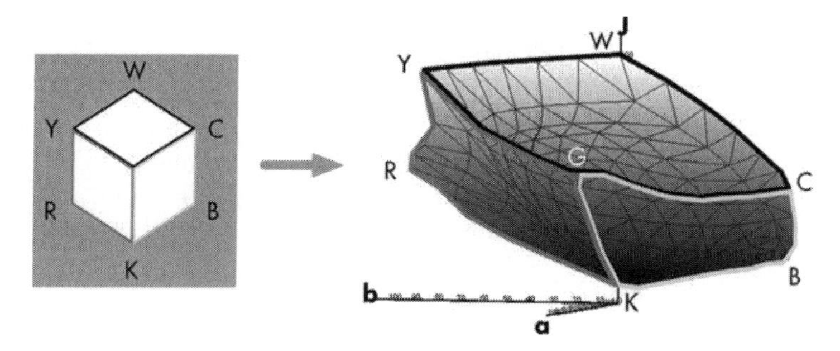

Figure 8.6 Herzog's gamulyt – realizing that medium gamuts tend to have the shape of distorted cubes.

algorithm is able to encode a gamut surface using a particularly small number of parameter values.

While the above methods work well for describing certain medium gamuts, they are not suitable for dealing with image gamut boundaries or the gamut boundaries of other, general sets of colors (e.g. the set of spot colors used in a company's corporate palette). Furthermore, even in the context of medium gamuts alone, some of these methods only work for one type of medium (e.g. print), whereas others work for other types. As a consequence both of applicability only to medium gamuts and the lack of generality even there, it is generic methods that tend to be used more widely.

Generic GBD methods are applicable to the computation of gamuts for, at least in principle, arbitrary sets of color coordinates, and recent years have seen an explosion in their number. What will have to be borne in mind, though, is that the boundaries that each one of these methods generates may be more usable for some sets of colors than for others. In addition to the three methods that will be looked at in detail in this chapter – *convex hull* (Kress and Stevens, 1994), *alpha shapes* (Cholewo and Love, 1999) and *segment maxima* (Morovič and Luo, 1997, 2000) – there are a number of others that each have their own merits and preferred contexts of application.

8.2.1 Convex hull

The convex hull of a set of points is a formal version of the idea of putting an elastic band around a point set. In two dimensions, the convex hull is the smallest convex polygon that contains a given set of points (3D: smallest convex polyhedron; *n*D: smallest convex polytope). A convex polytope then is one that contains all *convex combinations* of its vertices, where a convex combination c is simply a weighted sum of the form

$$c = \sum_{j=1}^{k} w_j p_j \tag{8.1}$$

where k is the number of the polytope's vertices, w_j are nonnegative weights that sum to one and p_j are the n-dimensional vectors of the polytope's vertices.

The simplest method of determining which of a set of points form the vertices of its convex hull in two dimensions (XY) is the so-called *Jarvis March* (Figure 8.7) or *gift-wrapping* (Jarvis, 1973):

1. Choose the set member with the smallest y coordinate, p_0 (guaranteed to be a vertex of the convex hull).
2. The next member of the convex hull, i.e. p_j, is that point from among the set that has the smallest counter-clockwise angle relative to the line connecting the previous two points (for $j = 1$ it is relative to the horizontal axis). The pair of points p_j and p_{j-1} result in a line that has the entire set of points to the left of it.
3. Continue until $p_j = p_0$.

The aim of describing the above method was just to illustrate a particular, simple way of finding the convex hull of a set. In practice, however, other methods are used due to

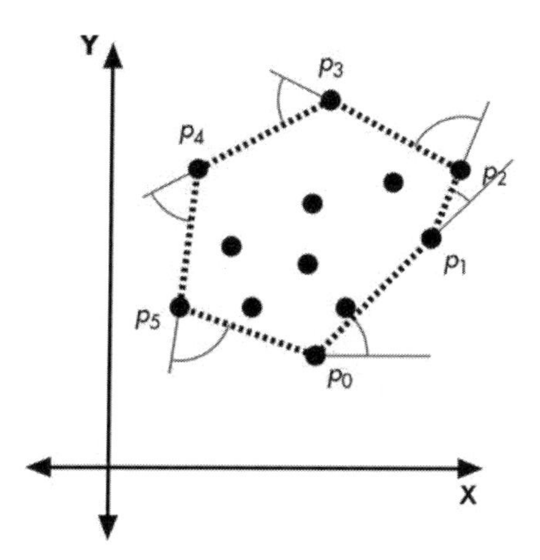

Figure 8.7 Determining the convex hull (dashed line) using the Jarvis march.

greater efficiency, and a particularly popular one is the *Qhull* algorithm (Barber *et al.*, 1996; www.qhull.org).

Using convex hulls as GBDs (Kress and Stevens, 1994) has serious limitations, though, as it is an approach that results in 'loose' boundaries that in many cases can cause the problems discussed in Section 8.1.3, i.e. uncontrolled clipping or excessive compression. For those gamuts that are indeed convex (e.g. the gamuts of additive systems in *XYZ*) or close to being so (e.g. display gamuts even in CIELAB), this method of gamut boundary description works; however, in general it fails, since there are often gamut concavities. For example, Figure 8.8 shows the projection of samples from a print's gamut onto the CIELAB a^*L^* plane alongside the convex hull computed for them. What is apparent from this figure is that there are significant regions close to the

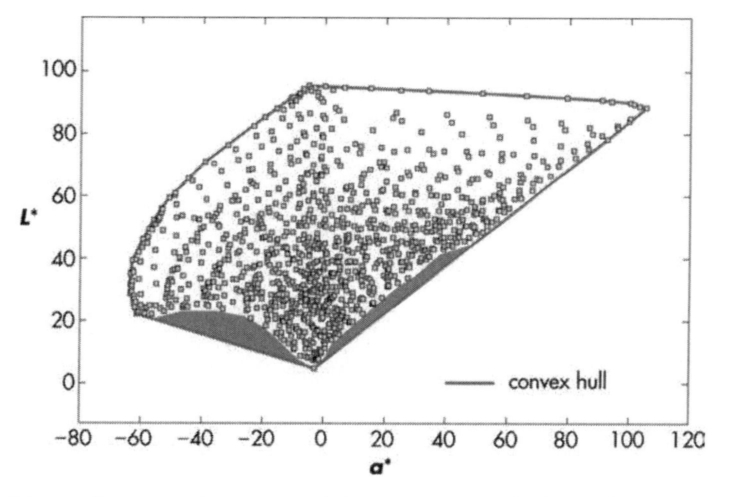

Figure 8.8 Print a^*L^* samples (squares) and their convex hull (solid line).

bottom boundary (shown as gray) that have no samples in them at all and that are likely not to be achievable in the print they represent.

Owing to this challenge of robustly and reliably describing the nonconvex shapes of gamuts, the majority of work on GBDs – aside from efforts to speed up the process or reduce the amount of data needed for encoding the result – focuses on how to provide a gamut boundary description that is more accurate and does not have a convexity constraint imposed on it.

One approach, in fact, uses convex hull computation, but modifies it to allow for concavities to be preserved. Balasubramanian and Dalal (1997) have proposed a *modified convex hull* approach which consists in a preprocessing of the data that makes it more convex, followed by convex hull computation and then the undoing of the preprocessing to bring the data back to its original shape. The preprocessing is akin to an inflation of the data, which brings some of the concavities' surface onto the convex hull and, therefore, allows for its inclusion in the final boundary.

The form of the preprocessing is, for example, to take CIELAB orthogonal coordinates, convert them to spherical coordinates (Section 7.1.1; Equations (7.3)–(7.5)) with the center chosen as some point inside the set to which it is applied (e.g. the mean, or a fixed point at $LAB = [50, 0, 0]$) and then altering the radius value of each point as follows:

$$r' = \alpha(r/\mathrm{r}_{\mathrm{MAX}})^{\gamma} \tag{8.2}$$

where r is the original radius, r_{MAX} is the largest radius of the set, α and γ are parameters that allow for different degrees of 'inflation' and r' is the modified radius. Taking the data from Figure 8.8 and applying this modification with $\alpha = r_{\mathrm{MAX}}$ and $\gamma = 0.2$ (a value that gave good results in a recent study (Bakke *et al.*, 2006)) is shown in Figure 8.9. Then,

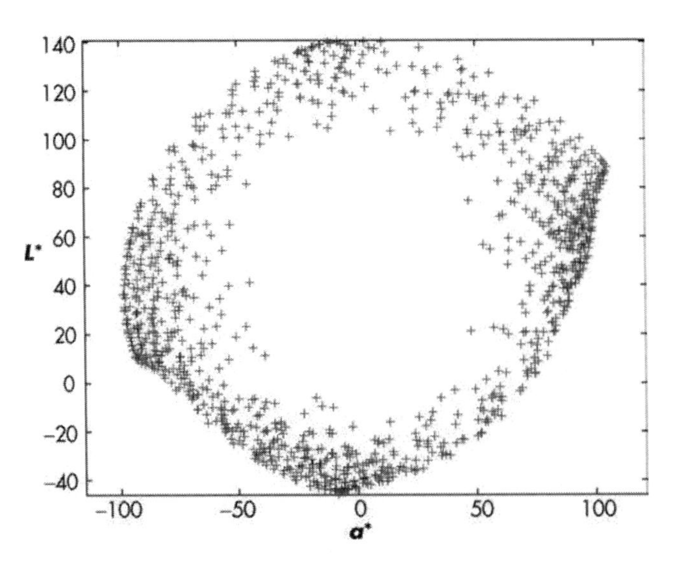

Figure 8.9 Applying the Balasubramanian and Dalal transformation to the data from Figure 8.8 for $\alpha = r_{\mathrm{MAX}}$ and $\gamma = 0.2$.

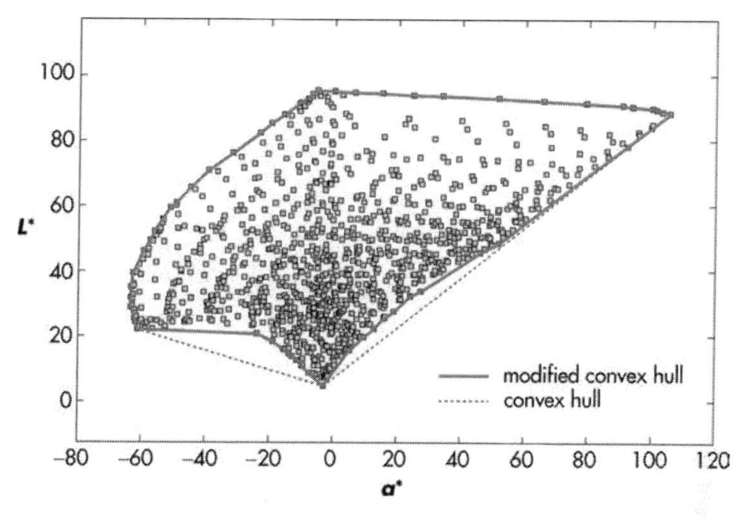

Figure 8.10 The modified convex hull for $\alpha = r_{\mathrm{MAX}}$ and $\gamma = 0.2$.

computing the convex hull in the inflated representation and transforming its vertices back via the inverse of Equation (8.2) gives the boundary in Figure 8.10.

As can be seen, the result is a much improved gamut boundary for the given data set, and this method of gamut boundary description has indeed got very good potential for being used in practice. A key challenge, though, is the need to decide the values of two parameters, as incorrect choices can lead to undesirable results that either over- or under-predict the gamut boundary (Figure 8.11).

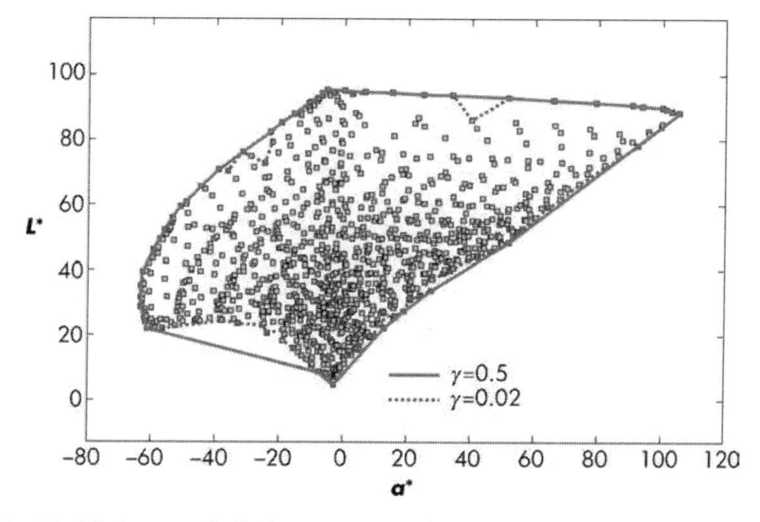

Figure 8.11 Modified convex hulls for $\alpha = r_{\mathrm{MAX}}$ and $\gamma_1 = 0.02$ and $\gamma_2 = 0.5$.

8.2.2 Alpha shapes

The use of alpha shapes (and resulting alpha hulls) was proposed by Cholewo and Love (1999) as a way of describing color gamut boundaries, and this method has enjoyed popularity since. The intention behind the definition of alpha shapes is precisely to offer a way of better describing the shape represented by a set of points; and, to acknowledge the ambiguity also discussed in Section 8.1 here, alpha shapes provide a parameterized shape description. Given a set of points, there is a whole family of corresponding alpha shapes, rather than just one (as was the case with convex hulls).

Intuitively, the idea of an alpha shape of a set of points S can be understood as follows (Edelsbrunner and Mücke, 1994: 45):

> Think of R^3 filled with styrofoam and the points of S made of more solid material, such as rock. Now imagine a spherical eraser with radius α. It . . . carves out styrofoam at all positions where it does not enclose any of the sprinkled rocks, that is, points of S. The resulting object will be called the α-hull . . ..

As can be seen, the *alpha hull* of S is dependent not only on S, but also on the choice of α: if α is infinity, then the result is the convex hull; and if it is sufficiently small, then it is S itself. Values of α in-between allow for a choice of 'resolution' (curvature of permissible concavity), and when used for gamut boundary description they could be related to the properties of the sample from which a boundary is to be derived. Note that for any value of α, the resulting polygon is a subset of the Delaunay tessellation (Section 7.5) of the entire data set (i.e. it can be obtained by removing triangles or tetrahedra from the data set's Delaunay tessellated convex hull). The example in Figure 8.12 shows a set of points, its convex hull and two alpha hulls for different values of α.

Computing an alpha shape boundary for the test data used before (Figure 8.13) shows that it avoids the overestimation of the convex hull. Hence, using it with a GMA would result in a more controlled result. Also shown in this figure are the circles with radius 10 that were at the locations that determined the boundary (i.e. only circles outside the boundary were empty of points, and the circles shown were the closest ones to the boundary that were empty).

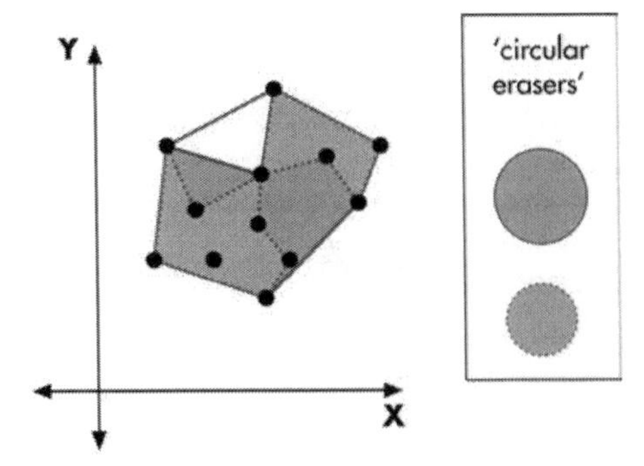

Figure 8.12 Alpha shapes for different values of α. The 'circular eraser,' i.e. sphere with radius α, for each of the two alpha shapes is shown, as is the convex hull (gray solid line with white fill).

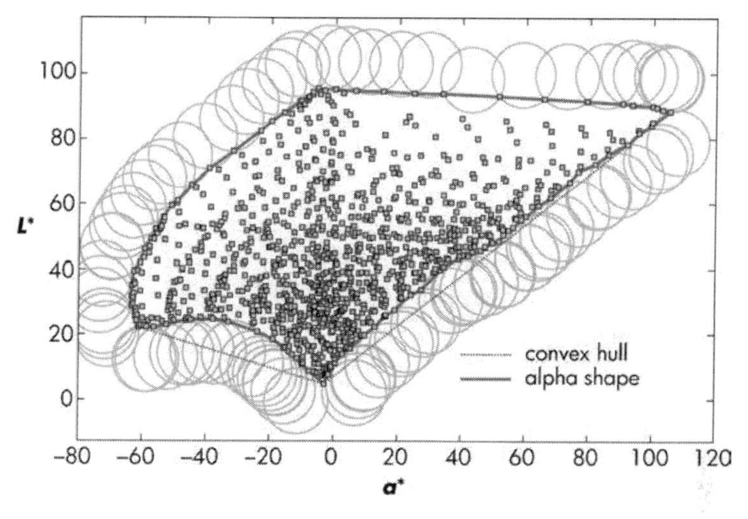

Figure 8.13 Alpha shape gamut boundary for $\alpha = 10$ (circles have radius of 10 and are the closest empty ones to the boundary).

The major challenge with alpha shapes, though, is the choice of α, as a value that is too large runs the risk of overestimation (like convex hulls do) and a value that is too small can result in false concavities. Changing the value of α from 10 (Figure 8.13) to 7 (where the units here are those of CIELAB, i.e. 1 is meant to correspond to a JND) results exactly in such problems (Figure 8.14).

What is required, therefore, is either a careful determination of α based on the data for which a gamut boundary is needed or the use of techniques where α (or a conceptual equivalent) is determined dynamically. For alpha shapes directly, such dynamic determination

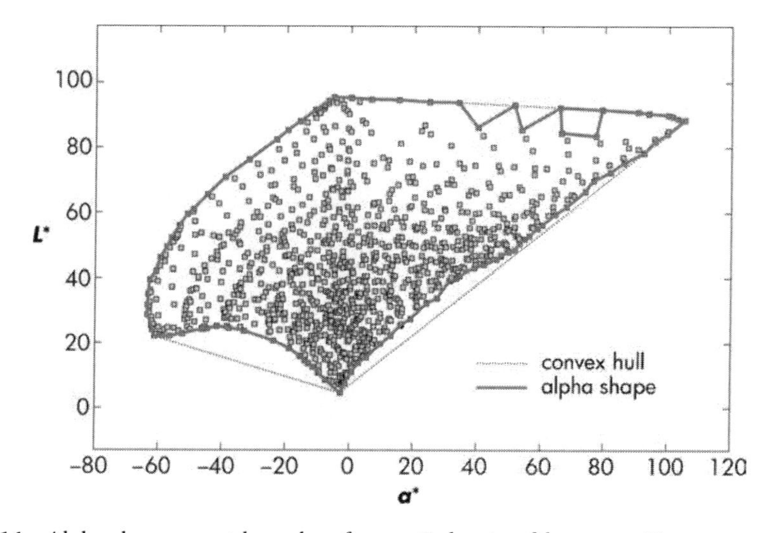

Figure 8.14 Alpha shape gamut boundary for $\alpha = 7$ showing false concavities.

of α (as opposed to α having a fixed value) – called *conformal alpha shape filtration* – has been proposed by Cazals *et al.* (2006). The *adaptively sampled distance field* (Frisken *et al.*, 2000) and *complex flow* approaches (Giesen *et al.*, 2005) are also along these lines.

Another related approach worth mentioning is Maltz's (2005) *expanding spheres tetrahedralization*, which, in addition to constructing a nonconvex, alpha-shape-like boundary around a set of points, also tetrahedralizes them (this is the three-dimensional equivalent of triangulation (Section 7.5)). The value then of the tetrahedralization is that it allows for interpolation, which is a popular method of device characterization, and also that both the boundary and its tessellating tetrahedra are arrived at more quickly than if Delaunay tetrahedralization were followed by the removal of tetrahedra that lie outside the desired alpha hull.

8.2.3 Segment maxima

An alternative to convex hull or alpha shape techniques is a set of approaches that involve the *segmentation* of color space and then the selection of one color per color space segment to represent the gamut boundary. While the method that will be covered here in detail is the *segment maxima* algorithm (Morovič and Luo, 1997, 2000), it has several variants that have been proposed since its publication, which will be covered at the end of this section.

Using the *segment maxima* method, color space is divided into spherical segments, which is like taking an orange, splitting it into segments and further cutting each segment along lines that meet at the orange's centre (Figure 8.15). The gamut boundary of a set of colors is then described by a matrix containing the most extreme colors (i.e. colors with largest radii) for each *spherical segment* of color space and a method for generating a surface on their basis. Note that instead of only storing the maximum radius for each segment, it is the full

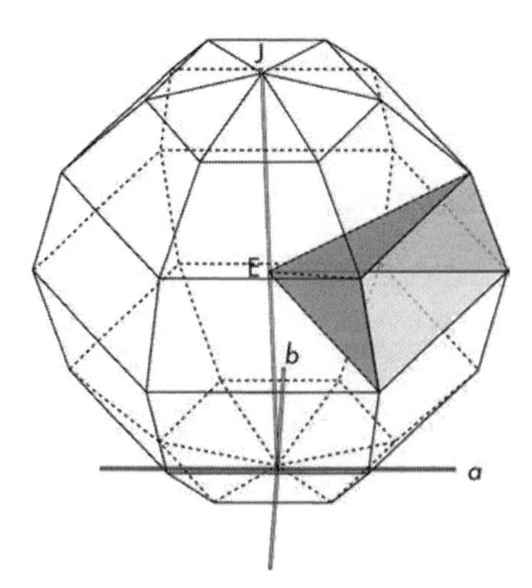

Figure 8.15 A sphere segmented in terms of spherical coordinates α and θ (only 6 × 6 segments – of which one is highlighted – are shown for the sake of clarity).

three-dimensional coordinates that are used. As a result, the segmentation is an uneven sampling that, nonetheless, has an underlying one-sample-per-segment structure that provides a host of benefits.

A segment maxima GBD matrix is calculated by first dividing color space into $m \times n$ segments (e.g. $m = n = 16$ is often used) according to α and θ (see Equations (7.3)–(7.5) for spherical coordinate notation). The number of segments to use depends on two factors: first, the accuracy of gamut boundary description needed, where accuracy potential increases with $m \times n$; second, a number of samples from the gamut to be described that is sufficient for the given choice of m and n. If only an insufficient number of samples are available, increasing m and/or n will not increase accuracy and is likely to result in false concavity artifacts. The choice of m and n is related to the choices of α for alpha shapes or α and γ for the modified convex hull method. As a rule of thumb, it is good to use a uniform sampling with no less than 40 (and ideally 60) samples per device color space dimension when describing device gamuts. Alternatively, fewer samples suffice (i.e. $6(n-1)^2$ instead of n^3, where n is the number of samples per dimension) if it is sufficient to the sample only the surface (e.g. as is the case with well-behaved dRGB controlled devices), fewer samples suffice.

To calculate a GBD for a set of colors (e.g. samples of an imaging device's gamut, an image or color palette) in terms of spherical angle segmentation, the following can be used:

1. Set up an empty $m \times n$ GBD matrix.
2. Transform each of the set of, for example, CIECAM02 *Jab* values for which a GBD is to be calculated into spherical coordinates (Equations (7.3)–(7.5)). Set the center E either to $Jab = [50, 0, 0]$ or the mean of the set's *Jab* values.
3. For each of the spherical coordinates from step 2, compute the zero-starting indices of the GBD matrix to which it belongs as follows:

$$\alpha_{\text{index}} = \max\{\text{floor}[\alpha/(360/m)], m-1\} \tag{8.3}$$

$$\theta_{\text{index}} = \max\{\text{floor}[\theta/(180/n)], n-1\} \tag{8.4}$$

where floor[ ] rounds a value down and max{ } gives the maximum of a pair of values. Note that the range of α is 360°, but θ only has a range of 180° as it is computed in slices of constant α. Next, do the following:

(i) If the $[\alpha_{\text{index}}, \theta_{\text{index}}]$ segment is empty, then store the current coordinates in it.

(ii) Else, if the radius of the current color is larger than the radius of the color stored for the segment, then store the current color for that segment. Note again that it is not only r that is stored for a given segment but the two spherical angles as well.

(iii) Otherwise, skip the current color.

4. Check whether any segment is left empty and, if that is the case, preferably increase sample size or interpolate a value for it based on neighboring occupied segments (for appropriately chosen sample density this is very rare for color imaging media, but it can often be the case when the gamuts of images or arbitrary color sets are calculated).

An important advantage of this method over other segmentation-based methods is that the GBD points obtained using it are actual colors from the set of samples used. The end

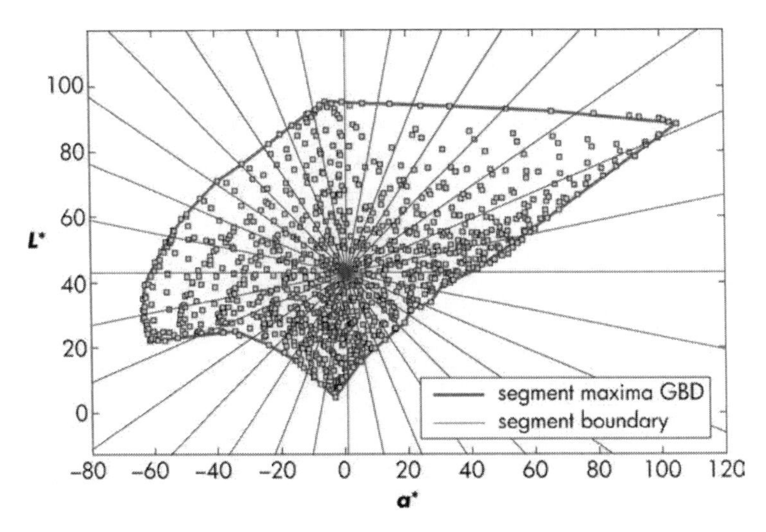

Figure 8.16 Segment maxima gamut boundary with 32 segments.

result is a set of points in color space where there is exactly one set of spherical coordinates in each of the $m \times n$ segments (Figure 8.16). As with previously discussed, fixed-parameter methods, here too the incorrect choice of m and n can lead to undesirable results (Figure 8.17).

The result of using the segment maxima method is initially a set of one-per-segment points. However, to provide a gamut boundary description it is also necessary to generate a surface on their basis, which can be done in at least the following two ways – thanks to the underlying one-per-segment property:

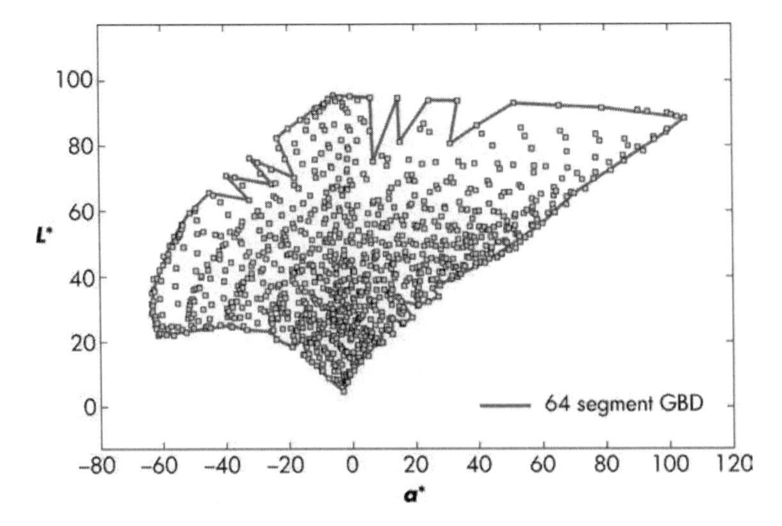

Figure 8.17 A 64-segment gamut boundary with false concavities.

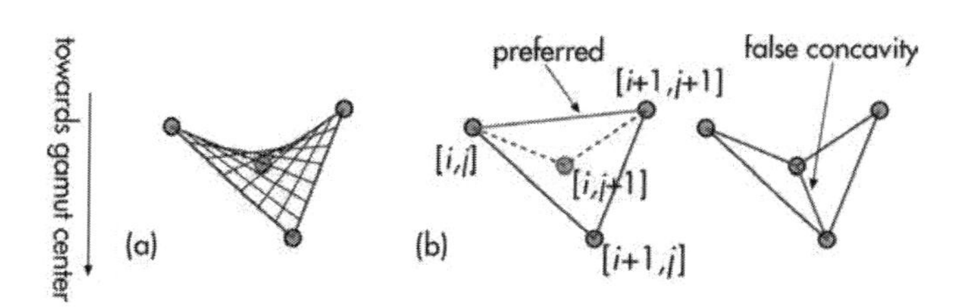

Figure 8.18 Generating surfaces from GBD points (circles): (a) using bilinear surfaces; (b) by triangulation.

1 . A *piece-wise bilinear* surface can be generated from the gamut boundary points such that each piece is delimited and generated by the lines connecting four GBD points with indices $[i, j]$, $[i+1, j]$, $[i, j+1]$, $[i+1, j+1]$ (Figure 8.18a). The advantage of this method is that it provides local smoothness, while a disadvantage can be an excessively concave surface for uneven gamuts.
2. A triangulated surface can be generated from the same sets of four points at a time as were used for the bilinear surface (this follows from the underlying structure of the data). This time, however, there are two alternative possible triangulations, as either $[i, j]$ and $[i+1, j+1]$ or $[i+1, j]$ and $[i, j+1]$ can form an edge. This ambiguity has been well explored by Pellegri and Schettini (2003) and there is a simple solution as to which one to choose: namely, the edge that is further away from the center is more likely to be a good gamut boundary, as the latter triangulation would result in a small local concavity. At least as far as device gamuts are concerned, these are unlikely (while concavities spanning larger parts of color space are common). Note that the bilinear surface here is always in-between the two possible triangulations.

A key benefit of the segment maxima approach is that it has an associated method – *flexible sequential line gamut boundary* (FSLGB) – for computing gamut boundaries along arbitrary lines of constant hue angle. FSLGB first finds the two-dimensional gamut boundary at the hue angle in which the mapping is to be carried out (Figure 8.19), which can be done as follows for a given color C:

1. Calculate the equation of the constant hue angle plane φ having the hue angle of C α_C.
2. For each θ level in the *segment maxima* GBD find the pair of hue-neighboring points from the GBD matrix of which one has a larger and one a smaller hue angle than α_C.
3. For each pair, calculate the intersection of the line connecting the two GBD points with φ.
4. In addition to these n points, calculate the points on the lightness (e.g. CIECAM02 J) axis where the surface defined by the GBD matrix intersects it. For the top of the lightness axis this can be done by considering only the n GBD points from segments having the largest θ values. Triangles are then formed between the point with the largest J and neighboring pairs of the other points. The intersection of each of these triangles and the lightness axis is calculated. and if it is within the triangle then it is the LGB point. An analogous procedure is used for finding the intersection of the gamut boundary with the bottom of the lightness axis.

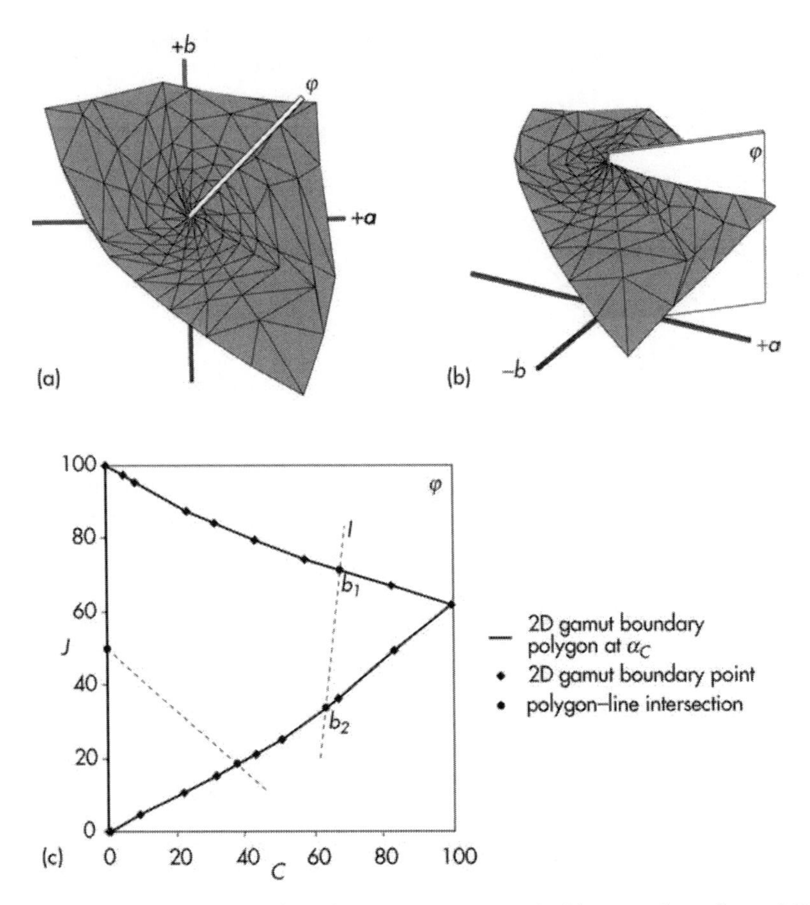

Figure 8.19 Overview of FSLGB algorithm in CIECAM97s: (a, b) gamut boundary of CRT and a plane of constant hue angle φ; (c) their intersection and the intersection of the resulting two-dimensional boundary polygon with a pair of lines (Morovič and Luo, 2000).

The resulting set of $n+2$ points form a polygon (Figure 8.19c) describing the gamut boundary for a given hue angle. The intersection of a given line l and this polygon can then be found using the following procedure:

1. For each pair of neighboring points in the polygon calculate the equation of the line determined by them.
2. For each of these $n+2$ lines calculate their intersection with l and if it lies between the two points from the polygon then it is an LGB point.

Depending on the shape of the gamut boundary, this procedure can result in varying numbers of LGB points for a given line, all of which constitute the LGB along a given line. It is up to the GMA to deal with cases where a larger number of LGB points than expected are calculated. Normally, the algorithm results in two points per line – the maximum and minimum along it (e.g. points b_1 and b_2 from Figure 8.19c).

While segment maxima divide color space in terms of spherical coordinates, another popular alternative is Braun and Fairchild's (1997) *mountain range* method, where segmentation is in the cylindrical lightness–hue space. Also, while segment maxima will result in an uneven distribution of one-per-segment points, the mountain range method only stores a chroma value in each lightness and hue segment and, as a consequence, requires many more segments for accurately describing device gamuts (especially their primaries and secondaries). For example, yellow can be particularly tricky, as it can have a very high lightness; if an insufficient number of lightness sectors are used, then parts of the gamut can be cut off. As a consequence, the original authors used a large number of gamut boundary points: 360 × 100 (i.e. unit steps in both lightness and hue angle).

Another alternative that involves evenly placed points in color space is the spherical segmentation method of Saito and Kotera (2000), which essentially trades off accuracy for a more compact gamut description (i.e. by storing only radii instead of full three-dimensional coordinates per segment). Note, though, that neither this method nor the mountain range method use points from the data set whose gamut is described, but instead result in a new set of points. As a consequence, both have particular difficulties in ensuring that especially extreme colors are accurately represented.

Finally, the most recent approach that involves segmentation is Shaw's (2006) *surface splines* method, where color space is divided into hue angle segments and each of them is characterized by a function (i.e. a spline) fitted to the points in the segment. A benefit of this approach is the smoothness of the resulting gamut that the method entails, and the author also claims computational speed advantages.

8.2.4 Gamut boundary descriptor algorithm summary

From reviewing the GBD algorithms discussed above it can be seen that, with the exception of the *convex hull* method, they can all be successfully used for general color gamut boundary description. With the exception of *alpha-shape*-like methods that dynamically adjust α (but that have not been tested for gamut boundary description to date), the other approaches all involve a parameter choice that can make or break the resulting GBD's success; therefore, care is needed when setting such values.

8.3 EVALUATING AND OPERATING ON GAMUT BOUNDARY DESCRIPTORS

For a gamut boundary computed using a GBD algorithm, it is often of interest to express how well it describes the gamut of a set of colors, to determine how it relates to other gamuts, to know its volume or to smooth it.

8.3.1 Evaluation

Although gamut boundary description is central to gamut mapping and it is important to know how accurate a GBD is (with inaccurate GBDs resulting in more *uncontrolled clipping* – Section 8.1.2), there is relatively little published work about how to evaluate GBD accuracy. Even what exists is restricted to the special case of knowing the boundary that a GBD is meant to describe. For example, Braun and Fairchild (1997: 150) evaluate the accuracy of their *mountain range* method as applied to the description of a CRT's gamut, predicted by the CRT's characterization model. Having computed the GBD from a set of samples generated using the model, they proceed to randomly generate a new test set of samples that are known to be on the gamut boundary (by having dRGB values from the surface of the dRGB cube). The accuracy of the GBD is then expressed by the color difference statistics between the model-generated test set and the corresponding GBD points – as the GBD uses L^*h_{ab}, sampling test colors are placed in correspondence with GBD surface points with the same L^* and h_{ab}. The same idea is essentially also used in later work by Green (2001) and Bakke *et al.* (2006).

What is implicit in the above is that these evaluation methods need to know the gamut boundary *a priori* to be able to express how well a GBD describes it, which clearly points to the difficulty of answering the general question: 'How accurately does a given GBD describe the gamut of a given set of points?' The roots of the problem are the continuity and concavity versus sampling density ambiguities discussed in Sections 8.1.2 and 8.1.3.

While a general solution to this problem might not be feasible, it is at least worth pointing out the flaws of a popular, intuitively attractive strategy, which is to check whether the whole color set is *inside* the gamut boundary computed for it. Even though this is desirable, it is by no means sufficient, as a very inaccurate gamut boundary can score well in terms of this. Take, for example, a gamut boundary that is a sphere with a radius of 200 units and its centre at $Jab = [50, 0, 0]$. This gamut boundary will have all possible colors inside it and, at the same time, be a very inaccurate description of any possible color gamut.

Looking at this unsuccessful strategy points to another way of seeing the ambiguities that plague GBD evaluation, namely the generation of colors that are not inside the given color set's gamut but are close to it. While it is trivial to generate colors that are far from a color set's gamut boundary, it is far more difficult to do so in proximity to the boundary. If it were possible to generate such a set, then a general gamut boundary evaluation would be possible, as it could revolve around testing whether colors known to be in gamut are also inside a GBD and those that are known to be outside are also outside the GBD. Instead of requiring knowledge of the gamut boundary, this method would only need IG and OOG color sets.

Even though a general solution of this problem will not be presented here, it can be pointed out that such IG and OOG color sets can be generated for densely sampled device gamuts along the lines suggested in Figure 8.2. Given a dense sampling of a device gamut, another sampling of the space surrounding the sampling could be generated in a way that would include only colors that are further away from IG colors than their closest other IG colors. The two sets could then be used to evaluate GBD accuracy. For samples of insufficient density this approach could lead to favoring methods with false concavities; therefore, it is not a robust solution.

The lack of a reliable way of evaluating GBD algorithms coupled with their reliance on parameter choices is the likely cause of there not being a clear winner among them and of a wide variety of them being used in practice.

8.3.2 Volume

An important property of color gamuts is their volume in color space, as it expresses, albeit in an indiscriminate way, how much of color space they contain. If gamut volume computation is performed in a visually uniform space that has JND units, which is approximately true for CIECAM02, then the resulting volume does relate to the perceived gamut of the corresponding color set (Heckaman and Fairchild, 2006a).

Given a GBD, there tend to be two ways of arriving at its volume. If computing the boundary also resulted in a tessellating tetrahedralization, then the volumes of these tetrahedra can be computed and summed. However, if all that is available is a triangulated surface, then it needs to be tetrahedralized first. As the sole purpose of such tetrahedralization is volume computation, a simple technique is to form tetrahedra with each of the surface triangles and a point inside the tetrahedron, which in most well-behaved cases is the mean of the GBD vertices. Given such a tetrahedralization, it is again down to computing tetrahedron volumes and summing them.

To compute the volume of a tetrahedron with vertices A, B, C and D in three dimensions X, Y and Z, the following equation can be used (O'Rourke, 2000: 22):

$$
\begin{aligned}
V &= \frac{1}{6}\begin{vmatrix} A_X & A_Y & A_Z & 1 \\ B_X & B_Y & B_Z & 1 \\ C_X & C_Y & C_Z & 1 \\ D_X & D_Y & D_Z & 1 \end{vmatrix} \\
&= \frac{1}{6}\begin{bmatrix} -(A_Z-D_Z)(B_Y-D_Y)(C_X-D_X)+(A_Y-D_Y)(B_Z-D_Z)(C_X-D_X) \\ +(A_Z-D_Z)(B_X-D_X)(C_Y-D_Y)+(A_X-D_X)(B_Z-D_Z)(C_Y-D_Y) \\ -(A_Y-D_Y)(B_X-D_X)(C_Z-D_Z)+(A_X-D_X)(B_Y-D_Y)(C_Z-D_Z) \end{bmatrix}
\end{aligned} \tag{8.5}
$$

where $|\mathbf{A}|$ denotes the determinant of matrix $\mathbf{A}$. Note that XYZ here stands for the three dimensions of any orthogonal color space such as CIECAM02 *Jab*, CIELAB $L^*a^*b^*$ or CIEXYZ. Once computed, gamut volume is referred to as being in *cubic color space units,* such as cubic CIELAB or cubic CIECAM02 *Jab* units, and it is customary to round these to thousands for device and medium gamuts. Such units will here be abbreviated as cLab and cJab respectively.

While volume is a fundamental property of a color gamut, it is agnostic of the fact that some colors are more important than others. This importance, in turn, can be absolute (e.g. memory colors) or relative (e.g. light, chromatic colors for printing systems that are meant for reproducing images originally present on displays). In any case, a reproduction gamut that covers important colors and has smaller volume is preferred over one that has larger volume but in areas of lesser importance.

It is worth noting, though, that a gamut's volume is not the same as the *number of distinguishable colors* within it, although the two are strongly related and, especially for gamuts of similar shape, the gamut with greater volume will also have the larger number of distinguishable colors. Essentially, the number of distinguishable colors would be the greatest number of unit-diameter spheres that could be packed into a gamut rather than its volume (assuming a perfect color appearance space). See the work of Wen (2006) for an intermediate approach to this issue, and consult Zong (1999) for more on sphere packing. In spite of this difference, the rest of this chapter will consider volume and the number of distinguishable colors are synonymous.

8.3.3 Intersections

Given two color gamuts, it is often of interest to know the boundary of their intersection; again, how it is computed depends on the GBD algorithm used. In the general case of having triangulated surfaces or tetrahedralized volumes, there are generic computer graphics techniques that can be used (e.g. Rabbitz, 1994), although these tend to focus on the case of convex polyhedra.

For gamut boundaries described using the *segment maxima* approach, intersection can be computed by setting up an empty GBD matrix and populating it as follows:

1. Generate a set of spherical angles consisting of an even sampling plus the spherical angles of both GBDs' vertices.
2. For each of the angles from step 1 calculate the coordinates of points on the two gamuts that have them and keep the one that has the smaller radius (Figure 8.20a).
3. Compute the segment maxima gamut boundary of the points kept in step 2.

Figure 8.20b also shows an example result of this computation when applied to the color gamuts of a display and a print.

A typical use of a gamut intersection is the computation of its volume and a subsequent comparison with the volumes of the gamuts whose intersection it is. This allows for a quantification of how much of one gamut is outside another gamut, which cannot be had

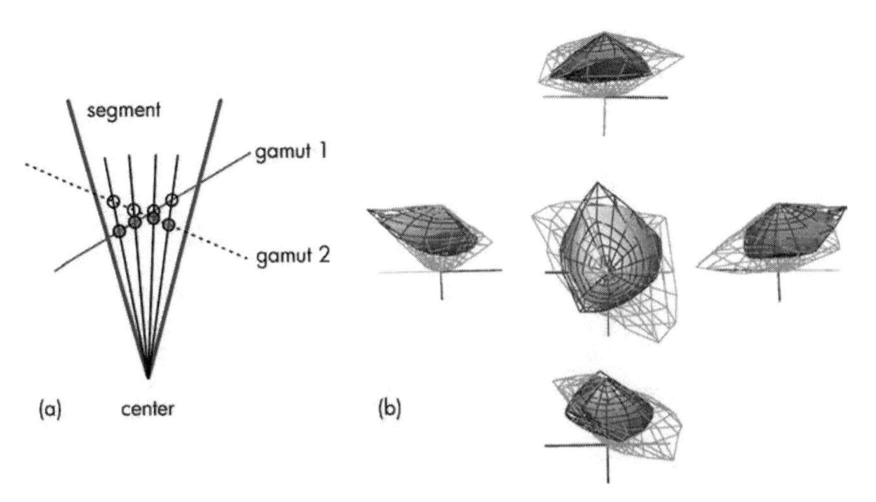

Figure 8.20 Intersecting two-segment maxima GBDs: (a) samples in a particular segment; (b) example of applying the computation to display and print gamuts.

just from computing the two gamuts' volumes (gamuts can have identical volume and limited or even no overlap).

Intersection gamuts are also useful for generating colors that can be matched in all of the gamuts they are derived from. This, in turn, is useful when a colorimetric match is desired between multiple media, as it allows for the restriction of all to their intersection gamut. In practice, this is a fairly niche application, though, as intersection gamuts tend to be too limited in general and, therefore, lead to low overall color reproduction quality.

8.3.4 Smoothing

The behavior of imaging devices, whose output is often measured, varies over time and is prone to noise, which leads to color gamuts sometimes having rough surfaces. Using such gamuts can, in turn, result in artifacts in color reproduction. For example, imagine a transition in a source gamut along the line in color space shown in Figure 8.21. Given rough source and destination gamuts and a simple algorithm that scales chroma to preserve the source chroma's relationship to the gamut boundary, the destination transition will exhibit fluctuations and is likely to result in visible banding or contouring. If, instead, the gamut boundaries are smoothed, then the resulting mapping also becomes smoother. Therefore, it is often desirable to smooth gamut boundaries before using them in gamut mapping.

Smoothness can then be achieved either directly, by applying a transformation to GBD colors or indirectly by controlling the sampling that underlies most GBD algorithms. An example of the first, direct approach is the *mean filtering* used by Braun and Fairchild (1997) that replaces each GBD point by the average of GBD points from a local neighborhood. For example, in the case of their mountain range method, each chroma value would be replaced by the average of $n \times n$ values from the GBD matrix, where n is odd and centered on the value to be smoothed. The larger n is, the smoother the result will be, but it needs to be borne in mind that smoothness is likely to reduce accuracy and the

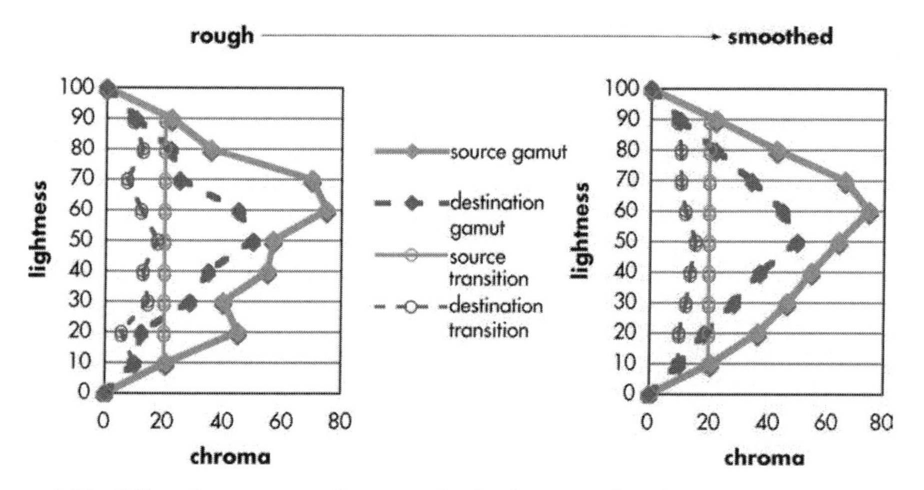

Figure 8.21 Effect of gamut smoothness on simple chroma-scaling GMA.

two need to be traded off. An alternative to such filtering is also only to replace a GBD point by the local mean if that mean has higher chroma. This is because direct mean filtering will reduce the color gamut at the primaries and secondaries of a medium, which is often not desirable. If explicit filtering is used, it is worth following a gamut mapping to the smoothed boundary by a second gamut clipping stage that ensures that all mapped colors are inside the actual, nonsmoothed gamut to avoid *uncontrolled clipping* (Section 8.1.2).

Smoothness can indirectly also be increased by the choice of GBD computation parameters, since they effectively change sampling density and, therefore, the potential for representing rough surfaces. In the *alpha shapes* case, for example, an α of infinity gives the convex hull, which has no surface roughness, and the greater α is, the smoother the roughest surface it can describe. In the *mountain range* and *segment maxima* methods it is the size of the GBD matrix that controls sampling, where smaller matrices have less potential for storing roughness. Finally, the α parameter of the *modified convex hull* method also has this effect, and the closer it is to unity the less roughness it allows for, as it is closer to the unmodified convex hull.

8.4 EXAMPLES OF SALIENT COLOR GAMUTS

Examples will be given next of the color gamuts of different imaging devices and media, as well as other color sets of interest, followed by an illustration of the effect of different viewing conditions and viewing modes (i.e. simultaneous versus sequential) on gamut size and shape. All the gamuts shown here have been computed using the alpha shapes method with a fixed $\alpha = 20$, are orthographically projected in CIECAM02 and are for a D50 white point.

Note that these examples are simply choices of single representatives of otherwise varied populations. Their purpose is solely to give an idea of the shapes of different kinds of color gamuts, rather than to provide a comparison between how large the gamuts of different imaging technologies are. As will be illustrated later, the color gamuts of stimulus sets and, therefore, also of imaging media are strongly viewing condition and mode dependent.

8.4.1 Object color solid

Unlike the rest of the gamuts that will be shown here, the *object color solid* (OCS; Schrödinger, 1920) is a theoretical one that shows the range of all possible surface colors (i.e. colors resulting from surfaces reflecting or transmitting light – Section 8.1.4). Note that while the OCS is the gamut of all possible surface colors when expressed in spectral reflectance or transmittance terms, once tristimulus and derived color space values are computed for it, the resulting colors are those that the OCS would have under a chosen illuminant and its gamut changes as a function of illuminant. Colorimetrically, therefore, there are as many OCSs as there are illuminants.

If the question were to compute the gamut of all possible color experiences that surfaces can elicit, then it would have to be the union of the OCS gamut under all possible

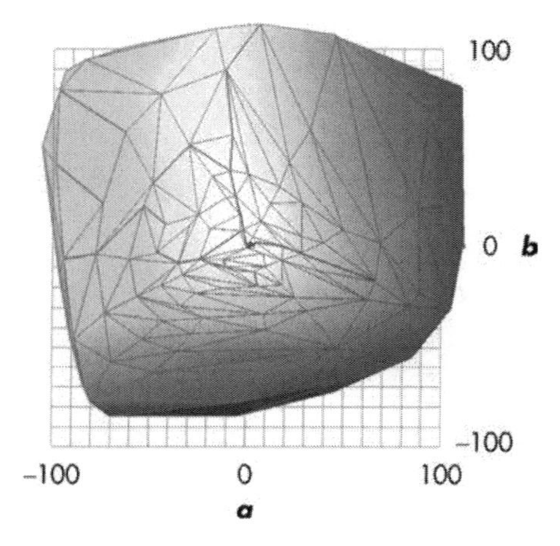

Figure 8.22 The OCS under D50.

illuminants (and mixed illuminant environments). If such a computation were to be performed (if it were possible), then the result would represent the gamut of all possible surface colors. The difficulty here is that, unlike those of surface reflectances, the bounds of light source spectral powers are not easily determinable and the same also holds for the SPDs of self-luminous objects. Also note that adaptation plays a controlling role in the range of perceived colors, and this has recently been shown to enable the generation of complex stimuli for which color experiences exceed even those for the *spectrum locus* (i.e. stimuli, each of which has power only in a narrow wavelength interval, e.g. lasers, that span the visible spectrum). By constraining the luminance of the white point (i.e. the most luminous achromatic stimulus) on a high dynamic-range display, Heckaman and Fairchild (2006b) have been able to generate images for which observers have reported colorfulnesses significantly exceeding those of the spectrum locus, which was previously believed to represent the limits of possible colors (MacAdam, 1935).

Figure 8.22 then shows the D50 OCS, whose gamut volume is 1 939 000 cJab units. Note that, as predicted by CIECAM02, the maximum chromas that surfaces can achieve are similar across the range of hues, with the largest ones being in the red–yellow–green region. In following figures, the OCS' outline will be shown for comparison.

8.4.2 Measured object colors

The next gamut to consider is that of a database of around 50 000 measurements taken from a wide variety of surfaces, called the *standard object color spectra* (SOCS; ISO, 2003) – which represents the properties of actual objects found in our environment – both natural (e.g. leaves, flowers, rocks) and artificial (e.g. plastics, textiles). The significance of this gamut (Figure 8.23), with a volume of 863 000 cJab units, is that it shows the likely colors of objects that might need to be reproduced using a given color reproduction device. Of note is also the gamut of 'real surface colors' by Pointer (1980), which is an earlier attempt at delimiting the gamut of typical colors in our environment and which is broadly similar to SOCS in gamut terms (Inui *et al.*, 2004).

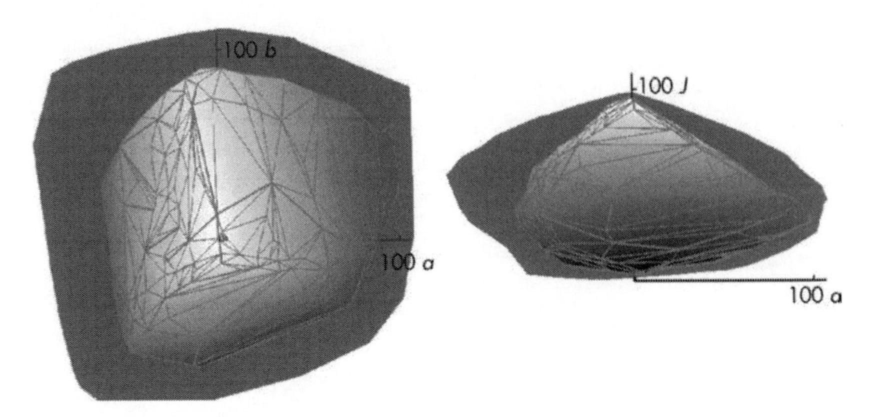

Figure 8.23 The SOCS gamut under D50.

8.4.3 Pantone solid colors

In the graphic arts the *de facto* standard for specifying colors for graphical elements like text, logos, etc. are the PANTONE® Guides, which are swatch books of pages each containing seven colors. Each color is specified as a mixture of some of 15 basic inks and the swatch books are available printed on different papers. Measuring the 1114 color patches of the *PANTONE® Formula Guide coated* gives a color gamut with a volume of 803 000 cJab units. As can be seen from Figure 8.24, it provides a good sample of all possible surface colors, whereby it is essentially a chroma-scaled version of the OCS.

8.4.4 Print

Looking at the print gamuts shown in Figure 8.25 we can still see a resemblance to the OCS gamut's shape, as they are yet another instance of subtractive color reproduction. To

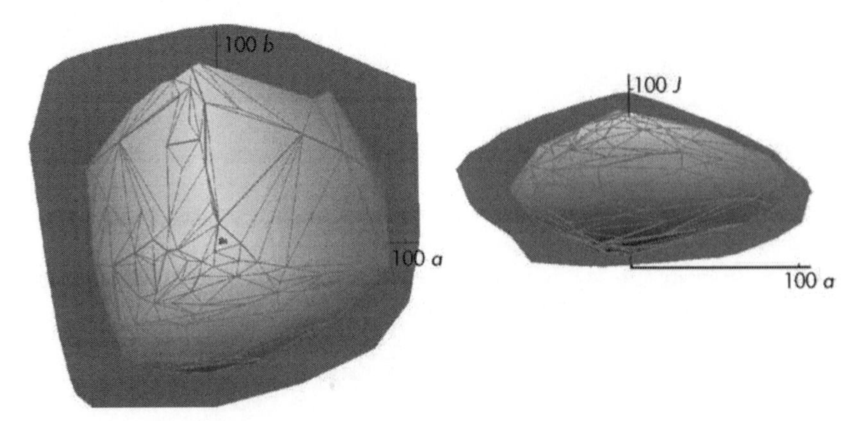

Figure 8.24 The gamut of the 'PANTONE® Formula Guide coated' under D50. Reproduced by permission of Pantone, Inc.

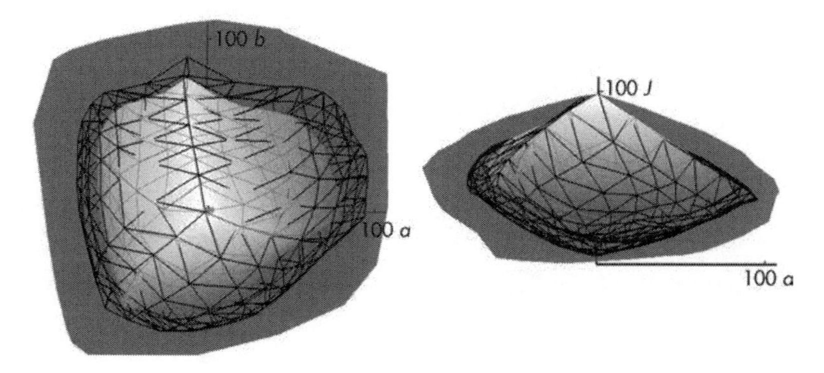

Figure 8.25 The gamuts of prints made on two papers: coated (black mesh) and uncoated (solid).

illustrate the importance of the substrate on which prints are made, the figure shows the color gamuts of prints made using a single printer but two different papers: a glossy, coated one (with a gamut volume of 796 000 cJab) and a matte, uncoated one (436 000 cJab).

8.4.5 Projected motion picture film

The final subtractive reproduction example (Figure 8.26) is that of a motion picture film stock projected in a cinema with dim ambient light that has a gamut volume of 731 000 cJab units. As can be seen, this gamut covers much more space in the chromatic greens than the other measured color gamuts and also gives access to some high chromas at low lightnesses.

8.4.6 Display

Moving along to technologies that use additive rather than subtractive color reproduction, Figure 8.27 shows the color gamut of a display seen in a dark room that has a gamut volume of 854 000 cJab units. As can be seen, this display has a very different gamut shape

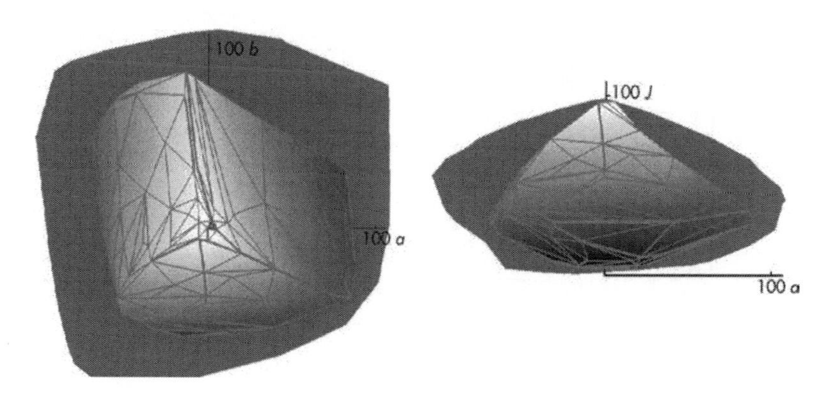

Figure 8.26 A gamut of projected motion picture film stock.

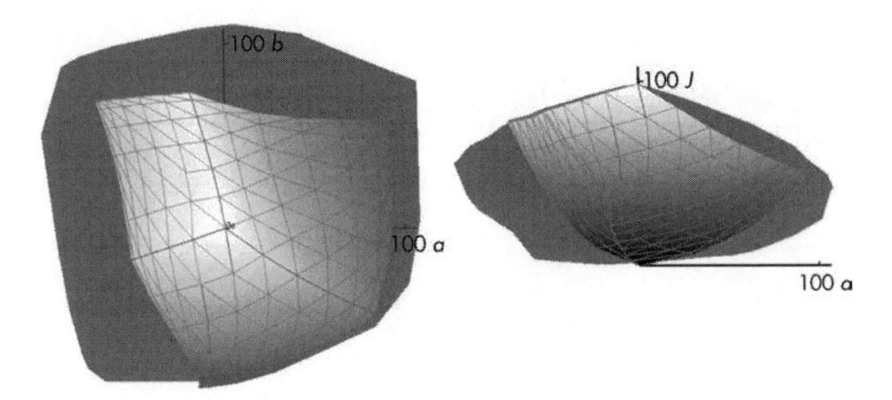

Figure 8.27 The gamut of a display with a D50 white point.

compared with the subtractive gamuts shown above. Also of note is the fact that part of its gamut falls outside of the OCS, and this is due to it being self-luminous rather than a surface that filters light incident on it or passing through it. This is also an illustration of the fact that the OCS is not the gamut of all possible colors, but only of surface colors (and under specific viewing conditions). The difference in shape between subtractive (e.g. print) and additive (e.g. display) gamuts is a key source of difficulty for finding a robust gamut mapping solution.

8.4.7 Digital projection

Next, the gamut of a digitally projected medium (i.e. one using an LCD-based projector) is shown in Figure 8.28 for two levels of ambient illumination (Morovič and Sun, 2000b). Here, ambient illumination affects the result dramatically (gamut volume differs by a

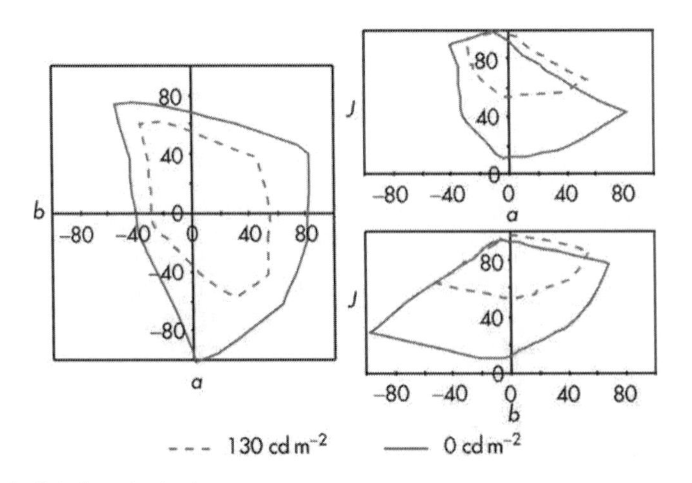

Figure 8.28 A digital projection's gamut.

factor of six), since the projection is the result of reflection from a surface. If there is other light (e.g. ambient illumination) reflected from that same surface then it is added to it, which in the case of white-looking light sources results both in a lightening and a desaturation. Being a linear addition in luminance terms, it then results in a reduction of lightness range, since the latter is a nonlinear transformation of the former, where a given amount of added luminance results in greater lightening for dark than light stimuli. The gamuts in this particular example vary from around 100 000 (for a typically lit room) to around 600 000 cJab units (for a dark room).

8.4.8 Scanner and digital cameras

Finally, as far as cameras are concerned, their gamuts span the entire OCS when their settings are correct (less some loss due to quantization and repeatability errors (Morovič and Morovič, 2003)). To see what happens when a camera is not set up correctly for a given scene, Figure 8.29 shows the results for two different incorrect aperture and exposure combinations. Here, the solid gamut has a volume of 1 139 000 cJab and the mesh a volume of 1 057 000 cJab units.

8.4.9 Effect of viewing conditions

In addition to factors like the ones already illustrated in the above figures (i.e. print substrate, camera settings), the most important external factors that affect color gamuts are the viewing conditions under which stimuli are seen and whether those stimuli are seen simultaneously or in separate locations and/or at different times. To illustrate how big a difference viewing conditions make, Figure 8.30 shows the color gamuts of two displays, i.e. a projection and a print under different levels of ambient illumination, when each of these media is seen in isolation. Note that the first row of plots shows projections onto the *ab* plane in CIECAM97s2 (Li *et al.*, 1999) – a precursor of CIECAM02 – and the

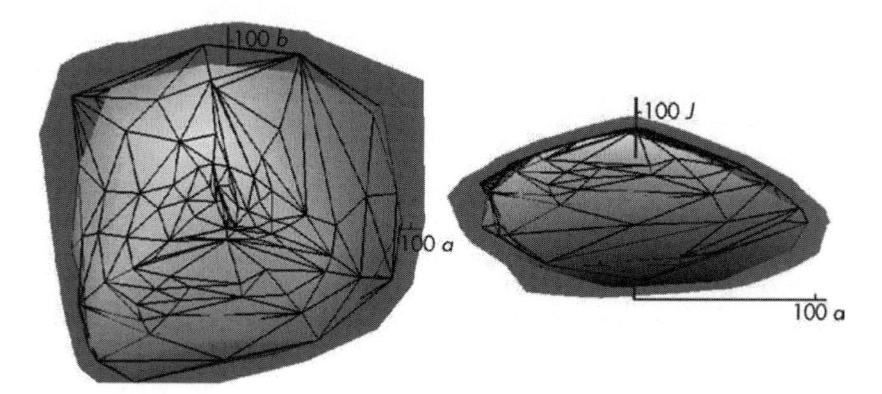

Figure 8.29 The gamut of a digital camera with two sets of incorrect settings, one shown as a shaded solid and the other as a mesh.

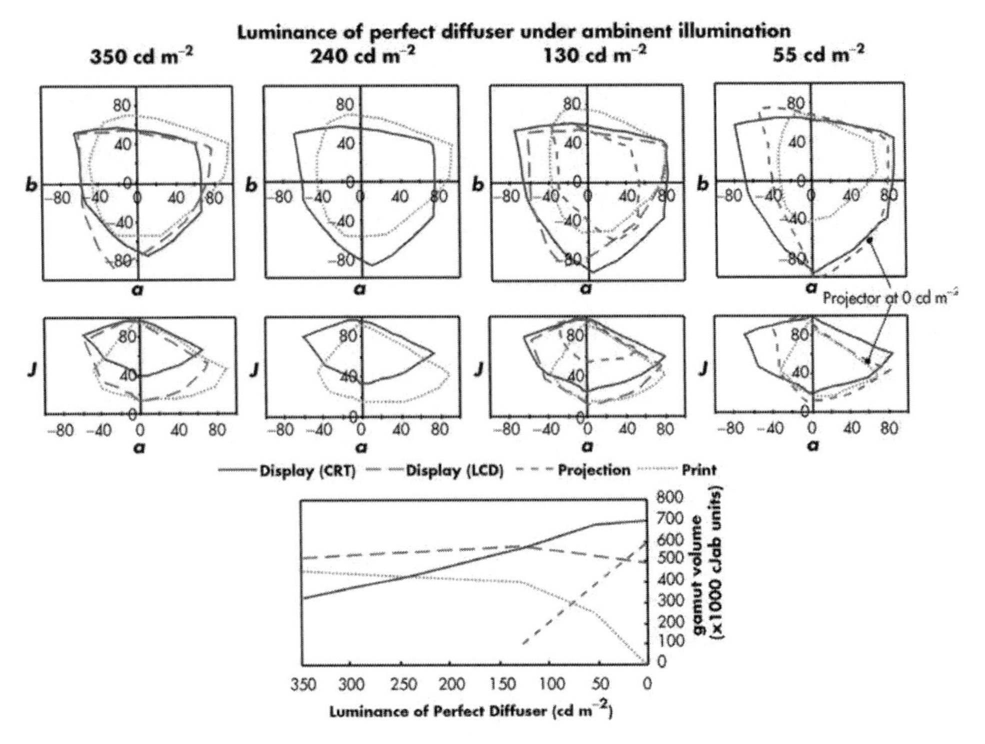

Figure 8.30 Effect of viewing conditions on medium color gamut for media seen in isolation.

second row shows their intersections with the *aJ* plane. At the bottom of the figure we can then see the gamut volumes under these viewing conditions.

One aspect of the above that stands out is that different media are affected very differently by changes in ambient illumination. While some (e.g. LCD displays) stay almost unaffected, others (e.g. CRT displays or projection) can change gamut volume by a factor of two to six even across the relatively limited ambient illumination range shown in the figure. Furthermore, media can have either positively (print) or negatively (glossy displays, projection) correlated gamut volumes with the level of ambient illumination, which also means that one gamut can be larger than another under a certain level and at a different level it can be the other way around. This is a further reason why it is not possible to talk about the gamuts of imaging devices or media without specifying viewing conditions.

To understand the influence of the viewing mode, Figure 8.31 shows the gamuts of the print and CRT display gamuts from Figure 8.30 for simultaneous viewing. The main consequence of this is that it will be the two media together that control the one adapted white of the viewing environment, which in turn will depend on the luminances of the two media's white points.

Gamut differences are even larger for simultaneous viewing, as the print's white point starts out with being of lower luminance than the display's under the low illumination level and then becomes a lot higher as ambient illumination is turned up. Since there is only a single white point for simultaneous viewing, the white point of the less luminous media does not have a lightness of 100, which also reduces the entire gamut. Across this range,

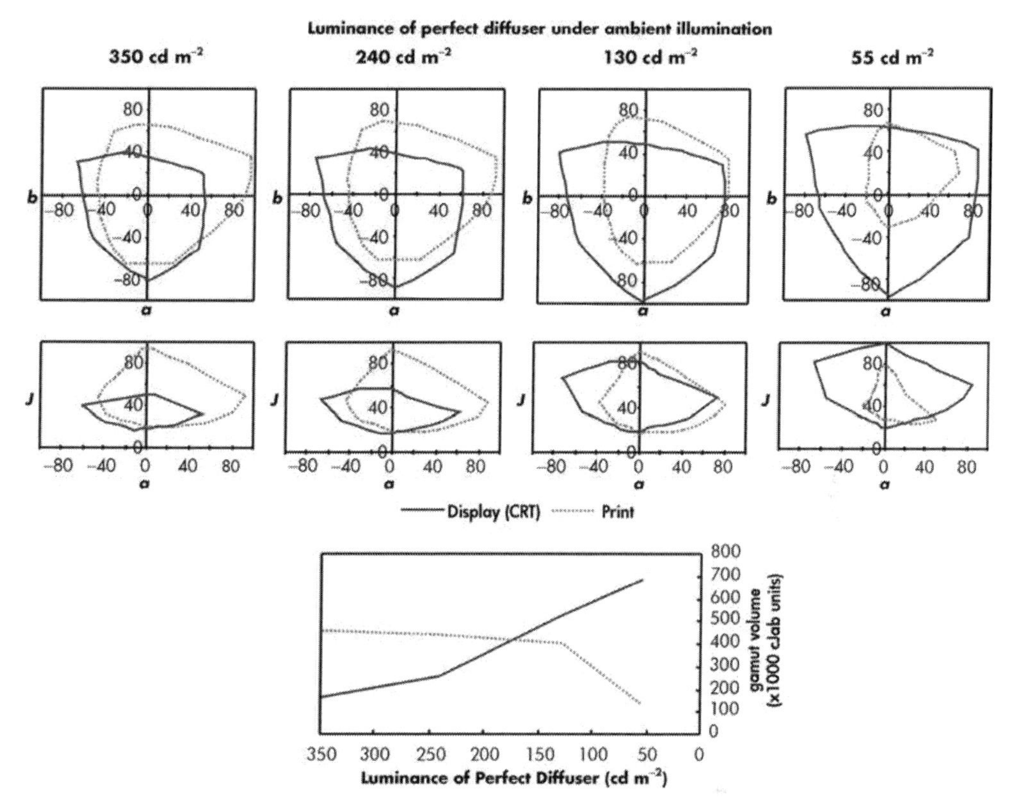

Figure 8.31 Effect of viewing conditions on medium color gamut for media seen simultaneously.

the lowest luminance case corresponds to looking at a print next to a display in a dim room, whereas the highest luminance case is that of having print and CRT display in a brightly lit room.

Having such highly variable gamut relationships that are a function of viewing conditions is a real challenge for gamut mapping and controlled color reproduction in general.

8.5 IMAGE GAMUTS

So far, we have only considered the description of the gamuts of imaging devices and media and arbitrary color sets. Applying the above techniques to images raises some additional issues.

The calculation of the gamut of all the colors in an image is a straightforward task and one that can be achieved using either specifically developed approaches (e.g. Kotera and Saito, 2003; Giesen *et al.*, 2005, 2006a) or even the methods discussed above. However, it is another matter altogether to calculate the perceived gamut of an image, in other words a gamut that describes the range of colors that an image is perceived to have rather than simply the gamut of all of its pixels (Morovič and Sun, 2000a). This distinction needs to be made, as images can contain pixels whose individual colors will either not be perceived (e.g. if they are below the human visual system's spatial acuity threshold) or which will not fully contribute to the image's perceived gamut (e.g. if they are very infrequent in an image).

Imagine, for example, an image that contains one pixel each with the following colors from a medium's gamut boundary: red, green, blue, cyan, magenta and yellow. Looking at the image, these pixels might not even be distinguishable in it, but computing the gamut of all the image's pixels would suggest that it is much larger than how it is perceived (Figure 8.32).

While there are no established methods for reliably determining the perceived gamuts of images, a number of approaches have been suggested previously and they all consist of excluding some of an image's pixels from gamut calculation (Morovič and Sun, 2000a).

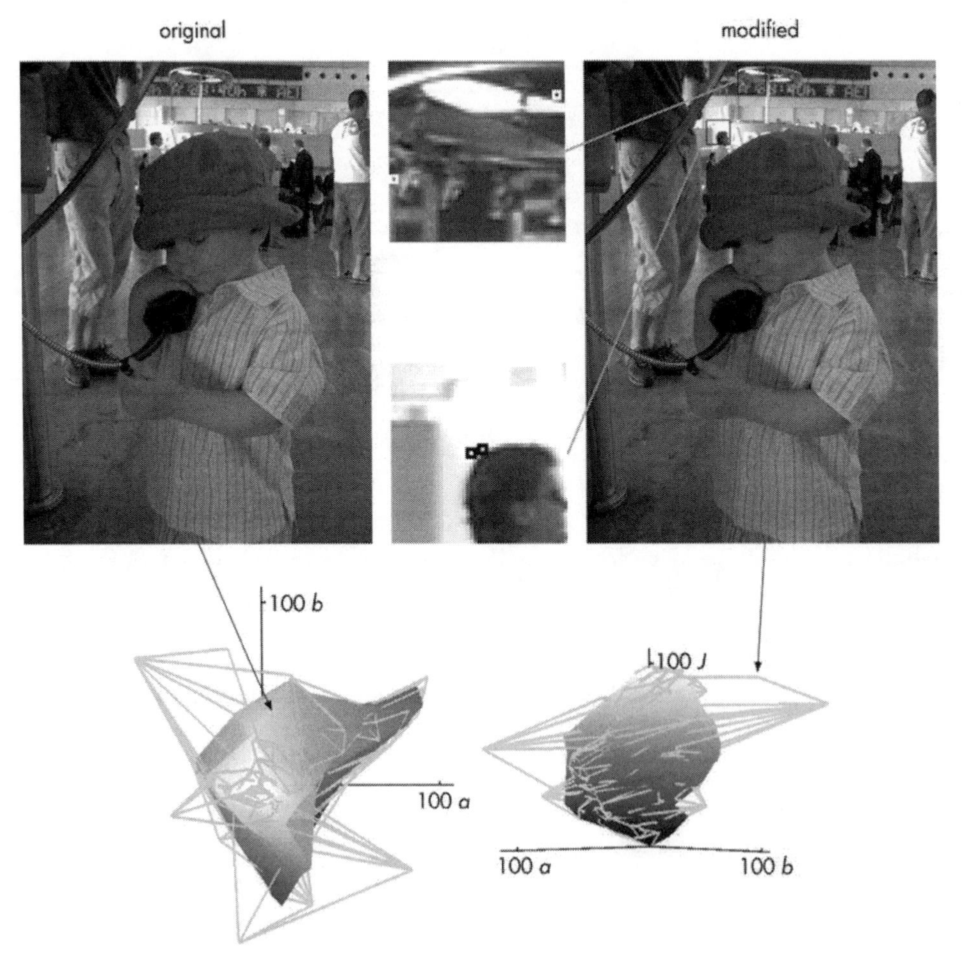

Figure 8.32 An image (top left), its gamut (solid), a modified version of that image that has six changed pixels (top right) and the modified image's gamut (mesh).

The first of these is to exclude those colors in an image that occur only rarely, i.e. whose frequency is below a certain threshold. Pixels in high spatial frequency areas should also be excluded, as they are likely to play a lesser part in how the image's color range is perceived. However, further research is needed for developing a robust and psychovisually justified image gamut description method.

8.6 SUMMARY

The main takeaways from this chapter should be the ambiguity of gamut boundary description – involving a struggle between representing concavities rather than sampling gaps, an overview of techniques for computing gamut boundaries, an idea of what the gamut boundaries of salient color sets and imaging media look like, how all gamuts depend on viewing conditions, and the additional considerations of image gamut boundary description.

ACKNOWLEDGEMENT

A portion of text in paragraph 8.2 has been reproduced by permission of the International Commission of Illumination (CIE).

Plate 1: 'Ski' test image used in CIE's Guidelines for the Evaluation of Gamut Mapping

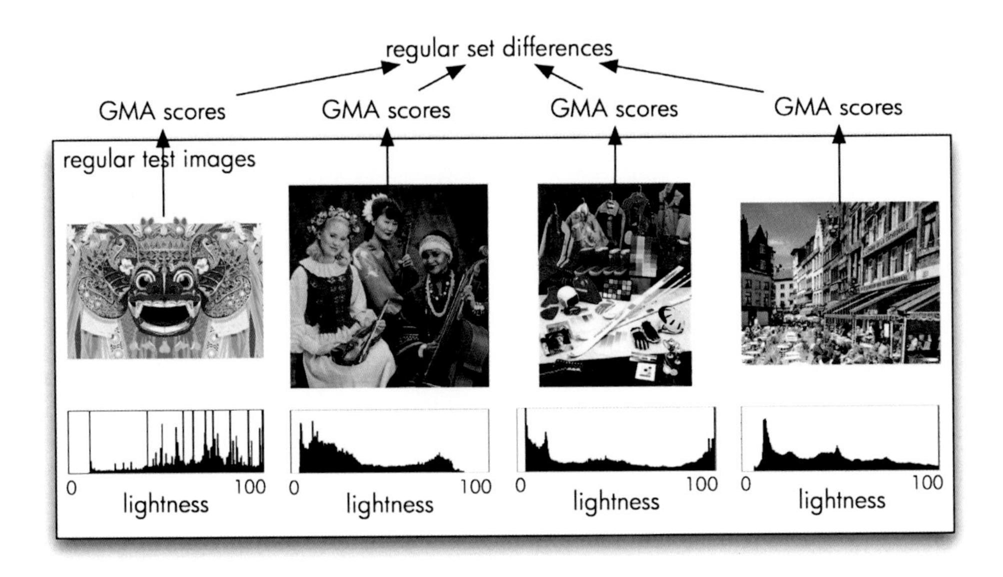

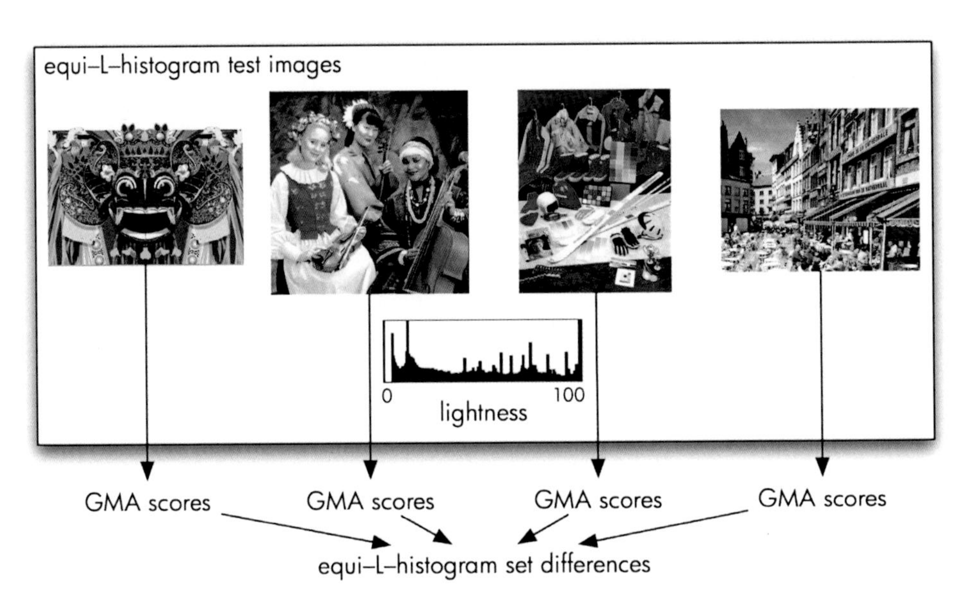

Plate 2: A regular set of test images and an artificial set, where all images have the same lightness histogram

Plate 3: Source image printed on coated (glossy) paper

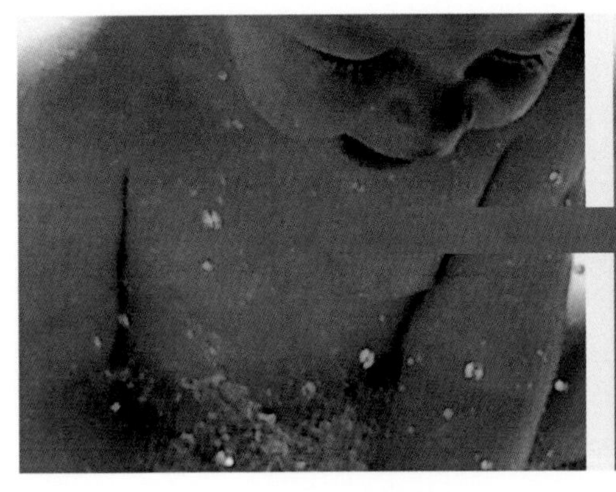
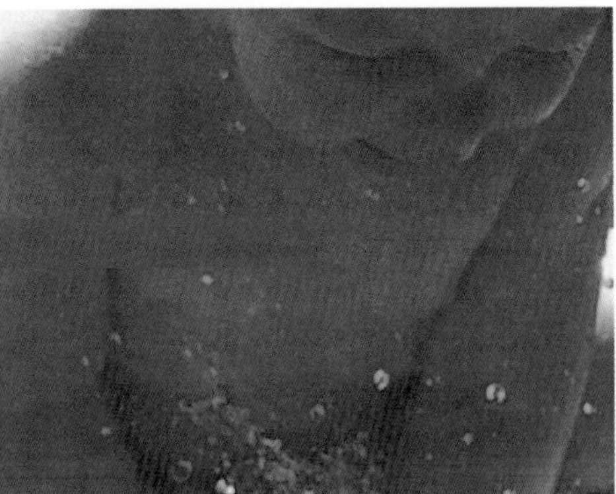

(a)

(b)

Plate 4: (a) Apparent change in color as a result of gamut mapping: detail from original (left) and reproduction (right) with uniformly colored bar going across to ease comparison. (b) An original – reproduced in Color Plate 5 using a clipping in two color spaces

Plate 5: Minimum color difference clipped reproductions of image in Color Plate 4b in CIELAB (top) and CIECAM02 (bottom)

Plate 6: Reproduction obtained using minimum color difference clipping in CIECAM02

Plate 7: Reproduction obtained using minimum color difference gamut clipping in CIELAB

Plate 8: Reproduction obtained using minimum ΔE_{WT} gamut clipping in CIECAM02 with [kJ, kC, kH]=[1,2.6,1.3]

Plate 9: Reproduction obtained using Black Point Compensation followed by minimum ΔE_{WT} gamut clipping in CIECAM02 with $[k_J, k_C, k_H]=[1,2.6,1.3]$

Plate 10: Reproduction obtained using SCLIP in CIECAM02

Plate 11: Reproduction obtained using LLIN in CIECAM02

Plate 12: Reproduction obtained using CLLIN in CIECAM02

Plate 13: Reproduction obtained using SLIN in CIECAM02

Plate 14: Reproduction obtained using CARISMA in CIECAM02 (using only mapping to intersection between lightness axis and line connecting cusps)

Plate 15: Reproduction obtained using Balasubramanian et al. spatial algorithm (filter size = 15x15; no initial lightness mapping; hue–preserving minimum ΔE)

Plate 16: Reproduction obtained using Morovic and Wang spatial GMA in CIECAM02

9

A Case Study: Minimum Color Difference Gamut Clipping

With about two-thirds of this book under our belts, let us now apply the theory presented so far to the specific case of reproducing an original print made on glossy, coated paper using a print on plain, uncoated paper using a minimum color difference gamut difference clipping algorithm. The key reason for picking this combination of original (source) and reproduction (destination) media is that it can be contained in a printed book even though other source–destination combinations pose greater challenges. Nonetheless, even here there is a significant gamut difference to be compensated for, and the basic process is the same in all cases. What will be presented in this chapter, therefore, is a step-by-step overview of starting with an original image, passing it through a sequence of transformations and ending up in its reproduction with a different color gamut.

9.1 THE ORIGINAL

The true original in our scenario is the experience a viewer has when looking at the stimulus shown in Color Plate 3 under ISO 3664 (ISO, 2000) P2 viewing conditions, i.e. the print being illuminated with a D50 simulator at 500 lx, which are also the ICC profile connection space's viewing conditions. Even though this, a viewer's experience, is the starting point and reproductions will be compared with it in its native domain, we have no direct access to it.

9.1.1 Original appearance predicted from measurements

The next best thing, albeit distinct from the actual original experience, is its prediction in a color appearance space such as CIECAM02. This can be obtained by taking telespectroradiometric measurements at regularly spaced locations across the stimulus and using them as inputs to CIECAM02 alongside viewing condition parameters that express how the stimulus is viewed. The result is brightness, colorfulness (or lightness, chroma) and hue predictions for the spatial samples, which is an attempt at estimating how a viewer would describe these attributes for each of the sampled locations – if it were only the stimulus at those locations that was viewed against a uniform background. Even though this is not the source image itself, and even though it is not even a prediction of the stimulus' parts' appearance as seen in the stimulus as a whole, this is probably as close to a quantitative expression of the source image as we can get today. An alternative to the use of a CAM is to employ an *image CAM* such as iCAM (Fairchild and Johnson, 2004) to make such predictions, as it would also attempt to take into account intra-image effects like simultaneous contrast, etc. Such models, however, are at an earlier stage of development still than CAMs.

9.1.2 Original image in device-dependent color space

Instead of either the experience itself or color appearance predictions from stimulus measurements, the initial description that is typically available of a source image in a digital workflow is an image that for each pixel (picture element) has a set of device-dependent RGB or CMYK values, i.e. instructions meant for the device on which the original stimulus was produced. To go from this image description (which by itself has no visual meaning) to an approximation of what the color appearance predicted from direct measurements would be, it is possible to use information about the printing system on which the original was printed by having the RGB or CMYK data sent to it.

In our scenario, the original is an offset-lithographic print made in accordance with the ISO 12647-2 standard (ISO, 2004b) on coated paper, and an ICC profile can be used to approximate the measurements that would be obtained for any CMYK combination printed in this way.

As an example we will take the pixel from the original image that is at the center of the crosshairs shown in the detail in Figure 9.1, which has CMYK percentage values of [4, 96, 72, 14]. Using the ICC profile, a prediction can be made of the CIELAB values that would be obtained if this CMYK combination were printed and measured under the ICC PCS' conditions, i.e. $L^*a^*b^* = [43,\ 58,\ 29]$. Color appearance attributes can then be predicted using CIECAM02, giving $Jab = [35,\ 69,\ 27]$, which will be the starting point for gamut mapping this particular original pixel.

This printed medium has the color gamut shown orthographically projected in CIECAM02 in Figure 9.2. The gamut was computed using the segment maxima algorithm (Section 8.2.3) using 16×16 spherical angle segments and with the centre of the spherical space at $Jab = [50,\ 0,\ 0]$. Using this technique means that, for each of the 16×16 spherical subdivisions of color space, one of the samples from which the gamut is computed is chosen as a gamut boundary vertex. The gamut then approximates the

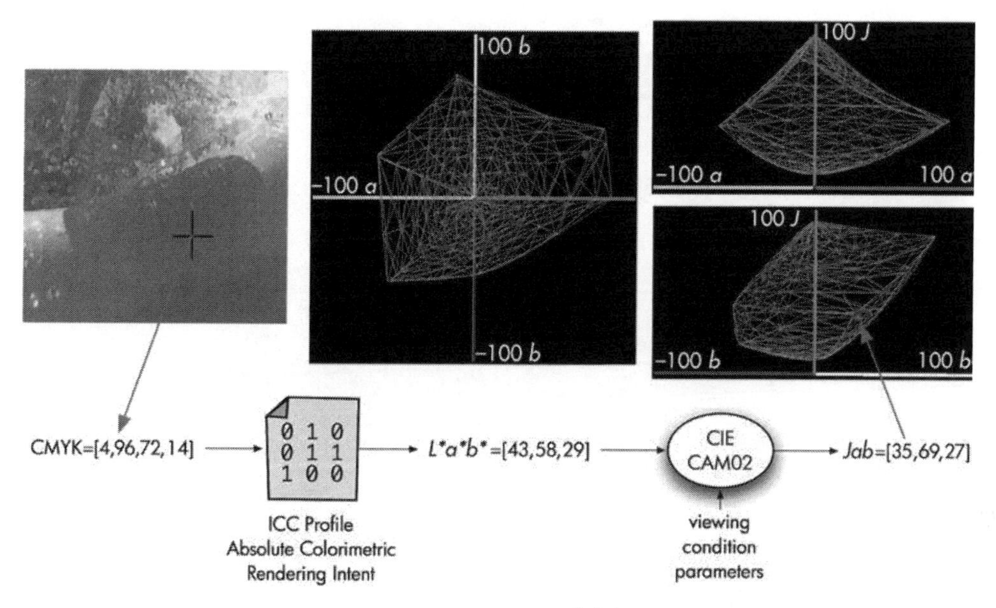

Figure 9.1 Sample from original CMYK image, its $L^*a^*b^*$ values and CIECAM02 *Jab* values (original medium gamut shown as mesh).

region in color space that encloses the colors resulting from printing all possible CMYK combinations using the chosen printing setup. As the reproduction is meant to look like the original when the two are seen side by side in this scenario, it is absolute colorimetry that will be used (i.e. when computing color appearance attribute, the adapted white will be the perfect diffuser).

The source image only uses some of all CMYK combinations (Figure 9.3); as a result, its color gamut is only a subset of the printing setup's gamut (Figure 9.4). What can be seen here is that, while the image does contain colors in the reds and magentas that are close to the source medium's gamut boundary, in the greens and cyans there is more room in the source gamut than the image uses.

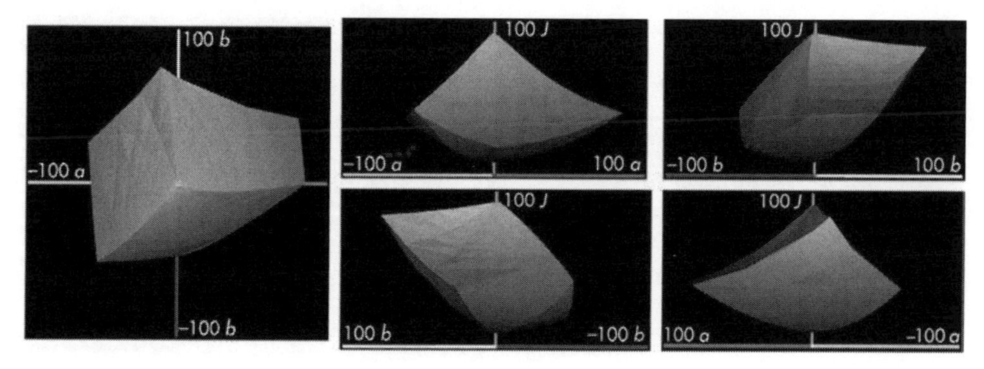

Figure 9.2 The source medium gamut (CIECAM02, D50).

Figure 9.3 CMKY (top left clockwise) planes of source image shown side by side (black indicates a large value and white a small one in each channel).

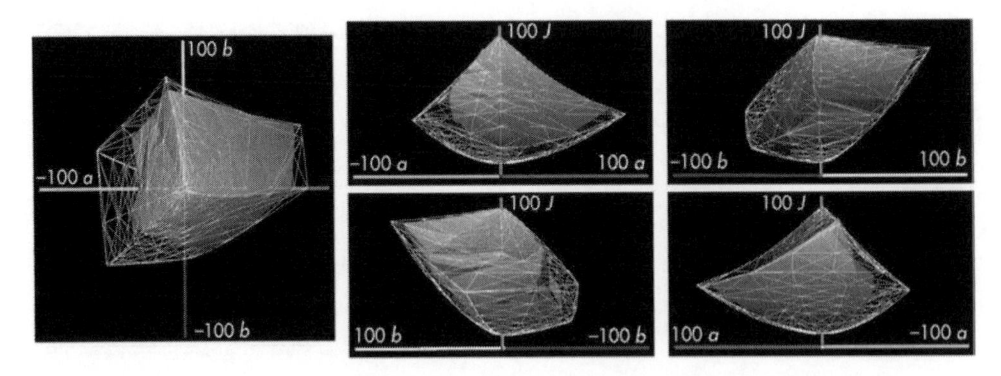

Figure 9.4 The source image gamut (solid) in the source medium's gamut (white mesh).

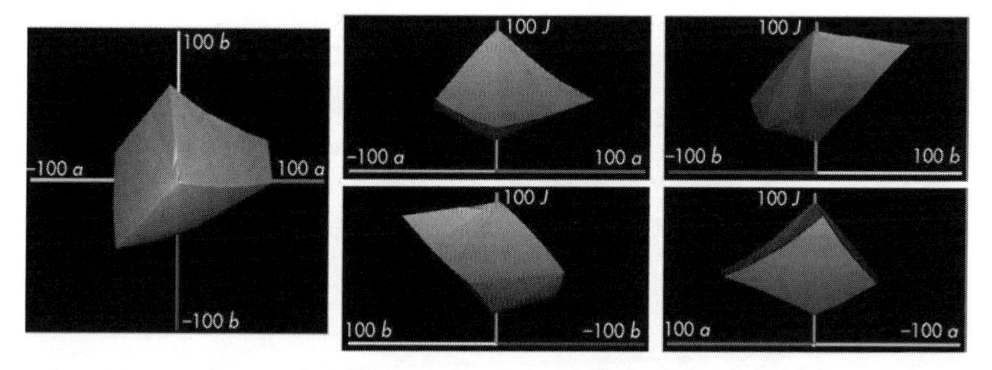

Figure 9.5 The destination medium gamut.

9.2 THE DESTINATION GAMUT

Given the above original image printed on coated paper, we will look here at its reproduction on uncoated paper – again following the ISO 12647-2 standard (ISO, 2004b). The gamut of this, destination medium is shown in Figure 9.5, and Figure 9.6 then shows how it relates to the gamut of the original image that will be reproduced in it. There are clearly significant parts of the original image's gamut that are outside the destination gamut, with a lack of chroma in the blue via red to greenish–yellow hues and also a lack of dark colors around the hue circle being most obvious.

While Figure 9.6 is indicative of where the challenges can be expected, it is also useful to see where in the original image there are colors that are outside the destination gamut (Figure 9.7). This figure also shows how far source colors are out of gamut, where white denotes IG colors and black a color that is 21 cJab units away from the nearest destination gamut boundary color.

We can see from Figure 9.7 that it will be the flowers at the top of the image, the shadows and parts of the paddling pool that will have to be moved most to fit into the destination

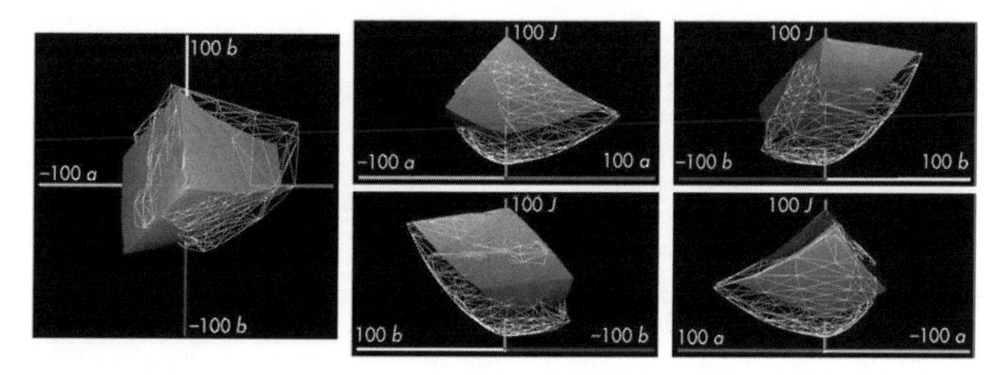

Figure 9.6 Comparison of source image gamut (mesh) and destination gamut (solid).

Figure 9.7 Distance from destination gamut for original image pixels.

gamut, which also agrees with the source image and destination medium gamut comparison, since this content is either dark or in the hues where differences are greatest.

9.3 MINIMUM COLOR DIFFERENCE GAMUT MAPPING

With the stage set for gamut mapping, let us now consider how the source image can be transformed into a destination image in CIECAM02 where each destination pixel's color is as close to its original color as the destination gamut allows. To achieve this aim, each IG color is kept unchanged and each OOG color is replaced by the color that has the smallest Euclidean distance to it on the destination gamut boundary. The earliest mention of this algorithm is in Sara's (1984) PhD thesis, where Euclidean distance is minimized in CIELAB.

9.3.1 A minimum color difference clipping algorithm

One way of implementing this algorithm, given a segment maxima GBD, is to do the following for each source color (Morovič and Sun, 2001).

Determine whether the source color is inside the destination gamut. This can be done by computing the destination gamut boundary at the source color's hue (Section 8.2.1) and seeing whether the intersection between the destination gamut and the line defined by the source color and the center of the spherical coordinate system has a greater radius than the source color. If it does, then the source color is inside the destination gamut and the destination color will equal it; and if not, then the following steps need to be followed:

1. For each triangle tessellating the gamut surface (and having vertices from neighboring GBD matrix members) do the following:
   (a) calculate the plane (and the plane's normal) determined by it;
   (b) intersect the plane and the line determined by the source color and the plane's normal;
   (c) if the intersection is inside the gamut surface triangle, then it is a candidate for having minimum color difference from the source color.

   Note that for three points from a GBD to be neighbors they need to be stored in members of the GBD matrix that are next to each other horizontally, vertically or diagonally (e.g. points with GBD matrix indices $[i, j]$, $[i+1, j]$, $[i, j+1]$ are a valid choice and $[i-1, j]$, $[i, j]$, $[i+1, j]$ are not as $[i-1, j]$ and $[i+1, j]$ are not neighbors). Also bear in mind that α is circular and that, when data are organized in a way where the columns relate to α and the rows to θ, the last column of the matrix is a neighbor of the first column.
2. For each of the tessellating triangles' edges do the following:
   (a) calculate the line determined by the pair of points delimiting it and a vector pair defining a plane orthogonal to the line;
   (b) intersect the plane determined by the vectors from step 2a and the source color with the line from step 2a;
   (c) if the intersection is inside the edge, then it is a candidate for being closest to the source color.
3. For each GBD vertex, calculate the Euclidean distance between it and the source color. The vertex with the minimum of these distances is the final candidate for being closest to the source color.
4. Choose the candidate with the smallest Euclidean distance from the source color from among the candidates from steps 1c, 2c and 3.

Looking at the above algorithm, it can be seen that various sets of original colors will be mapped onto single reproduction colors. More specifically, a given reproduction color from step 1 would be chosen for all the original colors that are along the line that is orthogonal to the plane of the triangle in which that reproduction color is located. Hence, a one-dimensional region of color space is mapped onto a zero-dimensional one. Reproduction color candidates considered in step 2 each correspond to two-dimensional regions of the original gamut, and three-dimensional parts of the original gamut are mapped onto the colors from step 3. What also follows from this is that different parts of color space will be subjected to different degrees of loss of variation (Figure 9.8). To see the effect on an actual gamut shape, Figure 9.9 shows those regions from a constant-hue section of a source gamut that would be mapped onto single points on a destination gamut

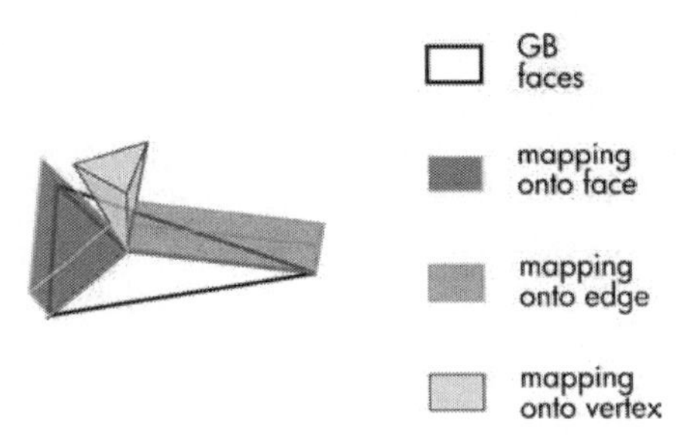

Figure 9.8 Types of variation loss due to minimum color difference gamut clipping.

as grayed out. This means that any variation in a source image in the grayed-out regions would be completely lost in the reproduction.

To illustrate how a gamut surface looks to a minimum color difference algorithm, Figure 9.10 shows the destination gamut's vertices as spheres whose lightness expresses their difference from the sample color used in Section 9.1.2.

Finally, color difference minimizing gamut clipping algorithms also need to consider the fact that, for concave destination gamuts, there is often more then one color on the destination gamut boundary that has the same minimal color difference from the source color. The reason for this is that spheres (which represent the set of all colors at a given distance from a source color) can touch a concave surface at more than one point (Figure 9.11). A GMA then needs to make a choice from among the equivalent alternatives; it is particularly important to do this in a systematic way (as opposed to randomly), as discontinuity artifacts may otherwise result.

9.3.2 The gamut mapped image

Applying the above minimum color difference gamut clipping directly to each pixel of our source image results in the color appearances in the destination gamut shown in Color

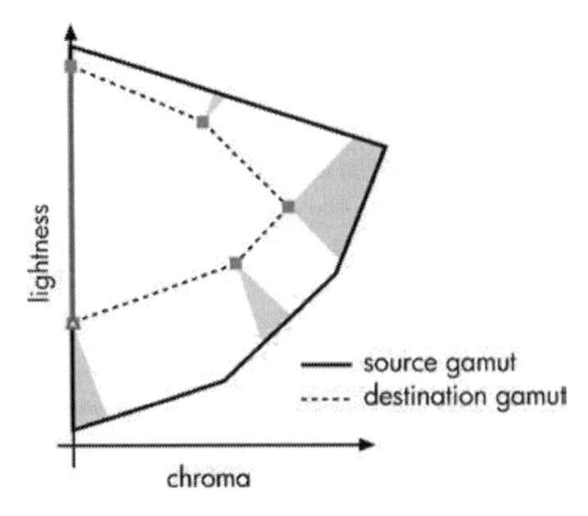

Figure 9.9 Regions of source colors mapped onto single destination gamut boundary colors shown in gray in a plane of constant hue angle.

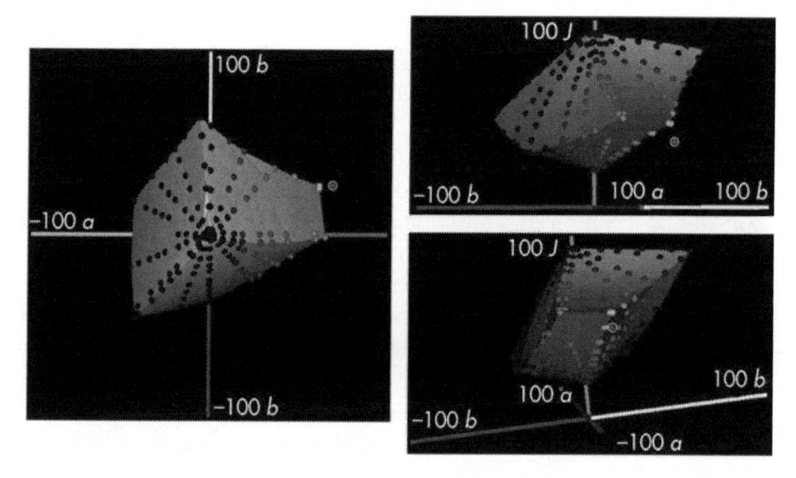

Figure 9.10 An OOG color (sphere with white border) and its color difference with each destination gamut boundary vertex indicated by the lightness of the sphere at each vertex (white represents zero difference and black is a difference of ≥80 units).

Plate 6. Note that this is not the same as using the algorithm to populate the data structures of, for example, an ICC profile and then applying the profile to the image's pixels. The difference is that, here (and throughout the following chapter), each image pixel color is gamut mapped explicitly using a chosen algorithm, whereas in the case of an ICC profile the gamut mapping is applied to a uniform sampling in either CIELAB or CIEXYZ and the colors of individual pixels are processed by interpolation among the gamut mapped outputs for the uniform color space locations. As a consequence, it is much more likely for artifacts to arise when gamut mapping (and especially gamut clipping) is applied directly, as using ICC profiles (or LUTs) effectively provides smoothing as a consequence of sampling.

What is most immediately noticeable when the gamut mapped image in Color Plate 6 and the corresponding original image in Color Plate 3 are compared is that:

1. the reproduction has *less contrast*;
2. the reproduction's colors look *less colorful*;

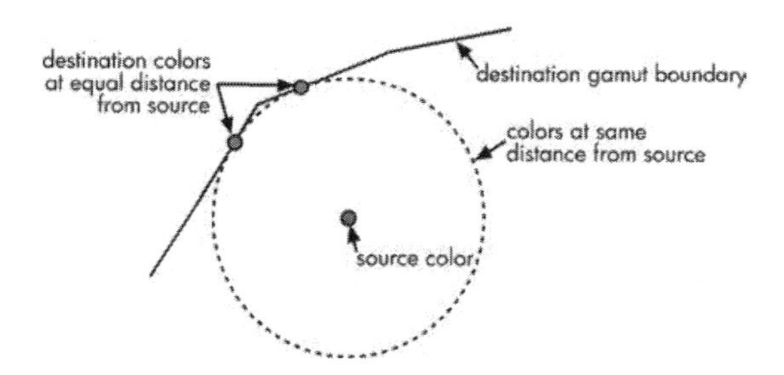

Figure 9.11 Multiple minimum color difference points on a concave surface (shown in two dimensions).

3. there is *loss of detail* in some parts of the image;
4. even some of the image's parts that are in the *intersection* of the two gamuts (and for which there is little of no difference shown in Figure 9.8) *look different* between source and reproduction.

Let us next look at each of these aspects in turn.

First, *loss of contrast* is a direct consequence of going from a larger to a smaller color gamut simply because the contrast between two stimuli is their difference; as a smaller gamut allows for smaller differences than a larger gamut, a reduction of gamut also implies a reduction in the potential for contrast. Looking at the pair of original and reproduction images used here, it can be observed that the difference between the lightest and darkest parts of the original is significantly greater than the corresponding difference in the reproduction, even though in both cases the lightest and darkest image parts are at the corresponding gamuts' boundaries. Having said this, it is essential to bear in mind that such an argument only applies to contrast *potential* rather than the actual contrast of a reproduction, and we will see in Chapter 10 that there is a lot of room within a given destination gamut for reproductions to have varying degrees of contrast. What can certainly be observed even from looking at this single reproduction is that its reduced contrast is the result of both the reduction in contrast potential in the smaller destination gamut and the choice of GMA used. Minimum color difference clipping here will tend to preserve as much overall contrast as is possible, since it clips OOG colors to the gamut boundary, thereby resulting in more contrast than compression techniques would provide in general.

Second, *loss of colorfulness* again leads to analogous considerations, where smaller gamuts tend to allow for less colorfulness but the actual colorfulness is again the result of colorfulness potential in the destination gamut and the choice of gamut mapping used to get there. Minimum color difference clipping here provides an even balance between changes to lightness, chroma and metric hue predictions (where *metric hue difference* is the part of color difference not accounted for by lightness and chroma differences, i.e. this is no hue *angle* difference). As a consequence, this reproduction will not be the one that maximizes colorfulness and there is likely to be more colorfulness potential in the destination gamut. To realize that potential, more of the original's lightness and hue would have to be given up.

Third, *loss of detail* is closely related to loss of contrast, as detail can be considered to be local contrast of specific spatial frequency properties and analogous considerations apply. Minimum color difference gamut clipping impacts detail differently, though, from overall contrast. The gamut clipping algorithm maximizes detail preservation in image parts that are entirely inside the destination gamut and results in strong detail loss for OOG image parts. This is because some volumes in color space are mapped onto destination gamut vertices, which means that variation in all of lightness, chroma and hue is reduced onto a single destination color. Therefore, a whole pyramid-like region below the darkest point of the destination gamut (and with its tip at that point) is mapped onto the darkest destination color. As a consequence, any variation in the source image that is in the part of color space shown as a pyramid in Figure 9.12 is lost.

An example of such loss of detail is shown in Figure 9.13, where a row of 50 pixels from the original is taken and its lightness, chroma and hue angle predictions are shown as a

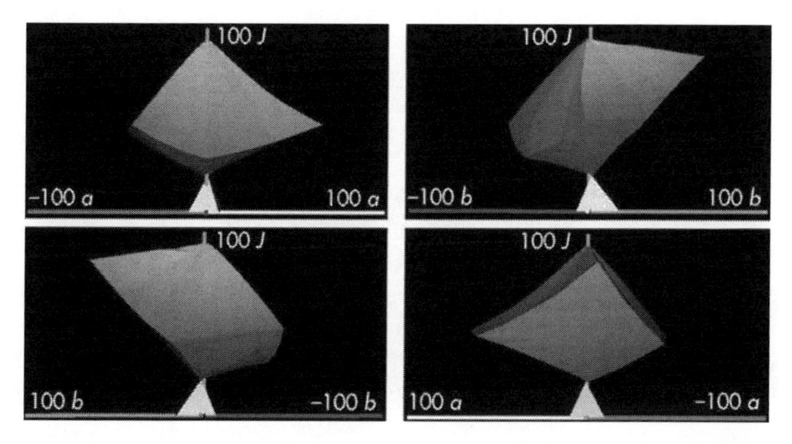

Figure 9.12 Colors mapped onto destination gamut's darkest point shown as a light pyramid.

solid line in the figure. Here, it can be seen that there is variation in all three appearance attributes. Gamut mapping these pixels' colors then gives the appearance attributes shown by a dashed line. Here (with the exception of some, probably imperceptible, variation in lightness), all of the original colors are mapped onto the same reproduction color and the detail from the original is lost. Such loss of variation is commonly referred to as *blocking*.

Fourth, there is also some *change in appearance for IG colors* (i.e. source colors that are already inside the destination gamut). For example, the center of the baby's chest in the original image is inside the destination gamut (as can also be seen from Figure 9.7); however, there is a different color experience to be had from the original and its

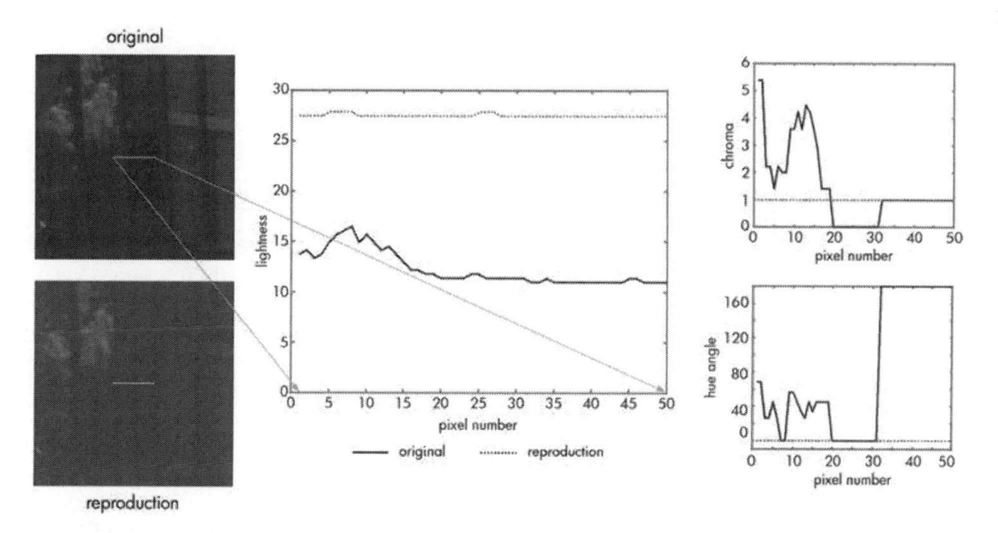

Figure 9.13 Detail from original image and its appearance attributes before (solid line) and after (dashed line) gamut mapping.

reproduction of that part of the image. The most natural assumption to make is that even this IG original color was changed in the process. Looking at the color predictions for both original and reproduction shows though that this part of the original has not changed (Color Plate 4a). The reason for even this IG color *looking* different is simultaneous contrast (Section 2.7.5). While in the original there are darker parts in the neighborhood of this color, gamut clipping to a smaller gamut has resulted in these darker parts being less dark, as a consequence of which the contrast between the IG color and them is reduced, making the IG color appear lighter than it looked in the original.

Note that minimum color difference clipping is also often used as the final stage in gamut compression algorithms, whereby preceding stages are set up without the necessity to result in all colors being IG and remaining (typically small) differences are left to the minimum color difference clipping algorithm (e.g. Stone *et al.*, 1988; Taylor *et al.*, 1989).

9.4 THE DESTINATION IMAGE

Given the gamut mapped image in CIECAM02 *Jab*, which is the immediate output of applying the minimum color difference gamut clipping algorithm to the original's *Jab* values, we need to first compute corresponding CIELABs. These are then the inputs to the destination medium's ICC profile, finally resulting in a set of CMYKs for the destination device at each image pixel. For the sample pixel from the source image used in Section 9.1.2, the gamut mapped *Jab*s are [45, 61, 27], which are equivalent to $L^*a^*b^*$ values of [52, 55, 30] and to CMYKs for ISO uncoated printing of [1, 98, 95, 0] (Figure 9.14).

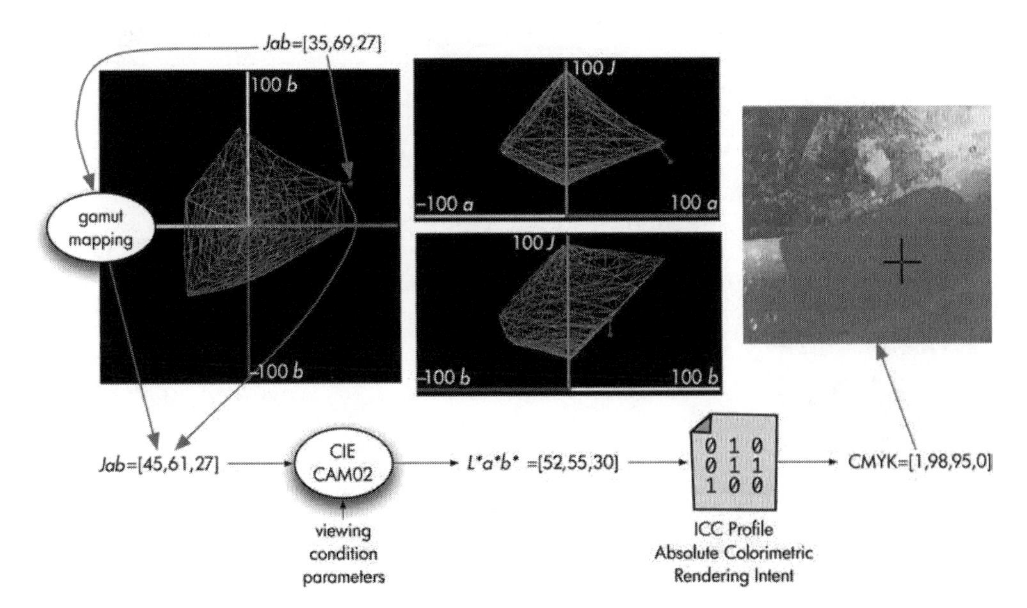

Figure 9.14 Gamut-mapped CIECAM02 *Jab* values, $L^*a^*b^*$ values and CMYK values of sample from original image (reproduction medium gamut shown as mesh). This is a continuation of Figure 9.1.

9.5 EFFECT OF ALTERNATIVES

In addition to the minimum color difference clipping shown in Section 9.3, where the gamut mapping space is CIECAM02, it is also worth looking at the effect of alternatives. Performing the same mapping algorithm in a different color space, i.e. CIELAB, will be looked at first, and the use of different weights for the three appearance attributes of lightness, chroma and hue will be considered next. In addition to clipping to a color that has predetermined properties (i.e. where gamut mapping equates to *finding* the point on the destination gamut that has a minimum value in terms of some difference metric – be it Euclidean, hue preserving or weighted in some other way), the clipping of source colors along lines determined by them and some point on the lightness axis will also be presented (i.e. here, gamut mapping is the *intersection* between a line and the destination gamut boundary).

9.5.1 Minimum color difference in CIELAB

Color Plate 7 shows the application of exactly the same algorithm as in Section 9.3, but in a different color space, i.e. CIELAB. Here, the destination gamut boundary is computed in CIELAB and the mapping also takes place in that space. The most noticeable difference between doing the clipping in CIECAM02 versus CIELAB is the greater loss of variation in the shadows of the image when CIELAB is used. The reason for this is the difference in how lightness and chroma are predicted in the two spaces, whereby the nearest destination gamut color to, for example, a dark chromatic green in the top left part of the source image is a near-black in the CIELAB space, but a lighter but still chromatic green in CIECAM02. This is a clear example of how differently color spaces predict appearance attributes and relationships, and it also shows that, at least in this case, the CIECAM02 prediction is more useful in the context of color reproduction.

Looking at another example (source: Color Plate 4b; destination: Color Plate 5), we can also see that the hue predictor of the two spaces differs significantly and, while minimizing color difference in CIELAB results in a reddening of blue parts of the image (e.g. the rain jacket), the same algorithm in CIECAM02 maintains more of the original appearance.

9.5.2 Minimum weighted color difference in CIECAM02

Even though it might at first seem that the closest (at least color-by-color) reproduction will be obtained by minimizing the most accurate color difference metric, there is in fact more to making a reproduction as similar to an original as gamut differences allow. Since this might at first seem somewhat counterintuitive, let us take a closer look next (Morovič *et al.*, 2007).

The units of ideal color difference equations represent JNDs. Pairs of colors judged as only just distinguishable, therefore, are described by a color difference $\Delta E = 1$. Then the meaning of $\Delta E = 2$ units between a pair of colors is that there exists a third color on the line connecting them in an ideal, perceptually uniform color space that has $\Delta E = 1$ from both colors of the pair. In general, a $\Delta E = n$ means that there are $n - 1$ intermediate one-JND-spaced colors along the line between the color pair whose difference it is.

Given a color, there are then spheres in an ideal, uniform color space that have it as their center and whose every point has the same color difference from the central color (this color difference being the sphere's radius). For representations in nonideal color spaces (i.e. all current ones), the ideal sphere is replaced by another geometric figure that, furthermore, varies with location in the space. The key point here is that for every given color there are a multitude of other colors that are equally different from it.

Let us then ask the question: 'For a color *c* and the set *D* of all other colors that are a given distance from it, what will be the result of asking observers to pick the most similar member of *D* compared with *c*?' If similarity is the same as distance, then for a sufficiently large sample the result will be a uniform distribution of choices across the entire set *D*. If, however, color appearance attributes have different levels of importance, then observers' choices will concentrate in specific parts of *D*. A consequence of such differences in importance are then also the results of previous work on cross-gamut visual similarity, where colors that favor one dimension over another are chosen instead of colors having the smallest overall difference.

Having an accurate color difference metric, therefore, is not the end of the story, and simply applying it when having to choose how to reproduce an OOG color will not result in the most similar looking choice. Instead, it is necessary to understand the relative importance given to the three color appearance attributes when an observer is asked to judge similarity. For example, if an observer is presented with an original and the choice between two reproductions – one that matches lightness and differs by two units of chroma and the other that matches chroma and differs by to units of lightness – then it would be expected that each alternative would be chosen equally often (for a sufficiently large number of observers) if differences in each attribute had equal importance. Doing such a test, however, will tend to give much greater preference for the reproduction that maintains lightness than chroma, and this applies both to single colors and complex images.

The idea of applying different weights to differences in lightness, chroma and metric hue has a long tradition in gamut mapping, and the weighted color difference formula is typically the following (though other forms have also been used):

$$\Delta E_{\mathrm{WT}} = \sqrt{\left(\frac{\Delta L}{kL}\right)^2 + \left(\frac{\Delta C}{kC}\right)^2 + \left(\frac{\Delta H}{kH}\right)^2} \tag{9.1}$$

Here, the weights $[kL,\ kC,\ kH]$ divide each of ΔL, ΔC and ΔH before they are squared and summed. Having weights of, for example, $[kL,\ kC,\ kH] = [1,\ 2,\ 3]$ would result in chroma being given only half the weight that lightness receives and metric hue only one-third – as a result, this minimum weighted ΔE gamut clipping would tend to preserve more of the original's lightness than its chroma and more of its chroma than its hue. Note that, in some studies, weights are used to multiply rather than divide the differences in each attribute and that the two relate simply by being each other's inverses, i.e. multiplying weight $= 1/(\text{dividing weight})$ and vice versa.

Given such a weighted color difference formula, it is possible to set the weights so that they map original OOG colors to reproduction colors that look most similar to them, and such weights can be determined experimentally.

The first such attempt was the work of Katoh and Ito (1996), where original images on a CRT display were reproduced by applying minimum ΔE_{WT} clipping with different values for the weights. The resulting printed reproductions were shown to observers in a pair comparison experiment (Section 3.3.2), where they had to choose the reproduction that was most similar to the original. The authors found that, based on having 24 observers judge the reproductions of three computer-generated images, the reproductions that were most similar to the originals were obtained when the $[kL, kC, kH]$ coefficients were [1, 2, 1] or [1, 2, 2] in CIELAB, i.e. larger changes were acceptable in chroma than in hue and the smallest change was tolerated in lightness.

These results are also supported by the work of Ebner and Fairchild (1997), who had observers adjust the colors of simple, synthetic images in a restricted gamut when they were asked to give the most similar appearance to an OOG original. The way this was done was by having an image of the destination gamut's surface from which observers could pick the color that they thought was most similar to a given OOG color. The results again suggested that most change was acceptable to chroma, but in this case the observers were somewhat more sensitive to hue changes than lightness changes. Next, Wei and co-workers (Wei and Sun, 1997; Wei *et al.*, 1997) reported that $[kL, kC, kH] = [1, 3, 1.5]$ gave closest matches when applied again in CIELAB. A further study (Morovič *et al.*, 2007) showed that $[kJ, kC, kH] = [1, 2.6, 1.3]$ in CIECAM02 resulted in gamut mapping to the visually most similar colors based on having color choices made both by naïve and professional observers. These results were obtained in a similar way to the Ebner and Fairchild (1997) experiment, with the difference that choices were made from the print of a gamut's surface rather than its simulation on a display.

Minimizing a $[kJ, kC, kH] = [1, 2.6, 1.3]$ weighted color difference in CIECAM02 results in the reproduction shown in Color Plate 8, and the way to compute this gamut mapping is the same as described in Section 9.3.1 with the one difference that it is ΔE_{WT} that is used instead of Euclidean distance. Note that while minimum Euclidean distance mapping is akin to expanding a sphere with a source color as its center until it touches the destination gamut, using ΔE_{WT} is in effect the expansion of a distorted sphere, such as shown in Figure 9.15 (Morovič *et al.*, 2007).

Other color difference equations can also be used in this type of minimization clipping, and good results have been reported by Ito and Katoh (1999) for using ΔE_{94} (CIE, 1995) in CIELUV and ΔE_{BFD} (Luo and Rigg, 1987) in CIELAB.

The weighted color difference minimizing reproduction in Color Plate 8 preserves the appearance of some parts of the original image better than Euclidean distance minimization in CIECAM02 (or CIELAB). For example, the shadows in the baby's face or the boat are preserved much more than in the alternatives. The downside of this approach, though, is that in the region below the black point of the destination gamut (and in a region just above it too) the greater weight given to lightness means that a lot of the source gamut gets mapped to near the destination black point. This can be seen clearly in the shadow region in the top left part of the image, where straight minimum ΔE mapping gives better results. So, while giving more weight to lightness than chroma works well for many colors, it does fall down around the destination black.

An extension of such weighted color difference clipping is one that is preceded by a mapping of source lightnesses or luminances, e.g. by applying the Adobe BPC algorithm (Section 6.4). This is done by taking each source pixel's *Jab*, transforming it to *XYZ* and

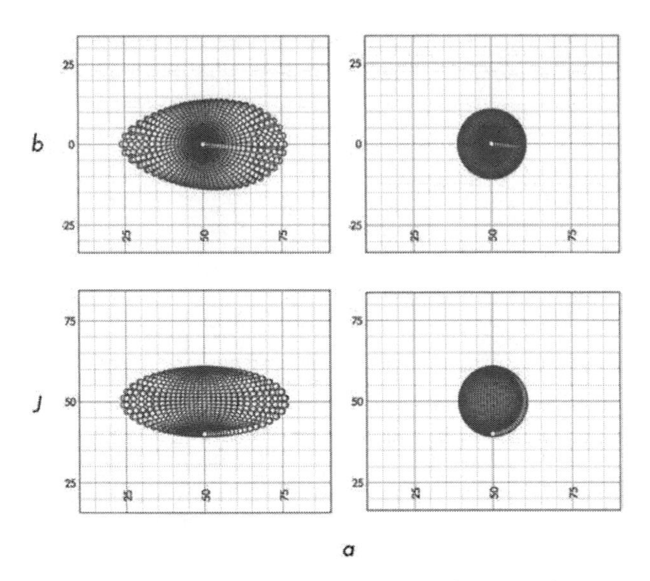

Figure 9.15 Set of points at 10 units difference from *Jab* = [50, 50, 0] using (left) weighted and (right) nonweighted Euclidean distance metrics.

applying the BPC algorithm to it (Equation (6.3)), which linearly scales the source relative luminance range onto the destination one. The resulting scaled *XYZ*s are then transformed back to *Jab* and gamut clipping is applied (Figure 9.16). The benefits of the BPC can be seen clearly in Color Plate 9, as it results in the preservation of a lot more shadow detail (e.g. see the top left part of the image) as well as a better preservation of overall image appearance (but not the appearance of individual pixels) than keeping IG colors

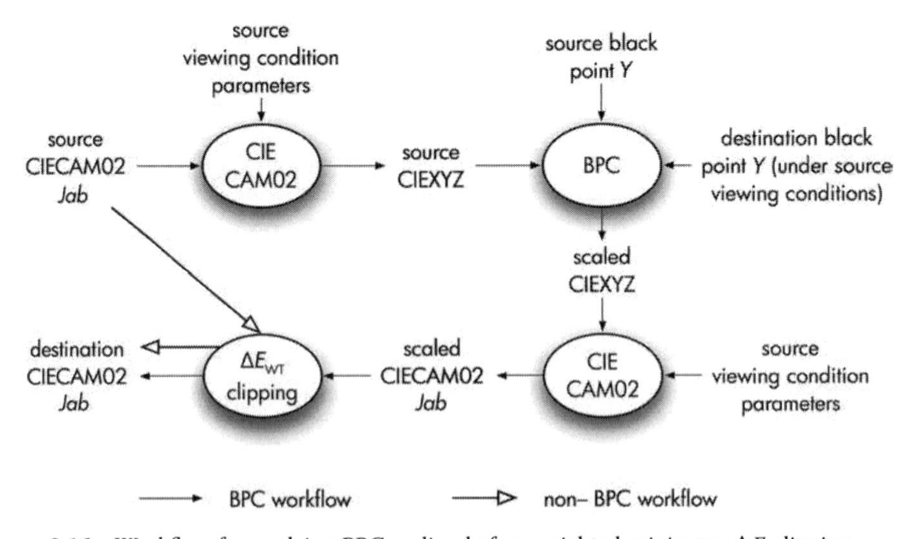

Figure 9.16 Workflow for applying BPC scaling before weighted minimum ΔE clipping.

unchanged results in doing. Notice, for example, the baby's chest, which (although IG) was changed by the BPC but which looks more like the original than in the other reproductions. The result is an appearance more similar to the original, since relationships among colors are better preserved.

9.5.3 Clipping towards point on lightness axis

An alternative to mapping onto destination gamut colors with a minimum value for a difference metric is to map along predetermined lines. The simplest of such approaches are the SCLIP and CCLIP algorithms.

SCLIP maps OOG colors onto the destination gamut boundary color that has the same α and θ as the source color when the centre of the spherical coordinate system is at *Jab* (or $L^*a^*b^*$) of [50, 0, 0]. In other words, it is a mapping towards the center of the lightness axis and was first proposed (in compression form) by Sara (1984). An alternative is CCLIP, which makes the mapping point on the lightness axis change as a function of hue, by making it match the lightness of the cusp at each hue (MacDonald and Morovič, 1995). The *cusp* at a given hue is the color of maximum chroma and is a feature of color gamuts that is often used in gamut mapping. The reason for the cusp being of importance is that it divides the lightness range at a hue into two parts, each of which have monotonic chroma changes on the destination gamut boundary with changes to lightness (for well-behaved device or medium gamuts).

These mappings also preserve hue angle and involve the intersection of the destination gamut boundary with a vector defined by the source color and a color on the lightness axis (Figure 9.17).

Looking at the reproduction in Color Plate 10, we can see that the SCLIP algorithm does much better than minimum ΔE clipping-based approaches in avoiding blocking artifacts. This is indeed because it only ever collapses variation along lines onto points,

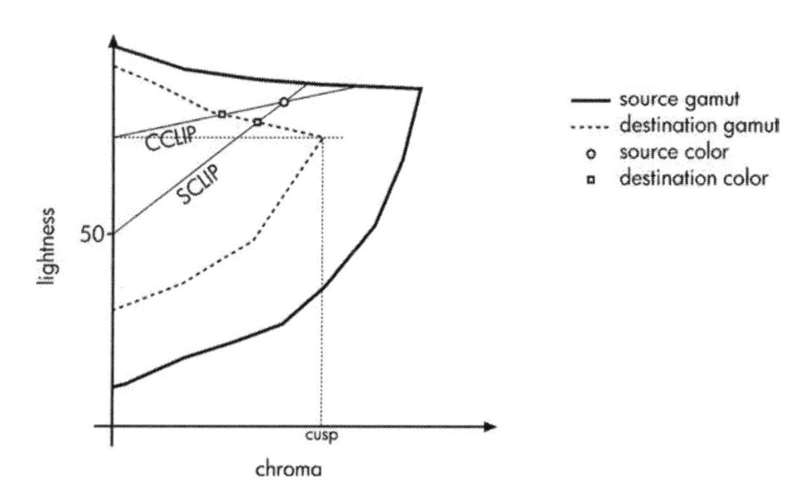

Figure 9.17 The SCLIP and CCLIP algorithms shown in a plane of constant hue angle.

whereas minimum ΔE clipping can do so for entire two- or three-dimensional regions from the source gamut. SCLIP, therefore, is less likely to compromise detail from a source image, and methods that map towards points on the lightness axis have been popular in gamut mapping studies. On the other hand, it is also clear that SCLIP performs worse in some respects – especially in terms of the chroma it provides in the reproduction, as this is maintained noticeably better in the minimum ΔE case for the source and destination gamuts considered here. Look, for example, at the paddling pool or the flowers in the background and compare how much chroma the two types of approach preserve.

Clipping can also be performed by keeping one or two of a color spaces' dimensions unchanged and minimizing distance along the remaining dimension(s) (Gentile *et al.*, 1990). Popular examples of this category are LCLIP, where lightness and hue angle are kept constant and only chroma is reduced for OOG colors until the destination gamut is reached (Sara, 1984), and HPMINDE, where hue is preserved and minimum distance to the destination boundary is then sought (Murch and Taylor, 1989; Taylor *et al.*, 1989).

While LCLIP is an approach that has been used repeatedly, it has a serious limitation in that it results in desaturation artifacts when the destination gamut boundary's chroma changes rapidly with lightness. Highly chromatic source colors (typically yellows or blues) are then mapped to destination colors with significantly less chroma but with the same lightness (Figure 9.18). A further artifact that LCLIP implies is a reversal of chroma whenever source and destination cusp lightnesses are not aligned (which is almost always the case). In Figure 9.18 we can see that while source color S_1 has slightly higher chroma that S_2, the gamut mapped version of S_1, i.e. D_1, has much lower chroma than D_2.

Gentile *et al.* (1990) have compared gamut clipping obtained by keeping various combinations of color attributes (and attribute pairs) constant while minimizing differences in the remaining attribute(s). Their small-scale pair comparison experiment, which involved mapping from a display's gamut to a synthetic destination gamut simulated on the same display, had lightness and hue preservation come out top in terms of similarity to

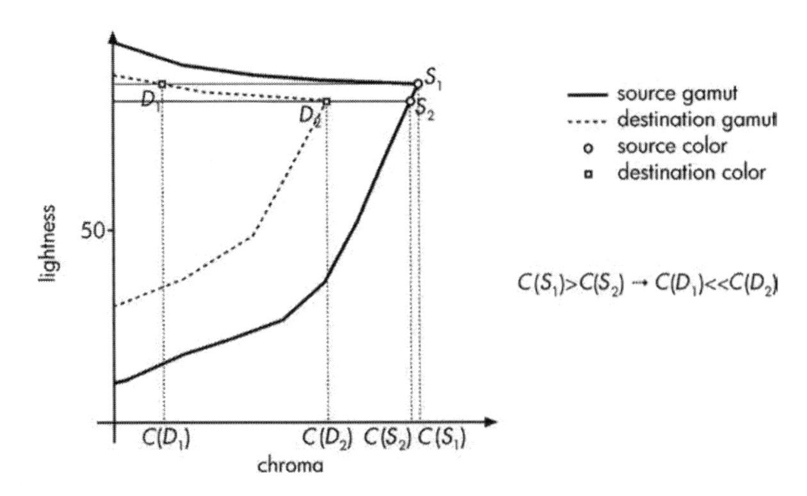

Figure 9.18 Clipping along lines of constant hue and lightness for 'steep' gamut boundaries.

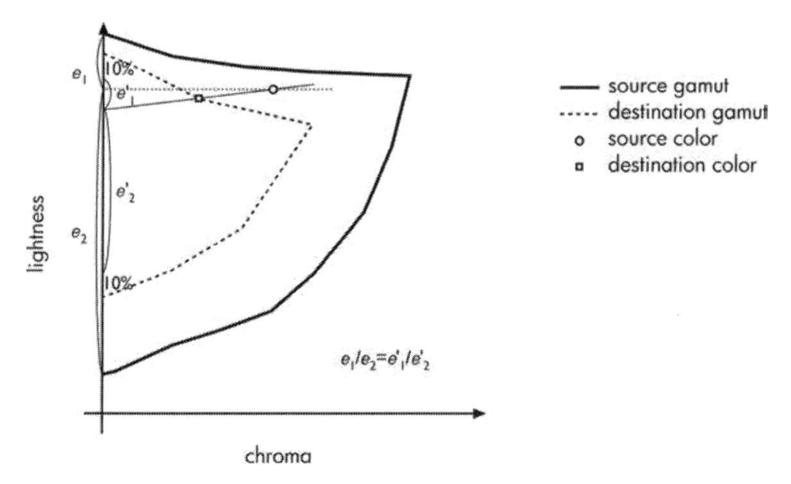

Figure 9.19 Marcu and Abe's gamut clipping algorithm.

originals. These results agree with the weights later found to give best minimum ΔE_{WT} clipping (Section 9.5.2) and are also supported by the findings of Pariser (1991).

A related gamut clipping technique was proposed by Marcu and Abe (1996), where the point on the lightness axis towards which clipping is performed depends on the source color's relationship to the source gamut's black and white points. Taking the lightness range of the source gamut and representing black as 0% and white as 100%, a source color will be at some percentage point in the source gamut's lightness range. Gamut clipping is then towards the percentage point in the destination gamut's lightness range on the lightness axis that is the same as in the source, after an offset has been applied to both destination black and white points towards the gamut's center. The offset is added to ensure that more of light and dark colors' chroma is preserved. If it were not used, then colors in those regions would be mapped close to the destination's black and white points, which could result in greater loss of detail than in the proposed algorithm (Figure 9.19).

Finally, colors can also be mapped along lines that have a fixed angle with the chroma axis (Ruetz, 1994), whereby that angle is applied counterclockwise above the lightness of the cusp and clockwise below it (also resulting in a region of twice the chosen angle that clips onto the destination gamut's cusp).

9.5.4 Interpolating among explicitly defined clipping vectors

An interesting alternative to either minimizing a metric or clipping towards predefined points on the lightness axis is a method proposed in Sara's (1984) PhD thesis, which consists of two stages:

1. Subjectively (manually) defining how 26 points (i.e. an even sampling of a device RGB cube's surface) are to be mapped onto a given destination gamut and then interpolating mapping vectors for a given source color from them.

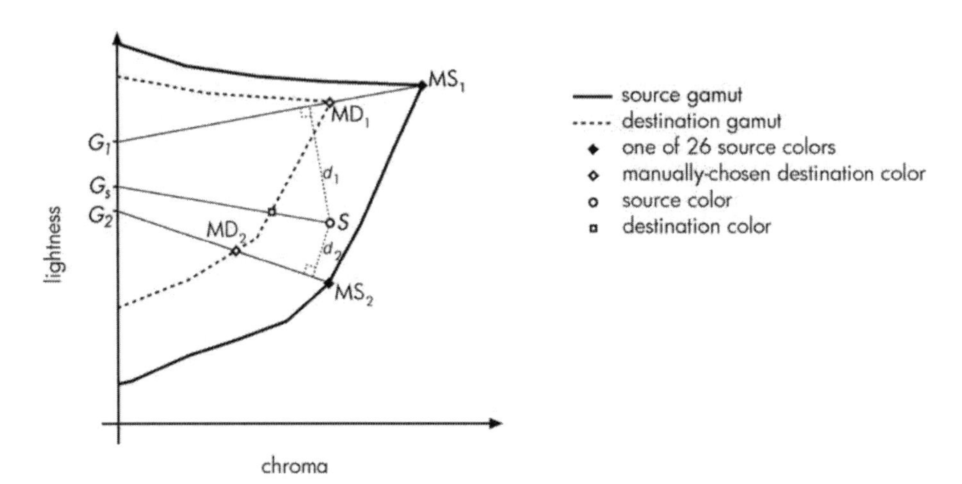

Figure 9.20 Sara's gamut clipping by interpolating among manually chosen clipping vectors.

2. For each of these 26 points, a center of gravity is defined as the intersection of the lightness axis and the line connecting the original and manually gamut-mapped colors. The center of gravity towards which a given source color is clipped is then calculated as a weighted average of the centers of gravity for the 26 points (Sara, 1984). Given the 26 centers of gravity $G_i(i \in \{1,\ 2,\ \ldots,\ 26)$, the centre of gravity G_S towards which a given source color S is to be clipped is computed as

$$G_S = \sum_{i=1}^{26} \frac{G_i}{d_i^2} \bigg/ \sum_{i=1}^{26} \frac{1}{d_i^2} \tag{9.2}$$

where d_i is the Euclidean distance between S and the closest point to it on each of the 26 lines that connect the center of gravity G_i with the corresponding source color for which it was manually determined (Figure 9.20).

9.5.5 Clipping along electrostatic field lines

Finally, van Gestel and Draaisma (2001) have proposed an intriguing gamut clipping algorithm that mimics the behavior of charged particles along electrostatic field lines. The idea here is to have a mapping where OOG colors are mapped onto the gamut surface with minimal loss of differences so that similar OOG colors end up being similar on the gamut surface. Since electrostatic field lines have this property, gamut mapping is 'translated' onto a physical model. Colors defining the destination gamut surface are represented as negatively charged point charges at their locations in a three-dimensional space, collectively forming an electrostatic field outside the gamut. OOG colors are then represented as positively charged particles that get moved along field lines defined by the destination gamut until they reach its surface. The result is a mapping that only collapses one-dimensional trajectories onto single colors and that results in a smooth transformation of the source onto the original.

9.6 SUMMARY

The aim of this chapter was first to provide a walk-through of how an image gets gamut mapped using the minimum color difference clipping algorithm and then to present some alternatives in terms of the color space in which to perform the clipping and the metric to minimize. Finally, gamut clipping algorithms that do not involve the use of color difference metrics were also discussed. While there is a fair amount of variety in gamut clipping, it is gamut compression where there is greater divergence, and this will be the subject of Chapter 10.

10
Survey of Gamut Mapping Algorithms

In many ways this chapter can be expected to be the most important one of this book, as it presents the details of a host of GMAs proposed in the last 30 years. It might also be expected that a neat evolution, chronicling progressive improvement in the quality of GMAs, will be laid out and that the reader will be left with a recommendation of what algorithm (or algorithm per desired reproduction property) to use that represents the state of the art.

Such expectations will, however, be disappointed, as the current state of gamut mapping is very far from an idyllic state – primarily due to the tremendous difficulty of answering the question: 'Is GMA *X* better than previous algorithms developed for the same desired reproduction property?' Unlike areas of color science like color difference prediction or color appearance modeling, there is no growing set of data that models need to predict, and judging whether a new GMA is better than previous ones would involve time-consuming psychovisual evaluation, comparing it against all previous algorithms. GMAs also need to work for different combinations of source and destination gamuts and for a myriad of images, which essentially makes even a representative evaluation of a GMA's performance near impossible.

What is often found is that the algorithm that works for one image does not work for another and the one that deals well with one pair of source and destination gamuts fails for another. How to attempt at least an improvement of this inhibited evolution due to evaluation constraints will be addressed in Chapter 12. For now it should suffice to say that this chapter will give an overview of some of the GMAs proposed to date and point to some general trends where there is consensus in the published literature. What will not be done is to point to best in class solutions, since these are certainly not known and probably not even there.

While the above is a pretty pessimistic picture, it has to be said that there is a positive aspect to gamut mapping research, which is the expanding understanding of what limits the broad applicability of algorithms proposed to date (Chapter 12). Such an understanding,

Color Gamut Mapping Ján Morovič

together with GMA evaluation that is consistent across studies and that can be aggregated, has the potential to lead to more robust solutions in the future.

With these caveats in mind, let us take a look at algorithms in the following categories: color-by-color reduction and expansion, spatial reduction, spectral mapping and algorithms for niche applications.

Note that, unless stated explicitly otherwise, *LCh* will refer here to lightness, chroma and hue predictors in general, rather than to those of a specific color space, and a and b will be the orthogonal equivalents of Ch (i.e. $a = C^* \cos(h)$ and $b = C^* \sin(h)$). In terms of notation, S will refer to the source, D to the destination, G will represent a gamut, min the minimum, max the maximum and $x(y)$ will mean x for given values of y. Hence, $G_{S\max C(L,h)}$ is the maximum chroma of a source gamut at a given value of L and h (i.e. the cusp at h) and D_C is a destination chroma.

10.1 COLOR-BY-COLOR REDUCTION

Since *clipping* GMAs were already covered in Chapter 9, it will be *compression* algorithms that will be looked at here. These can change not only source colors that are outside the destination gamut, but also those that are inside it – to distribute changes across more of the source gamut. The primary motivation here is to try to avoid the blocking issues inherent in clipping algorithms (Section 9.3.2).

10.1.1 Sequential, per-dimension mapping

In this first group, mapping consists of separate, sequential mappings between the ranges of a source and destination gamut in the dimensions of the gamut mapping color space, which in the vast majority of cases has lightness, chroma and hue predictors.

Initial lightness mapping

Given a source–destination gamut pair, it is almost invariably lightness that gets mapped first by applying a mapping to all source lightnesses based on the source and destination gamuts' lightness ranges.

The simplest such mapping is a linear one:

$$D_L = G_{D\max L} - (G_{S\max L} - S_L)\frac{G_{D\max L} - G_{D\min L}}{G_{S\max L} - G_{S\min L}} \tag{10.1}$$

Such scaling was first used by Buckley (1978), and while it has been reported to work well in several cases (e.g. Sara, 1984; Laihanen, 1987; Viggiano and Wang, 1992), in others it has led to an unacceptable loss in image contrast at the expense of lightness difference preservation. Looking at the effect of the mapping (Figure 10.1), we can see

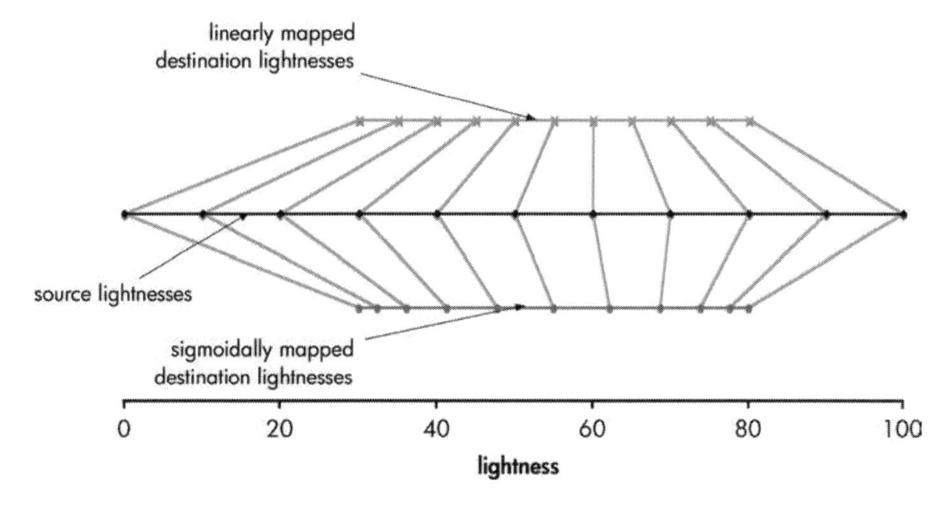

Figure 10.1 Linear versus sigmoidal lightness mapping.

that differences are preserved equally from across the source range, but that this implies a great reduction in contrast between, for example, the 10% and 90% points of the source lightness range, which in turn shows up as reduced image contrast for some images.

To address such loss of contrast, sigmoidal lightness compression had been proposed already by Buckley in his 1978 PhD thesis. Looking at Figure 10.1, it can be seen that this approach to mapping lightness results in detail being somewhat suppressed (but not entirely lost) to preserve more of the source's contrast potential. The idea of sigmoidal lightness mapping was revived more recently by Braun (1999), who implemented it in the following way:

1. LUT N is first computed on the basis of a discrete cumulative normal function:

$$N_i = \sum_{n=0}^{n=i} \frac{1}{\sqrt{2\pi}\Sigma} e^{-[(100n/m)-x_0]/2\Sigma^2} \tag{10.2}$$

   where x_0 and Σ are the mean and standard deviation of the normal distribution respectively, $i = 0, 1, 2, \ldots, m$ and m is the number of points used in the LUT (it is recommended for m to be at least 100). Hence, N_i is the value of the cumulative normal function for the i/m percentage point.
2. The m values in N are then scaled into the lightness range of the destination device. This gives LUT M, whose ith entry contains the destination lightness for the i/m percentage point of the source lightness range:

$$M_i = \frac{N_i - \min(N)}{\max(N) - \min(N)}(G_{D\max L} - G_{D\min L}) + G_{D\min L} \tag{10.3}$$

Table 10.1 Sigmoidal mapping parameter calculation.

G_{DminL} (CIE L^*)	5.0	10.0	15.0	20.0
x_0	53.7	56.8	58.2	60.6
Σ	43.0	40.0	35.0	34.5

3. Given a source lightness S_L, the percentage point p at which it is in the source lightness range is computed:

$$p = 100(S_L - G_{SminL})/(G_{SmaxL} - G_{SminL})p = 100 \times \frac{S_L - G_{S\,\mathrm{min}\,L}}{G_{S\,\mathrm{max}\,L} - G_{S\,\mathrm{min}\,L}} \tag{10.4}$$

and p is used for interpolating among the m values in M.

The x_0 and Σ parameters of the discrete cumulative normal function can be determined from the lightness of the reproduction gamut's black point by interpolating in Table 10.1, which was experimentally determined by Braun (1999). Figure 10.2 then shows sigmoidal lightness mappings for a number of parameter values.

To improve the quality of gamut mapping further, Braun (1999) also developed an image-dependent computation of the sigmoidal mapping's x_0 and Σ parameters. This was done by first conducting a psychovisual experiment in which participants were asked to adjust these parameters until a reproduction in a limited lightness range most closely matched an original with a larger lightness range. Looking at the experiment's results, Braun found a systematic link between image statistics and participant choices. More specifically, it is the lightness of the 75% point of the source image's cumulative lightness

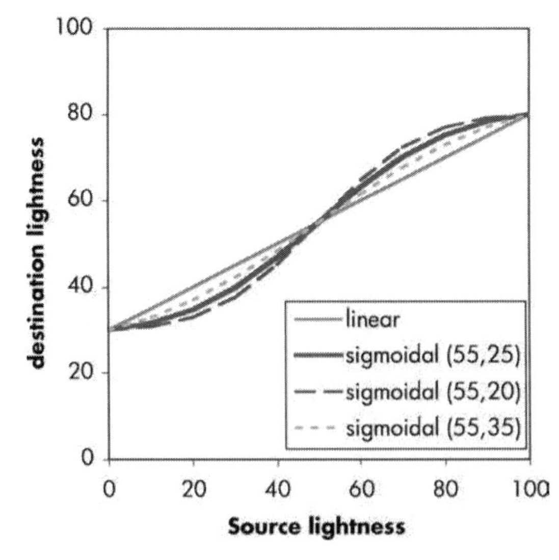

Figure 10.2 Sigmoidal lightness mapping for various (x_0, Σ) combinations compared with linear mapping.

histogram and the black point of the destination gamut that are used to determine the most appropriate x_0 and Σ values. For details of the computation, see Braun (1999: 122–125); and for an alternative to making gamut mapping dependent on a source image's lightness histogram, see the work of Chen and Kotera (2000).

Finally, an interesting approach to lightness range mapping has been proposed by Braun *et al.* (1999a) in the form of their *inverse–power–inverse* (IPI) method, which intends to preserve image contrast (note that it has also been referred to as *inverse gamma inverse* (IGI; Bala *et al.*, 2001)). The algorithm involves the following stages:

1. Transform source medium white point relative tristimulus values (normalized so that $Y = 1$ for the reference white) to a colorimetric RGB space (cRGB):

$$\begin{bmatrix} \mathrm{cR} \\ \mathrm{cG} \\ \mathrm{cB} \end{bmatrix} = \mathbf{M} \begin{bmatrix} X \\ Y \\ Z \end{bmatrix} \quad \text{where} \quad \mathbf{M} = \begin{bmatrix} 2.944 & -1.461 & -0.457 \\ -1.095 & 2.026 & 0.036 \\ 0.078 & -0.272 & 1.4552 \end{bmatrix} \tag{10.5}$$

2. Compute destination cRGBs as:

$$\mathrm{cD}_A = 1 - (1 - \mathrm{cS}_A)^{\gamma} \quad \text{for} \quad A \in \{R, G, B\} \tag{10.6}$$

   where γ is computed as follows to map 95% of the source Y range to 95% of the destination Y range:

$$\gamma = \frac{\log(1 - D_{Y_{95}})}{\log(1 - S_{Y_{95}})} \quad \text{where} \quad B_{Y_{95}} = (1 - B_{Y_{\min}})(1 - 0.95) + B_{Y_{\min}} \quad \text{for } B \in \{S, D\} \tag{10.7}$$

3. Compute destination XYZs as follows: $[X\ Y\ Z]^T = \mathbf{M}^{-1}[\mathrm{cR\ cG\ cB}]^{\mathrm{T}}$ where $\mathbf{M}^{-1}$ is the inverse of $\mathbf{M}$.
4. Transform destination XYZs into gamut mapping space and apply further mapping (e.g. mapping towards a point on the lightness axis).

The end result is a transformation of source lightness shown in Figure 10.3; and, since the change is applied to all three tristimulus values, there is also some change in chroma (an increase) and in hue (mainly for colors with higher chroma). Note also that, since this transformation does not map the source black onto the destination black, some of the lightness mapping is effected by the subsequent transformation. Based on the psychovisual evaluation performed by its authors, this algorithm performs particularly well when the desired reproduction property is preference, but it also worked well when subjective accuracy was judged.

Apart from a lightness mapping that is applied equally across color space, an alternative is to make that lightness mapping depend on the chroma of the source color as well. The reason for wanting to do this is that chroma-independent lightness mapping also tends to increase the lightnesses of chromatic source colors. These can, therefore, move into parts of color space where the destination has less available chroma, which is particularly the case

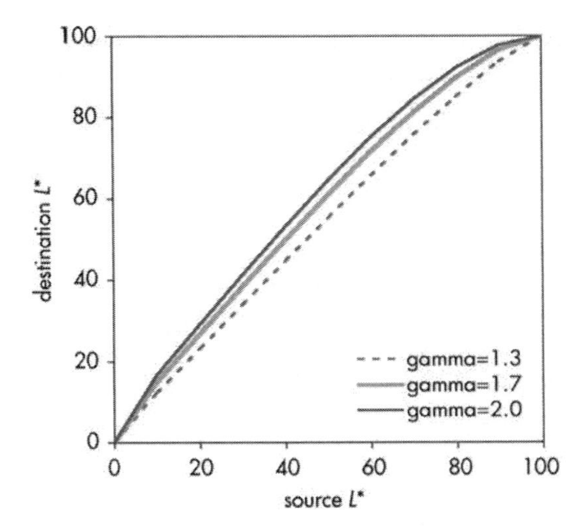

Figure 10.3 Inverse–power–inverse transformation's effect on CIE L^*.

when the source is a display and the destination a print. Since the former uses the additive and the latter the subtractive (combined with the additive) principle, secondaries in a display are lighter than the mean lightness of their primaries, while in a print they are darker. There is, therefore, often a great mismatch in the lightnesses of cusps (i.e. the most chromatic colors at a given hue), and a lightness increase for chromatic source colors is not desirable. To address this issue the GCUSP algorithm (Morovič, 1998) and later also SGCK (CIE, 2004d) have an initial lightness compression that is chroma dependent. The compression is applied fully only at and around the neutral axis and then drops off with chroma. Given a source lightness S_L and its fully lightness compressed version C_L (using any chosen compression approach, e.g. linear, sigmoidal, ...), the destination lightness is computed as

$$D_L = (1 - p_C)S_L + p_C C_L \tag{10.8}$$

where p_C is the chroma-dependent weighting factor (Figure 10.4):

$$p_C = 1 - \left(\frac{S_C^3}{S_C^3 + 5 \times 10^5}\right)^{1/2} \tag{10.9}$$

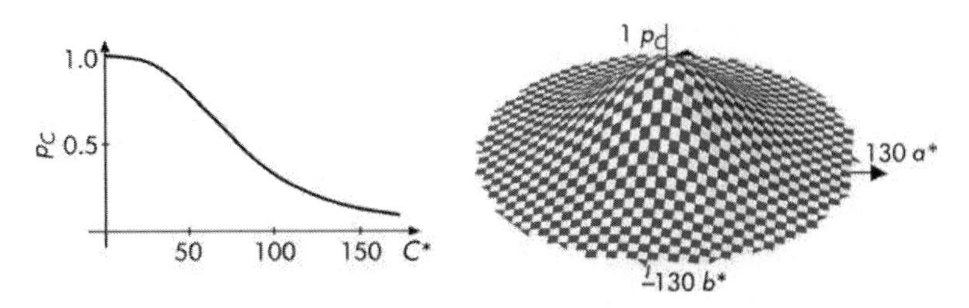

Figure 10.4 Lightness mapping percentage (pC) as a function of chroma (Morovič, 1998).

This type of chroma dependence was also used in one of the algorithms proposed by Braun *et al.* (1999a), where they used a piece-wise linear computation of p_C instead of the nonlinear one used here.

An alternative chroma-dependent lightness compression was later proposed by Green and Luo (2002) with the aim of mapping source and destination cusps onto each other without affecting the chosen lightness mapping for the neutral axis. This was achieved by first applying a chosen mapping to source lightness – S_L (i.e., linear mapping in their case – Equation 10.1) and by considering the result to be an intermediate stage – I_L. Destination lightness is the computed as follows:

$$D_L = I_L + (D_{L\text{cusp}(h)} - S_{L\text{cusp}(h)}) \frac{S_C}{S_{C\text{cusp}(h)}} \quad (10.10)$$

Following such cusp-dependent lightness, compression with lightness-preserving chroma compression results in the source cusp being mapped onto the destination cusp.

While these types of chroma-dependent lightness compression provide clear benefits in many cases, in some they can lead to lightness inversions that are objectionable, i.e. the reproduction of a lighter original is darker than the reproduction of a darker original. This type of behavior, whereby an algorithm works well in one context but badly in another, is a universal property of all GMAs so far and a major challenge in gamut mapping research.

Finally, alternatives to initial lightness mapping are also the mapping of ranges in other domains – this is typically done linearly and most notably in CIE *XYZ* (as is done in Adobe's BPC – Adobe, 2006; Section 6.4) and optical density (e.g. Buckley, 1978).

Chroma mapping following lightness mapping

Having altered the source colors' lightness predictors based on the source and destination gamuts' lightness ranges, the next step in sequential GMAs is to map chroma predictors. This can be done by applying a single scaling factor to all source chromas (e.g. Buckley, 1978), by determining that scaling factor as a function of hue angle (e.g. UGRA, 1995; Lammens *et al.*, 2005), or most commonly by determining the extent of compression along each line of constant lightness and hue angle separately (e.g. Johnson, 1979). Compression can then be done linearly (Equation (5.2)), using a piece-wise linear function with two or three segments (Section 5.5.2) or nonlinearly. An example of nonlinear mapping is the *knee* (or *soft-clipping*) function first proposed by Sara (1984: 110) and later used by several others (e.g. Stone and Wallace, 1991; Herzog and Müller, 1997; MacDonald *et al.*, 2002). The key point of this kind of mapping is to have a part of the source range that is changed very little or not at all and then to transition nonlinearly from that to a compression that ends up in the destination gamut. The percentage of the destination range in which source values are left unchanged is often reported when specifying knee functions, i.e. a 90% knee function will apply no change to source values that are in the first 90% of the destination range (starting from the IG point towards which mapping is applied) and then have nonlinear compression into the remaining 10%. In MacDonald *et al.*'s (2002) implementation, compression along a line (e.g. of constant

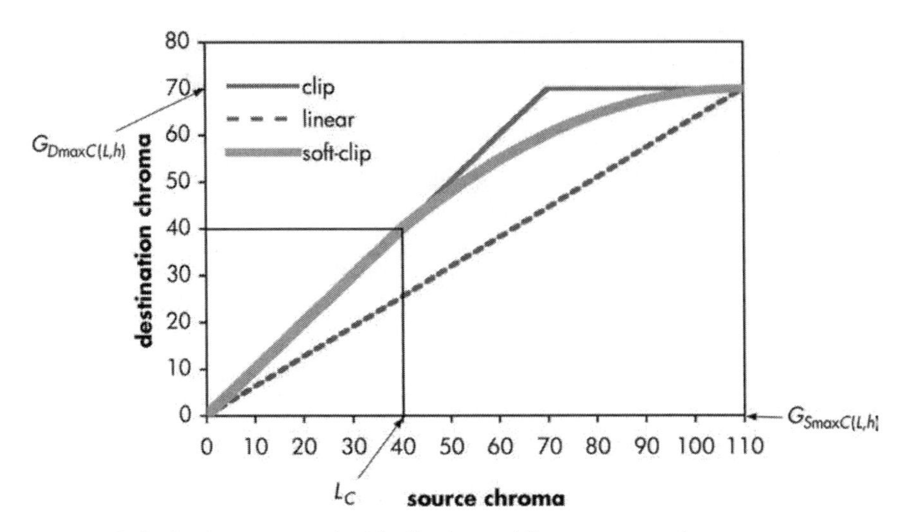

Figure 10.5 Soft-clipping compared with clipping and linear compression.

lightness and chroma in the case of the GMAs discussed here) is computed as follows (Figure 10.5):

$$D_C = \begin{cases} S_C; S_C \leq L_C \\ L_C + 4(G_{D\max C(L,h)} - L_C)(\zeta - \zeta^2); L_C < S_C \leq G_{S\max C(L,h)} \end{cases} \tag{10.11}$$

where $G_{D\max C(L,h)}$ is the maximum chroma of the destination gamut at the source color's lightness and hue (if the mapping was preceded by lightness mapping, then it is at the mapped lightness rather than the native lightness of the source color), L is the point along the mapping line up to which the mapping is to result in no change, and ζ is

$$\zeta = \frac{|S_C - L_C|}{2|G_{S\max C(L,h)} - L_C|} \tag{10.12}$$

Even though soft-clipping is typically used along lines of constant lightness and hue or other lines in planes of constant hue angle (Section 10.1.2), it has also been applied in some cases to initial lightness compression (e.g. Hoshino and Berns, 1993; Kang *et al.*, 2002) and has been reported to work well when lightness range differences are small but not when they are more significant.

While lightness and chroma are mapped in sequence, hue is kept unchanged in the vast majority of cases.

LLIN

Bringing all of this together, let us look at Johnson's (1979) linear lightness scaling followed by linear chroma scaling while maintaining hue (labeled LLIN in Morovič

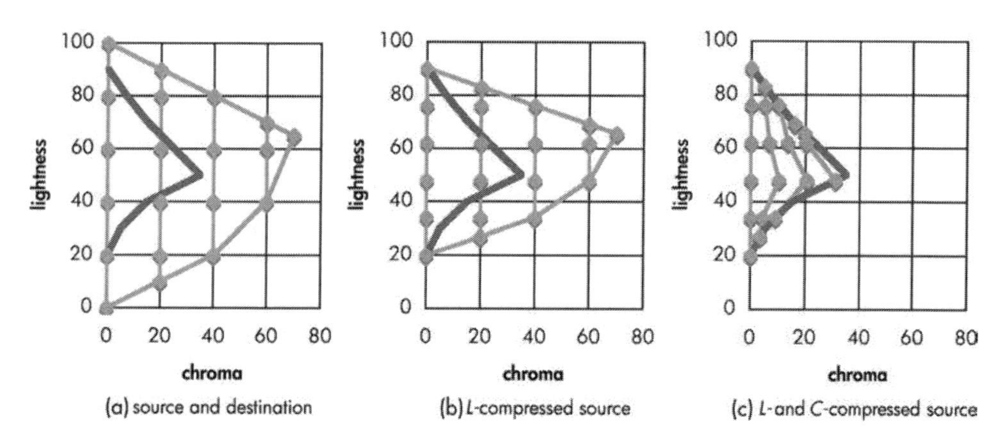

Figure 10.6 The LLIN algorithm: (a) source and destination gamuts; (b) lightness-mapped source gamut; (c) final gamut mapped result. Diamonds represent source colors and thick line shows destination gamut boundary.

(1998)). Given a source and destination gamut (shown in a plane of constant hue in Figure 10.6a), the first step is a mapping of lightness. The result is an intermediate state of the source data that has the same lightness range as the destination, but still the source chroma range (Figure 10.6b). This is further mapped along lines of constant lightness and hue so that along each such line the maximum source chroma is mapped onto the maximum destination chroma (using Equation (5.2)). Note that this chroma compression is done on the basis of the lightness-compressed source gamut rather than on the actual source gamut, since it is not source colors, but lightness-compressed source colors that are mapped. This is an important point to note in general, i.e. that mapping from a source to a destination can involve several stages and that at each stage the starting point may not be the native source colors, but some transformation of them. Therefore, it is an equally transformed source gamut that needs to be used for determining mapping parameters for subsequent stages.

Applying LLIN to the test image from Chapter 9 gives the reproduction in Color Plate 11. When this image is compared with the gamut clipped reproductions, it can be seen to preserve more detail than they do, but at the expense of some loss in contrast. In addition to detail, this reproduction also does a better job at preserving relationships within the image, but the loss of colorfulness and contrast makes it a weak choice even in this case where source and destination gamuts are of similar shape. In cases where the two gamuts have significantly different shapes, e.g. display and print, LLIN can result in very large drops in chroma where the source gamut is very steep. For example, in the case shown in Figure 10.7, LLIN will result in a highly chromatic source color being mapped to an almost neutral destination color. In spite of its obvious limitations, this algorithm was found to work well under some conditions though (e.g. Gentile *et al.*, 1990; Pariser, 1991).

Returning to the sample points mapped using LLIN in Figure 10.6, we can also see that while the initial lightness mapping does preserve relationships among the source samples (as the same transformation is applied to all of them), the chroma mapping introduces a significant distortion. For example, the source cusp (i.e. source color of greatest chroma) is

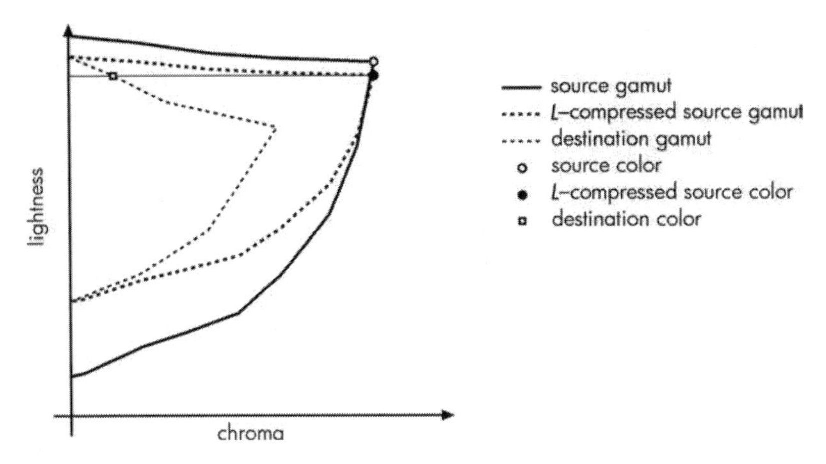

Figure 10.7 Large chroma drops due to LLIN.

no longer the most chromatic in the destination, which also implies a chroma inversion. While this is clearly not desirable, the question of what specifically is is not clear.

One might say with Evans (1943) that 'the rendition of some colors must not be better than of others.' However, such a principle leads to using a constant scaling factor for compressing a larger into a smaller gamut (Gordon *et al.*, 1987), which in turn results in a much greater change to an original than any other GMA ever proposed. In terms of the above example, this would be a far more drastic compression than LLIN and would also result in large parts of the already small destination gamut not being used (Figure 10.8). Preserving the original's lightness and chroma relationships would come at the cost of very low overall chroma and a much more limited lightness range. In general, such an approach is not desirable except for cases where gamut differences are either very small or where gamut shapes are so similar as to result in the vast majority of the destination gamut being

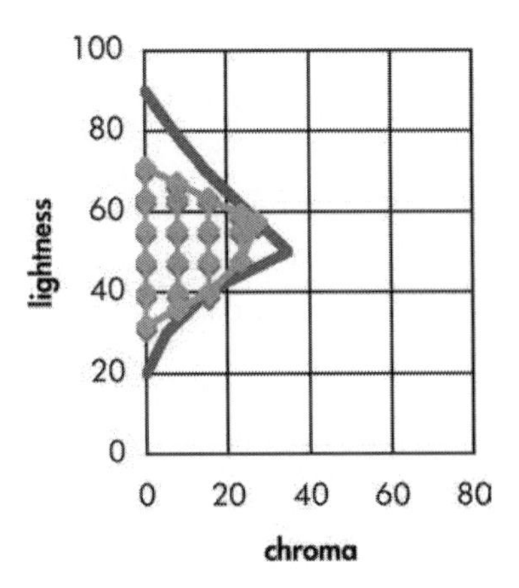

Figure 10.8 Uniform global compression.

used when a single, global scaling factor is applied. All in all, this *Evans consistency principle* is a good example of an idea that sounds reasonable, is almost certainly the right thing to do under some specific narrow range of conditions, but ultimately does not generalize to a broad range of gamut mapping scenarios. The road to gamut mapping is littered with many such good principles.

Initial chroma mapping

Staying with sequential GMAs, an alternative to mapping lightness before chroma is obviously to map chroma before lightness. One such approach, namely CLLIN (Morovič, 1998), starts in a plane of constant hue by finding the two cusps (i.e. source and destination) and by compressing source chromas by the ratio of the destination to source cusp chromas. The second step is then a lightness mapping along lines of constant chroma and hue whereby the chroma-compressed source range is mapped onto the destination range (Figure 10.9). Compression is done in the same way as the global lightness compression from Equation (10.1), but instead of the lightness range extremes being those of the whole gamut they are only those along the line that has the source's hue and its compressed chroma. The result of this algorithm is a mapping that preserves chroma relationships and uses the entire destination gamut – as a consequence, mapping the source cusp onto the destination cusp (Color Plate 12).

Hue shifting

While the vast majority of compression algorithms keep hue unchanged, Johnson (1992), by observing the work of color reproduction professionals, found that some change to hue can result in greater similarity between original and reproduction. This is because it can

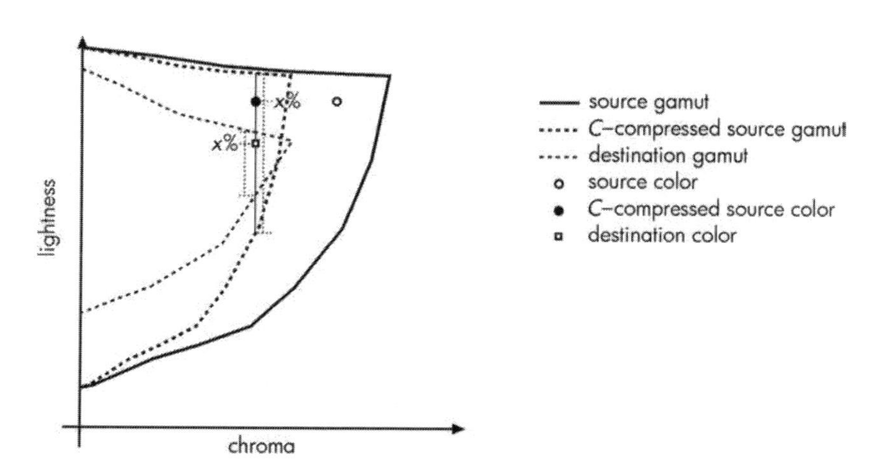

Figure 10.9 CLLIN – chroma compression followed by lightness mapping.

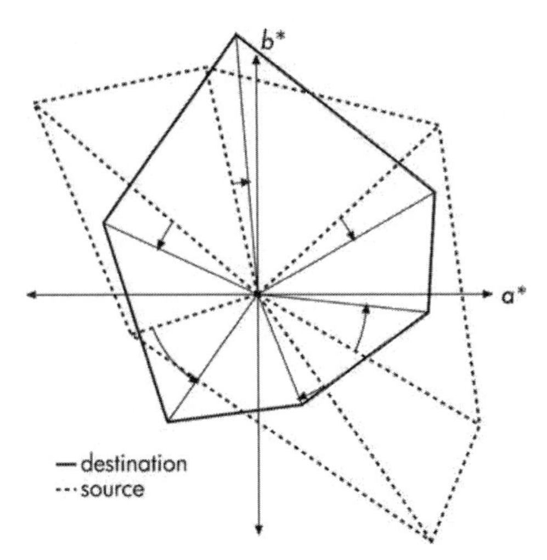

Figure 10.10 Hue shift determined by source and destination primaries and secondaries.

enable a better preservation of source lightnesses and/or chromas. Johnson saw that there was a consistent preference for moving source hues from the hues of the source gamut's primaries and secondaries towards those of the destination gamut when a photographic transparency is reproduced in print (Figure 10.10).

Given a source color, its hue is transformed as follows:

$$D_h = S_h + \frac{\text{huedif}(S_h, S_{h\text{Pc}}) \times \text{huedif}(D_{h\text{Pcc}}, S_h) + \text{huedif}(S_{h\text{Pcc}}, S_h) \times \text{huedif}(S_h, D_{h\text{Pc}})}{\text{huedif}(S_{h\text{Pcc}}, S_{h\text{Pc}})} p \tag{10.13}$$

where Pc is the primary or secondary that has a source hue closest to the source color's hue in the clockwise direction, Pcc is closest in the counterclockwise direction, the huedif (a, b) function gives the counterclockwise distance from hue angle b to hue angle a, and $p \in [0,\ 1]$ is the percentage of hue shift. In Johnson's work it was found that $p = 0.5$ most closely matched the choices made by professionals. Note that this function is used instead of direct subtraction to account for hue angle circularity (i.e. huedif(1, 359) $= 2$, whereas $1 - 359$ is -358).

Hue shifting of this kind, introduced in Johnson's (1992) CARISMA algorithm, was later used also in a number of other algorithms (e.g. Schläpfer, 1994; UniGMA – Morovič and Luo, 1999).

10.1.2 Simultaneous mapping in constant-hue planes

While there are a number of sequential GMAs, the bulk of color-by-color algorithms are ones that make changes in planes of constant hue angle (either keeping hue unchanged or after hue has been altered in the way described at the end of Section 10.1.1). Before looking at mapping strategies, it is worth taking a closer look at the gamuts between which mapping takes place.

What gamuts to map between

While many of the earlier gamut compression algorithms are defined between an original and a reproduction medium's gamuts, Stone *et al.* (1998) have proposed the use of the original image's gamut as the starting point. Indeed, this approach has been shown repeatedly to perform better than a mapping that starts from the source medium's gamut (Gentile *et al.*, 1990; Pariser, 1991; Hoshino and Berns, 1993; MacDonald and Morovič, 1995; Nakauchi *et al.*, 1996, 1999; Montag and Fairchild, 1997; Wei *et al.*, 1997; Kim *et al.*, 1998) and there are good reasons for such a result. Image-dependent gamut mapping – even just at the level of compressing the source image's rather than the source medium's gamut – has the potential only to make changes where they are necessary and only to the extent that a given image dictates. The flip side, though, is that the same source color in multiple images is likely to be reproduced by multiple destination colors. While this might not always be a problem, it would most certainly be that if it were a corporate identity color or a color occupying a prominent place in the images (e.g. scenes set in the same environment and that environment being reproduced in different ways for each image).

In addition to alternatives for the source gamut, there are also some choices to be made about the destination gamut. Sara (1984) had already recognized this and defined a *core* region within the destination gamut where no compression takes place (Figure 10.11). Compression is then applied between the parts of the source and destination gamuts that are outside the core region. What this essentially allows is a combination of the benefits of clipping and full compression algorithms. Instead of differences between source and destination being distributed across the entire destination gamut, a part of color space around the lightness axis is protected from change. Maintaining identity in that region contributes to the experience of greater similarity in the end, since the core region (if set appropriately) can contain important memory colors, such as some skin tones and neutrals. This core region concept is very popular in gamut mapping, and many have adopted it successfully (e.g. Ito and Katoh, 1995; Spaulding *et al.*, 1995; Katoh and Ito, 1999; MacDonald *et al.*, 2001; Lammens *et al.*, 2005).

There is also the question of whether to map only to the intersection of the source and destination gamuts or whether to map to the entire destination gamut (i.e. also involving

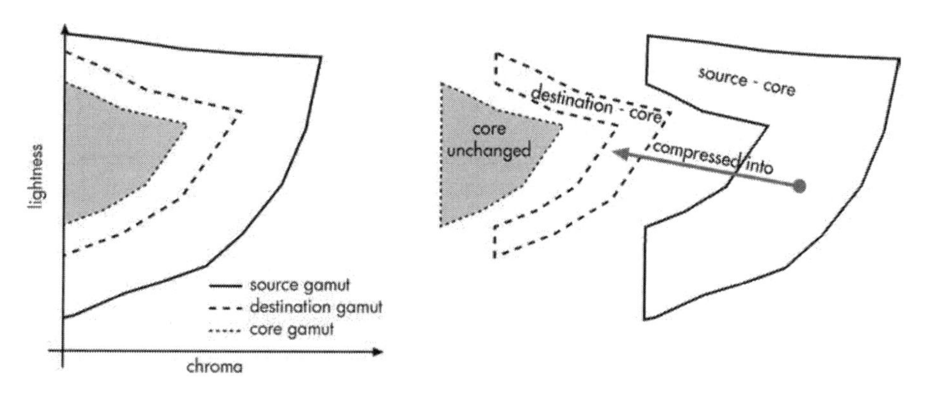

Figure 10.11 The core region – its content is kept unchanged and compression takes place outside it.

some expansion in regions where the destination gamut is larger than the source gamut). While most compression algorithms restrict themselves to the former (i.e. compression only), some do also expand into regions of the destination gamut outside the source gamut, and most algorithms lend themselves to use in either of these ways.

Finally, it is also an option to use one (e.g. a compression) algorithm to map to a smoothed (or convex) version of the destination gamut and then to use another (e.g. clipping) algorithm from this intermediate state to the actual, nonsmooth or concave gamut. Farup *et al.* (2004) proposed this strategy and reported improved results for it.

Mapping towards a single point in a constant-hue-angle plane

The most basic simultaneous mapping approach is one where source colors are compressed along lines towards a single point: the *focal point*. The simplest approach here is to have the focal point at the center of the lightness axis, which was first proposed in Sara's (1984) PhD thesis and then used in many subsequent studies (e.g. MacDonald and Morovič, 1995; MacDonald *et al.*, 1995). Not having an initial lightness compression, this effectively gives SLIN a piece-wise linear lightness mapping, with two segments that meet at $L = 50$.

Looking at an SLIN-mapped reproduction (Color Plate 13) we can see that it provides a result that is somewhere between LLIN and the clipping algorithms in terms of colorfulness while still having the detail preservation properties of LLIN. What has to be borne in mind yet again is that these are results for a print-to-print case where gamut shapes are relatively similar. Both SLIN and LLIN in fact perform very poorly for some other cases, like that of reproducing an image from a display in print, where gamuts have very different shapes. The SLIN approach has also been extended by having it preceded by a chroma compression that varies only with hue (Herzog and Müller, 1997) or by having it preceded by a global lightness compression (Laihanen, 1987).

A similar approach to SLIN is one where, instead of a single focal point for the entire gamut, the mapping uses focal points that vary with hue and are at the lightness of the cusp at each hue angle (MacDonald and Morovič, 1995). Such mapping (labeled CUSP) has the advantage of preserving more chroma than SLIN does and is used as a component of many complex algorithms (e.g. SGCK – CIE, 2004d).

Two of the component algorithms of Johnson's (1992) CARISMA algorithm (which will be revisited in full later) also have single focal points in planes of constant hue angle – after an initial lightness range mapping. The first has the focal point on the lightness axis where the line connecting the source and destination cusps intersects it (Figure 10.12a); and there is a later alternative to it by Han *et al.* (1999), who propose moving the focal point away from the lightness axis and towards the cusps along this line – thereby effectively increasing the chroma of the reproduction. The second has the focal point on the chroma axis that has half the chroma of the destination cusp at that hue (Figure 10.12b). In both cases the intersection of a mapping line with the source gamut is linearly mapped onto its intersection with the destination gamut. One thing that makes these two approaches different is that they do not aim to have universal applicability – and in fact they do not. For both mappings there are cases which do not result in a complete mapping of source onto destination and where they would result in artifacts. The reason for this in itself not

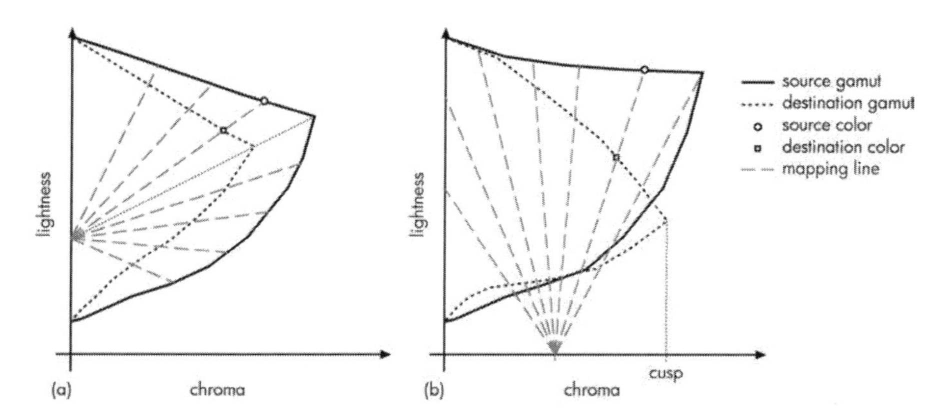

Figure 10.12 Two of the three component algorithms of Johnson's (1992) CARISMA.

being an issue is that these algorithms (with the addition of the CUSP algorithm) are used in tandem and the choice of which one to use is made after analyzing source and destination gamut properties. For an example of the CARISMA algorithm in use, see Color Plate 14, where only the method shown in Figure 10.12a is used (due to how the source and destination gamuts used here relate). As can be seen there, the reproduction is one where a lot of chroma is preserved at the expense of lightness.

Finally, Büring and Herzog (2002) proposed the interesting idea of having the focal point in the hue plane that is at 180° to the current one (Figure 10.13), as a response to existing single focal point methods sacrificing too much lightness. Moving the focal point into the extension of the current hue plane that has negative chroma addressed the issue but poses the challenge of how to select it without resulting in negative chroma for destination colors. Even though Büring and Herzog did not address this challenge, it led

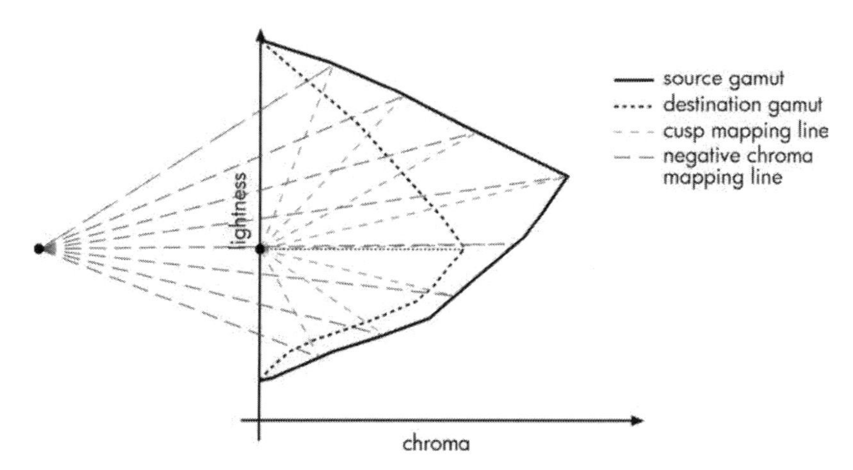

Figure 10.13 Mapping towards a point with negative chroma.

them to the definition of an algorithm that maps along curved lines, discussed in more detail later in this chapter.

Mapping towards multiple focal points in constant-hue-angle planes

Instead of mapping all colors of a given hue towards a single focal point, there are a number of algorithms that divide source colors into regions and apply different mappings to each, within a single plane of constant hue.

The first such algorithm is Ito and Katoh's (1995) four region compression that also includes an unchanged core gamut. Here, K is a point along the line between the lightness axis and the destination cusp that divides the source gamut into four regions (Figure 10.14):

A the core region, where source colors are kept unchanged;
B the region where source colors are compressed towards K;
C the shadow region, where source colors are compressed towards the destination white point on the lightness axis; and
D the highlight region, where colors are compressed towards the destination black point again on the lightness axis.

Neumann and Neumann (2004) later proposed making the core region (region A) a scaled version of the destination gamut (following MacDonald *et al.*'s (2002) approach) and having K at region A's cusp, mapping region D towards a point that has region A's lowest lightness and negative chroma (following Büring and Herzog's (2002) concept) and mapping region C towards a point with region A's highest lightness and again negative chroma.

A simpler approach was later proposed by Herzog and Müller (1997), where source colors with lightnesses above the destination cusp's lightness are mapped towards the

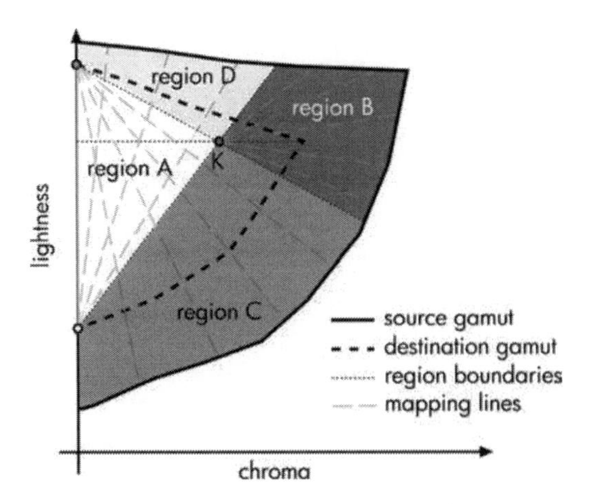

Figure 10.14 Ito and Katoh's four-region compression algorithm.

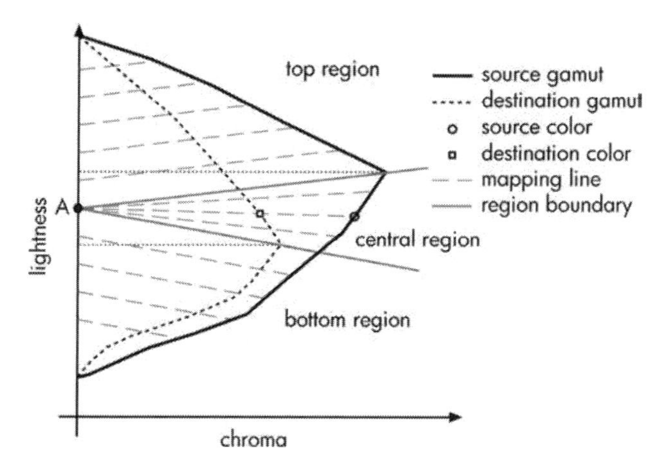

Figure 10.15 Lee *et al.*'s subdivision of the source gamut dependent on both source and destination cusp lightnesses.

destination cusp's lightness on the lightness axis and those with lightnesses below are mapped along constant lightness lines. Such a division of what to do for lighter versus darker colors is also found in the earlier work of Granger (1995), who used clipping for lighter colors and compression for darker ones.

Lee *et al.* (2000) devised a scheme that divides the source gamut into three regions depending on both the source and destination gamuts' cusps. This parameterization of gamut mapping and making its behavior adaptive to the relationship between the two gamuts is also a popular theme in gamut mapping – started by Johnson (1992) and later also used by Lammens *et al.* (2005). In Lee *et al.*'s algorithm, the parameterization is as follows: take the source and destination gamuts and divide the source gamut into three regions on the basis of the mean of the source and destination gamuts' lightnesses at the source color's hue angle (Figure 10.15) – this is referred to as the *anchor point* A. The central region is delimited by two lines: the first connects the source cusp with A and the second connects the destination cusp with A. Mapping in the central region is towards A, mapping in the top region is along lines parallel with the top boundary of the central region and, similarly, mapping in the bottom one uses the bottom boundary line.

Kang *et al.* (2002) proposed a related approach by having a mapping that divides the source gamut (in a plane of constant hue) into three lightness ranges, maps chroma along lines of constant lightness in the middle range and maps the top and bottom ranges towards the top and bottom boundaries on the lightness axis respectively (Figure 10.16). What is noteworthy about this algorithm (*GCA*) is that the region boundaries it uses are determined on the basis of a psychovisual experiment in which observers adjusted images in a restricted gamut until they looked as similar to larger gamut originals. Note that the compression along the various mapping lines in this algorithm is done in a nonlinear way (also derived from psychovisual experimental results) and that the three-region mapping described above is preceded by a piece-wise linear lightness range mapping.

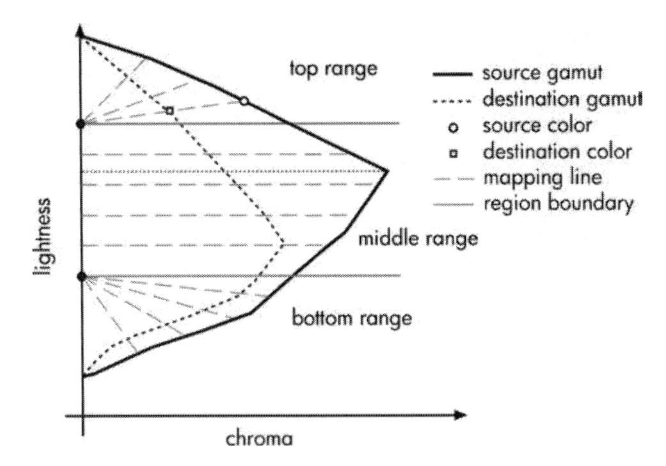

Figure 10.16 Three mapping regions of Kang *et al.*'s GCA algorithm.

Other constant-hue-angle plane algorithms

There are also a number of algorithms that depart from the idea of mapping towards a small number of specific points in planes of constant hue angle while still restricting change to be in those planes – an idea that was so popular in early GMAs and that has seen and still sees a great deal of development.

The first example here is the *TRIA* algorithm (Morovič, 1998), which derives from the realization that color gamuts in planes of constant hue angles are like triangles (which is particularly true for CIELAB). TRIA then proceeds by representing a source color as the linear combination of two vectors: the first being a vector from the lightest to the darkest source gamut color on the lightness axis and the second vector being one from the lightest source color to the source cusp at a given hue angle. Every source color is then uniquely identified by its hue angle and a pair of weights derived using a vector pair defined by the source gamut. The mapping algorithm then simply takes these source-derived weights and applies them to the reproduction gamut's vectors (Figure 10.17). In other words, a given

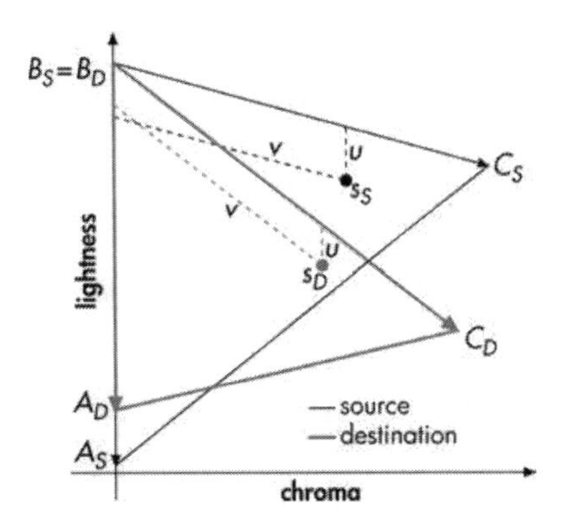

Figure 10.17 The TRIA algorithm.

source color first has weights computed for it using a pair of vectors from the source gamut at its hue angle and these weights are then applied to the vector pair from the destination gamut at the same hue angle. Finally, gamut clipping is applied, since the above mapping would only result in the source being mapped fully into the destination if the two gamuts were either triangular or their shapes so similar as to be a simple shearing (transvection) away from each other. The result is a mapping that guarantees no lightness or chroma inversions (i.e. if lightness/chroma was greater for one source color than another at a given hue angle, then that relationship will be maintained in the reproduction) and also maps the source cusp onto the destination cusp (this is significant, as the source's most chromatic color is reproduced with as much chroma as is available in the reproduction at that hue).

In terms of its performance, TRIA works well where algorithms that map cusps onto cusps (e.g. Green and Luo's (2002) CUSP2CUSP and Morovič's (1998) CLLIN) work well, which is when the lightness differences of cusps are not large. In those cases, TRIA results in a very smooth transformation. If that condition is not met, then this kind of approach results in objectionable changes that make light chromatic colors dark and dark ones light by following the relationship of source and destination cusps too closely.

The *TOPO* algorithm (MacDonald *et al.*, 2001) that also belongs to this category takes a very different, novel approach. It works by first setting up a core, no-change region, which is a compressed version of the destination gamut. Lines of compression are then determined in planes of constant hue angle based on linking points on the source and core boundaries that have the same relative distances along their respective boundaries (Figure 10.18). This is done by 'walking' along a boundary from the lightest to the darkest points on the lightness axis and representing each point as a percentage of the total distance. Colors are then mapped along these lines so that there is no change in the core region, and nonlinear compression between the source and destination gamuts is performed outside it (Equation (10.11)).

Büring and Herzog (2002) proposed a twist to the usual way of mapping colors along straight lines by introducing curvature to the mapping path (Figure 10.19), with the effect of having less lightness change for colors closer to the gamut boundary than for those further away from it. The proposed mapping has the largest chroma of the source $S_{C\max}$,

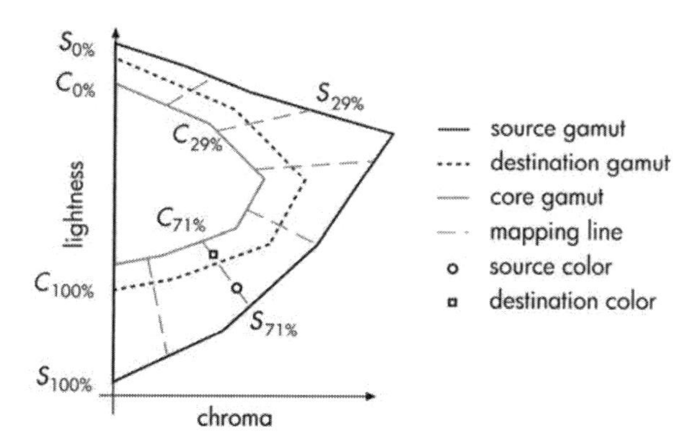

Figure 10.18 The TOPO algorithm.

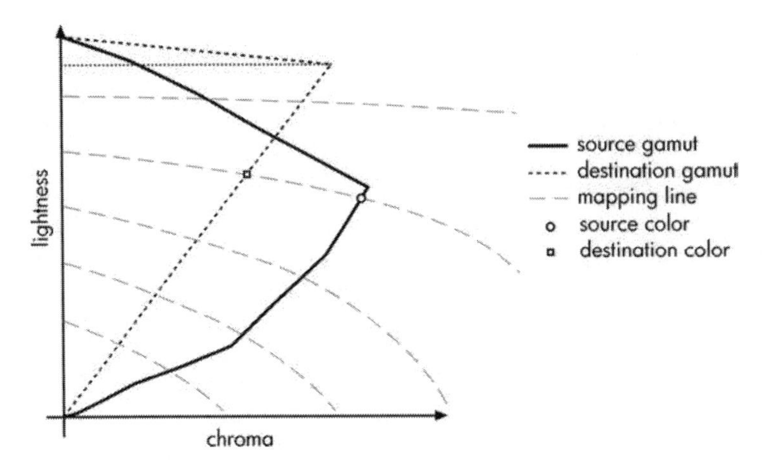

Figure 10.19 Mapping along curved lines.

the chosen percentage of soft-clipping λ and the degree of permissible lightness change percentage α (both with [0, 1] ranges) as parameters, and is computed as follows when the destination gamut is smaller than the source gamut at the source's lightness and hue:

$$\begin{aligned} D_L &= S_L + \alpha \frac{(D_{L\text{cusp}(S_h)} - S_L)(S_C - \lambda G_{D(S_L,S_h)})}{S_{C\max} - \lambda G_{D(S_L,S_h)}} \\ D_C &= \lambda G_{D(D_L,S_h)} + (1-\lambda) G_{D(D_L,S_h)} \frac{S_C - \lambda G_{D(D_L,S_h)}}{G_{S(S_L,S_h)} - \lambda G_{D(D_L,S_h)}} \end{aligned} \tag{10.14}$$

While this approach is indeed the first one where colors are mapped explicitly along curved paths, that effect is also achieved using the chroma-dependent lightness compression of GCUSP (Morovič, 1998) and SGCK (CIE, 2004d), as well as the lightness preprocessing proposed by Cui (2001). Cui found in a psychophysical experiment that reducing the CIE L^* of a stimulus while maintaining its C^* resulted in increased colorfulness. Since in gamut reduction the aim is also to preserve as much of the colorfulness of highly chromatic source colors as possible, his solution involved a chroma-dependent lightness reduction preprocessing step before other gamut mapping is applied. Source L^* first has ΔL^* subtracted from it, which depends on the ratio of the source C^* and the source gamut cusp's C^* at the source L^* and h^*_{ab} and which is computed as follows:

$$\Delta L^* = r \left(\frac{S_{C^*}}{G_{S(S_{L^*}, S_{h^*_{ab}})}} \right)^2 \tag{10.15}$$

The effect is a transformation that changes colors of equal source lightness to ones that lie on a curved path. As can be seen, Cui's idea is one of the few gamut mapping components that are specifically tied to a color space (namely CIELAB), whereas the majority of other algorithms are defined in terms of appearance attributes and it is only their implementation that involved attribute predictors.

Zolliker *et al.* (2005; Zolliker and Simon, 2006a) developed their *SGDA* (*smooth gamut deformation algorithm*) as an evolution of CARISMA by keeping its hue shift and addressing its cusp-to-cusp aim (also shared by Green and Luo's (2002) CUSP2CUSP algorithm) but performing the mapping nonlinearly (like Chen *et al.* (1999)) and between novel source and destination intervals. These intervals, which are mapped onto each other, are determined by having a focal point on the lightness axis for the source S_F that has the mean lightness of the source gamut's lightest and darkest color, i.e. $S_F = (G_{SminL} + G_{SmaxL})/2$. A line is then determined by a source color S and S_F, and where it intersects the source gamut is the source gamut boundary S_{Gsgda} point that will determine the mapping. Hence, the source color is located in the line segment between S_F and S_{Gsgda} and it will be mapped onto a line segment defined by D_F and D_{Gsgda}. Computing the first is straightforward, i.e. $D_F = (G_{DminL} + G_{DmaxL})/2$, and it is in the determination of D_{Gsgda} where the novelty of this method lies. Namely, D_{Gsgda} is that point on the destination boundary that (with D_F and G_{DmaxL}) delimits an area in the destination gamut that is of the same size relative to the total area in the constant-hue plane as is delimited by S_{Gsgda}, S_F and G_{SmaxL} in the source gamut. Computing where D_{Gsgda} is then gives a line segment in the destination gamut and the destination color is computed by mapping the position of the source color in its line segment non-linearly into the destination line segment (Figure 10.20). In some sense this is related to MacDonald *et al.*'s (2001) TOPO, where corresponding source–destination gamut boundary points are related by relative distance along the gamut boundary, whereas here it is relative area delimited in source and destination gamuts that provides the link. Both approaches are a clear attempt at making the mapping adaptive to gamut shape rather than only to gamut relationships along a single mapping path, which is what the majority of other methods do.

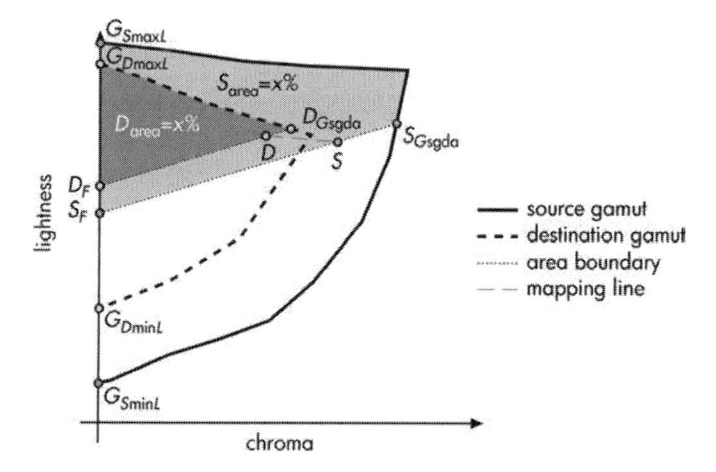

Figure 10.20 SGDA mapping in planes of constant hue angle.

Combining separate mapping algorithms across color space

What all the compression algorithms discussed so far have in common is that the same approach is used throughout color space and while this is a popular approach, several authors have felt the need to combine different mapping methods to gain improved color reproduction performance.

The first of these combined methods is Johnson's (1992) CARISMA, which uses three different mapping strategies depending on how the source and destination gamuts relate at a given hue angle. This algorithm, parts of which have already been mentioned here, and which is an attempt to replicate what color reproduction professionals did manually, consists of the following stages:

1. Map source lightnesses linearly so that the source lightness range is mapped onto the destination lightness range (Equation 10.1).
2. Determine the hue shift of the six primary and secondary colors between the two gamuts and translate the source medium's hues halfway towards the reproduction medium's hues (i.e. using $p = 0.5$ in Equation (10.13)).
3. Perform additional compression of L and C depending on the relative shapes of the gamut boundaries.
   (a) For each source primary and secondary compare the source gamut boundary at the source hue with the destination gamut boundary at the hue-shifted source hue in planes of constant hue angle. The result is then one of these three cases, each resulting in the choice of a different mapping direction:
      (i) Source gamut completely encloses destination gamut and intersection of line passing through two cusps and lightness axis is within destination lightness range. Mapping takes place towards that intersection point (Figure 10.12a).
      (ii) Source gamut completely encloses the destination gamut, but intersection of line passing through two cusps and lightness axis is outside destination lightness range. Mapping takes place towards point on lightness axis with same lightness as destination cusp (Figure 9.17).
      (iii) Destination gamut is not enclosed by source gamut. Mapping takes place towards point on chroma axis with half the chroma of the destination cusp (Figure 10.12b).
   (b) For each source color proceed to calculate the gamut-mapped reproduction with the methods used for its neighboring primary and secondary colors and interpolate between them using the angular differences between the given color and the closest primary and secondary colors as weights. That is, if the two angular differences are $\Delta\alpha_1$ and $\Delta\alpha_2$ and the two corresponding gamut mapped lightnesses are L_1 and L_2, then the resulting L is calculated as follows:

$$L = \frac{L_1 \Delta\alpha_2}{\Delta\alpha_1 + \Delta\alpha_2} + \frac{L_2 \Delta\alpha_1}{\Delta\alpha_1 + \Delta\alpha_2} \tag{10.16}$$

   C is interpolated analogously.

While this algorithm has a solid empirical basis and has been popular in gamut mapping research as a benchmark (Morovič, 1998; Chen *et al.*, 1999; Green and Luo, 2002; MacDonald *et al.*, 2002), it also suffers from lack of robustness due to being so closely

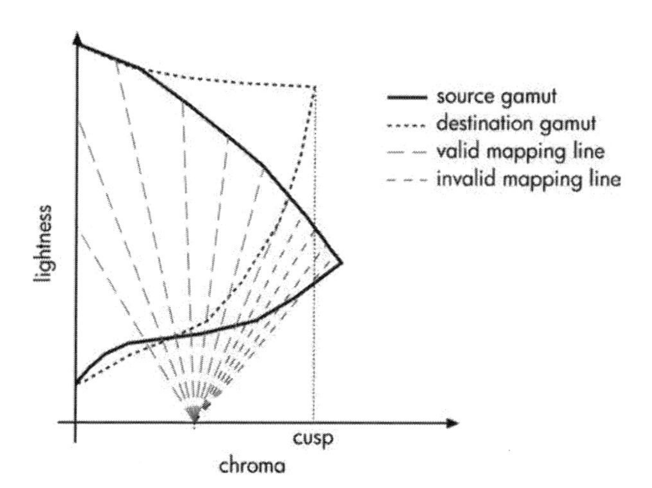

Figure 10.21 Pathological case when applying the third CARISMA mapping.

tuned to the photographic transparency to print case from which it was derived. In particular, the third mapping case of transforming colors towards a point on the chroma axis can result in pathological situations (Figure 10.21) where parts of the source gamut do not get mapped onto the destination gamut. Along similar lines to this algorithm are also Morovič's (1998) *UniGMA* or Zeng's (2001) idea of using different color spaces for gamut mapping different source colors.

Another algorithm that also operates in planes of constant hue angle is Wolski *et al.*'s (1994) approach, which is of interest both because it is rather intricate and because it uses saturation s instead of chroma:

1. Map source colors into the smallest rectangle containing the destination gamut by:
   (a) Soft-clipping saturation with a 95% cutoff, i.e. 95% of the destination range is left unchanged and the source region outside this 95% point is mapped onto the destination range's remaining 5%;
   (b) Soft-clipping lightness using 75% between $L = 50$ and maximum L as the cutoff and clipping colors with L below the destination black point, after the whole lightness range was shifted in an unspecified manner by between $-5L$ and $2L$ units depending on the target gamut.
2. Divide the source gamut into three regions, namely a cylinder around the neutral axis, an upper chromatic region ($L > 50$) and a lower chromatic region ($L < 50$).
3. Compress saturation and maintain lightness for the achromatic cylinder.
4. Maintain saturation and compress lightness for the upper chromatic region.
5. Map lightness and saturation simultaneously with a ratio of $\Delta L : \Delta s = 2 : 1$ for the lower chromatic region (although this ratio was found to vary a lot from image to image).

Another aspect that is particularly clear with this method (but is fairly widespread in gamut mapping in general) is that there are many details that the authors admit are quite fuzzy (e.g. the lightness shift, the size of the achromatic region) and that they have found to be quite image dependent (e.g. the $\Delta L : \Delta s$ ratio).

A related approach to having mapping chosen as a function of color space location is also its higher level extension where mapping algorithms are chosen as a function of image statistics (Sun and Zheng, 2004). Here, a source image is analyzed (e.g. in terms of its three-dimensional color histogram) and the result of the analysis determines which of a set of available algorithms to use for its gamut mapping.

10.1.3 Mapping in three dimensions

While the algorithms discussed in the preceding sections display a lot of variety in how a source is mapped into a destination gamut, they all operate in at most two dimensions at any one time; and even though the mapping is a three-dimensional to three-dimensional one, it is the result of a sequence of one- or two-dimensional transformations. Note that this is not to say that three-dimensional transformations are better than two-dimensional, but simply to draw attention to the fact that there are more degrees of freedom in the case of these algorithms.

Haneishi *et al.* (1993) provide a good example of an alternative approach that does result in true three-dimensional mapping from source to destination. They start from the realization that different categories of colors need different kinds of treatment (as is implicitly or explicitly acknowledged by authors of non-three-dimensional algorithms too, e.g. by the definition of a core region (Sara, 1984) or an achromatic cylinder (Wolski *et al.*, 1994)). The algorithm, therefore, starts with determining which of the following categories a given source color belongs to: skin tones, achromatic, red, green and blue and then determines separate transformations for each category.

Taking this idea further by making it more flexible is the *UltraColor* algorithm (Spaulding *et al.*, 1995), which poses gamut mapping as a morphing that is effected as follows:

1. Explicit definition of color gamut mapping functions for subsets of the original gamut (e.g. no change for neutrals, enhancement of memory colors by moving towards their archetypes, greater chroma preservation for saturated colors, no change in the core region).
2. Mapping of remaining colors (for which mapping was not defined explicitly) by using interpolation on the basis of explicitly mapped colors. This interpolation needs to be smooth and continuous.

Finally, the third related approach is the *spring–primary* algorithm (Zeng, 2006), where the morphing paradigm is described in terms of thinking about the source as having some key colors (e.g. primaries and secondaries, a few points in-between them plus black and white) joined by springs. The mapping into the destination then takes place by explicitly mapping the key colors and having the rest mapped by letting the 'springs' adjust to the new constellation. Essentially, this means using interpolation in polyhedra that tessellate the source gamut with vertices at the key colors; the result is a fast algorithm that allows for full three-dimensional changes to be applied in the source to destination transformation.

While the above three methods are truly three-dimensional, they are more like frameworks for mapping all colors once the mapping of some is defined explicitly (and typically in a way that needs to be done outside the framework), the next two approaches are directly about how to compress colors in three dimensions.

Following their extensive work on what color difference equations to minimize in gamut clipping (Chapter 9), Katoh and Ito (1999) moved on to propose an application of their findings to compression. The key idea – and here lies the ingenuity of their solution – is to pose compression as a clipping onto a scaled version of the destination gamut. Instead of clipping only OOG source colors onto the destination gamut, all source colors are clipped onto versions of the destination gamut scaled in proportion to those source colors' locations in the source gamut. The destination gamut, therefore, is not a shell, but instead an onion-like layered structure made up of scaled versions of the destination gamut, and the choice of which layer a color is clipped to (by minimizing a color difference equation) is determined by that source color. The end effect of such source-color-dependent clipping is in fact compression.

Their 'onion-peel' method proceeds as follows (Figure 10.22):

1. Map source lightness range into destination lightness range (e.g. using linear or sigmoidal mapping).
2. Define a core region as a scaled version of the destination gamut. Lightness-mapped source colors in this region will not be changed further.
3. For every lightness-mapped source color:
   (a) Compute chroma differences at its lightness and hue between it and the core gamut m and the source gamut n.
   (b) Compute the scaled destination gamut layer onto which clipping will be performed by finding the scaling factor for the destination gamut that will place the layer at a chroma distance from the core gamut of x along the lightness-mapped source color's lightness and hue line so that $x : y = m : n$, where y is the layer's distance from the destination gamut boundary along the same line.

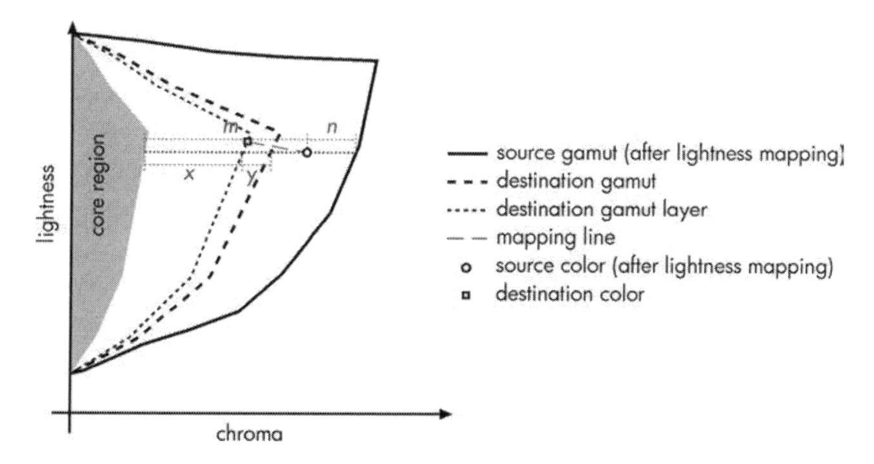

Figure 10.22 Katoh and Ito's (1999) 'onion peel' gamut compression algorithm.

(c) The destination color is the color with minimum color difference from the lightness-mapped source color on the destination gamut layer from step 3b. Note that 'color difference' does not necessarily mean Euclidean distance in the gamut mapping space, but can be some other (weighted) distance metric.

Finally, the last algorithm discussed here is Motomura's (1999, 2000, 2002) *categorical color mapping* approach, which presents a very particular and promising solution to gamut mapping that revolves around color naming. The basic idea here is to say that the key requirement for successful gamut mapping is the preservation of color names (as far as that is possible). Colors that were dark blue in the original also need to be dark blue in the reproduction and bright yellows need to retain that label (categorization) in the reproduction. While this requirement is well acknowledged in the literature, Motomura takes it beyond the realm of wishful thinking and makes it the processing criterion of his algorithm.

The first stage in this solution is to conduct a color naming experiment on samples from both the source and destination media. Here, samples (color patches), with known device color values (e.g. dRGB or dCMYK) are shown to a group of observers who are asked to categorize each sample into one of 11 categories, i.e. white, black, gray, red, brown, pink, orange, yellow, green, blue and purple, and this is done for both media. The end result is a set of 11 'average' colors – one per color category.

Having psychophysically determined which colors belong to which category and where each category has its mean in the source medium, a source color can be characterized by its set of *Mahalanobis distances* (Wikipedia, 2007a) from each of the color category centers. This distance metric uses an ellipsoid defined by the distribution of colors belonging to a category for judging how a new color relates to the category, which is a much more accurate way of determining category membership than a simple Euclidean distance from its mean would be. Not only does the Mahalanobis distance allow for a category shape to be different from a sphere, but it also considers what region in color space was occupied by colors judged to be in a category in addition to where that category's mean color is. Note that the three achromatic categories are grouped in Motomura's implementation and, therefore, it is nine such distances that are used. The Mahalanobis distance between a color's *Lab* vector $\mathbf{t} = [L,\ a,\ b]$ and the mean *Lab* vector $\mathbf{m}$ of one of the categories is computed as follows:

$$\Delta = (\mathbf{t} - \mathbf{m})\mathbf{S}^{-1}(\mathbf{t} - \mathbf{m})^{\mathrm{T}} \tag{10.17}$$

where $\mathbf{x}^{\mathrm{T}}$ is the transpose of matrix $\mathbf{x}$, $\mathbf{x}^{-1}$ is its inverse and $\mathbf{S}$ is the covariance matrix of the category whose average color is $\mathbf{m}$. $\mathbf{S}$ is computed as follows:

$$\mathbf{S} = \begin{bmatrix} s_L^2 & s_L s_a & s_L s_b \\ s_L s_a & s_a^2 & s_a s_b \\ s_L s_b & s_a s_b & s_b^2 \end{bmatrix} \tag{10.18}$$

Here, s_L is the standard deviation of lightnesses, categorized in this category in the psychophysical experiment, and s_a and s_b are analogously the standard deviations of the *a* and *b* values of colors in this category.

Given a source color and its Mahalanobis distance vector $\mathbf{S_M} = [\Delta_1, \Delta_2, \ldots, \Delta_9]$ containing distances from the color category means in the source, the next step is to compute the destination color by finding that point in the destination which has the same set of distances from the destination category means. However, since distances in the source are greater (as we are looking at the gamut reduction case here) than those in the destination, it is first necessary to scale the source distances before trying to match them in the destination. This is done by comparing the Mahalanobis distances between category means in the source and the destination and by computing a destination distance vector as follows:

$$\mathbf{D_M} = \boldsymbol{\phi}\mathbf{S_M} \tag{10.19}$$

Here, $\boldsymbol{\phi}$ is a matrix that is computed as follows:

$$\boldsymbol{\phi} = \begin{bmatrix} v_{1,1} & v_{1,2} & \cdots & v_{1,9} \\ v_{2,1} & v_{2,2} & \cdots & v_{2,9} \\ \vdots & \vdots & \ddots & \vdots \\ v_{9,1} & v_{9,2} & \cdots & v_{9,9} \end{bmatrix} \quad \text{where} \quad v_{i,j} = \begin{cases} \dfrac{\Delta(i,j)_D}{\Delta(i,j)_S}, i \neq j \\ 0, i = j \end{cases} \tag{10.20}$$

and where $\Delta(i, j)$ is the Mahalanobis distance of the mean color of category j from the mean of category i and using i's covariance matrix. Motomura further discussed a weighting that can be applied to give more importance to small values in $\mathbf{D_M}$, since these are of greater importance (they are to category means that are dominant).

Given a $\mathbf{D_M}$ vector of distances for the destination, a search algorithm is used to find the *Lab* values in the destination that have the most similar set of Mahalanobis distances to the destination color category means to $\mathbf{D_M}$, and that color is the destination color. The kind of mapping that this algorithm performs in three dimensions is illustrated in Figure 10.23 using a two-dimensional example involving five categories.

Note that the above is only an overview of the basic elements of Motomura's method and the reader is advised to consult the details in his journal papers (e.g. Motomura, 2001).

10.1.4 Summary of color-by-color reduction algorithms

This category of GMA has received by far the most attention to date and it is where the bulk of published work belongs. From the review presented above it is apparent that there is

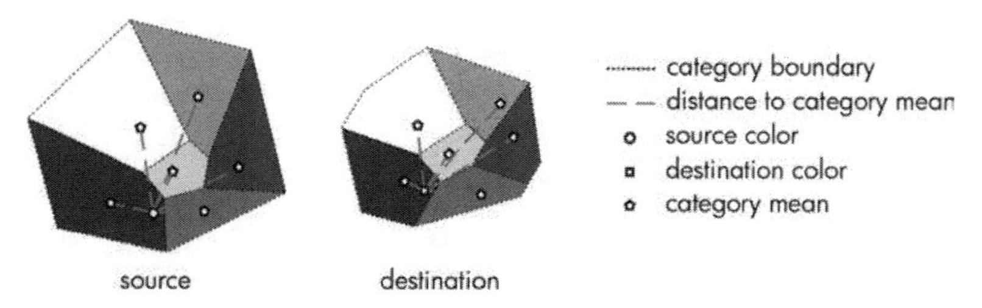

Figure 10.23 Illustration of the categorical color mapping concept. Uniform areas indicate regions where distance to one of the category means is smallest using the Mahalanobis distance.

tremendous variety in the proposed approaches. What further complicates the situation is that it is very difficult to recognize what are likely to be successful color-by-color reduction approaches for any given application. The reason for this lies in the fact that the vast majority of algorithms have been either evaluated only by their authors and (almost without exception) been found to outperform other algorithms that they were compared with, or they were evaluated by more than one group of researchers, who – as a result of differences in the evaluation conditions – have found contradictory results. This is a fundamental challenge of gamut mapping development and will be addressed in Chapter 12 in more detail.

Nonetheless, it would be overly cautious not to point to some of the recurring themes in these algorithms and their published evaluations, which include:

1. *Making gamut mapping more adaptive to source and destination gamuts.* While early approaches mapped along lines and only considered how source and destination relate along that line (e.g. LLIN: Sara, 1984), more recent methods try to make the mapping track gamut shape more closely (e.g. TOPO: Macdonald *et al.*, 2001; SGDA: Zolliker and Simon, 2006a).
2. *Sacrificing some detail preservation to preserve more contrast.* This has been recognized since the first gamut compression papers, and even though there were a number of approaches that proposed linear lightness compression, there is strong consensus that this is typically not a good solution. Instead, approaches like sigmoidal (Buckley, 1978; Braun, 1999) or IPI (Braun *et al.*, 1999a) lightness compression or linear luminance compression, as used in Adobe's *BPC* (Adobe, 2006), emerge as more robust.
3. *Maintaining a core region of the source gamut where no change is made.* This concept had already been proposed by Sara (1984) and has found many followers (e.g. Ito and Katoh, 1995; Spaulding *et al.*, 1995; MacDonald *et al.*, 2001; Lammens *et al.*, 2005).
4. *Mapping nonlinearly along mapping paths.* Instead of linearly mapping the source range onto the destination range, greater success is reported from methods that make less change closer to the origin of the mapping path (which tends to be on the lightness axis or somewhere else in the destination gamut) and apply greater compression close to the gamut boundary. This can be achieved either by piece-wise linear (i.e. locally linear) or directly nonlinear means (e.g. Sara, 1984; Stone and Wallace, 1991; MacDonald *et al.*, 2002). This can, in fact, be an indirect way of having a core gamut where no change takes place.
5. *Mapping from source image rather than medium gamut.* That this results in more accurate reproductions has been observed repeatedly and is also intuitive (e.g. Gentile *et al.*, 1990; Montag and Fairchild, 1997; Wei *et al.*, 1997; Kim *et al.*, 1998).
6. *Categorizing colors.* While this is much less strong a trend, it is an area that has received a lot of attention, ranging from doing different things for different categories (e.g. Haneishi *et al.*, 1993) to having the entire mapping determined by them (Motomura, 2001).

10.2 COLOR-BY-COLOR EXPANSION

Gamut reduction undoubtedly has the greatest urgency, since it is required when a destination gamut is even just partly smaller than a source gamut; therefore, it has received most attention to date. With the rise of imaging technologies that allow for the generation of increasing color gamuts, the opposite of reduction – *expansion* or *extension* – has also started to gain relevance. In particular, where there is significant legacy content bounded by smaller gamuts than are given access to by newer devices, there is a need for making use of the additional addressable color space to deliver a more pleasing experience. Gamut expansion is, in fact, exclusively applicable to varieties of *preferred color reproduction* (i.e. where what matters is how pleasing the reproduction is, irrespective of the appearance of the original) and is a form of *image enhancement.* Currently, the most dominant context where gamut expansion is applied is the use of wide gamut displays for high-definition television (HDTV), where television sets are able to produce gamuts that far exceed those of the legacy content displayed on them (e.g. Kim *et al.*, 2004).

Being a form of image enhancement (akin to methods like sharpening and contrast enhancement) brings with it two consequences:

1. *Likely to have optimum.* Expanding an image's gamut as much as possible is unlikely to give the most preferred result (e.g. Yendrikhovskij, 1998), just like increasing its contrast would not either. A key struggle here is between naturalness (or plausibility) and expansion.
2. *Likely to be image dependent.* This can be expected primarily because the aim of gamut expansion is to take a source image and change its gamut to maximize its preference, given the available color space in the destination. The optimal gamut of different images is then likely to be different, e.g. the optimal gamut of an image with objects having pastel colors will be different from the optimal gamut of an image depicting fireworks.

In the limited literature about gamut expansion, algorithms come in two flavors: GMAs designed specifically for expansion and reduction GMAs applied to expansion cases.

10.2.1 Expansion-specific algorithms

The first expansion-specific algorithm is Hoshino's (1991) mapping of printed images to the color gamut of HDTV. Here, lightness mapping is carried out using an unspecified, nonlinear tone reproduction curve. Source chroma is then mapped along lines of constant lightness and hue angle, which can be either an expansion or a reduction depending on which of the gamuts has greater chroma at that hue and mapped lightness combination. A hue shift is also applied for some color regions (i.e. cyan and blue) to make the degree of expansion more similar between neighboring areas. Essentially, this is quite similar in structure to the LLIN reduction algorithm, with the exception that lightness mapping is nonlinear and that there is some hue shift. A later revision of this algorithm involves a proportional lightness increase for source colors that have had their chroma increased, which is done to maintain more of the source's naturalness (Hoshino, 1994).

Kang *et al.* (2001) describe a study where they performed a psychovisual experiment in which a group of observers were given images with limited gamuts and asked to expand them manually. The resulting data showed that observers made changes in line with Hoshino's (1991) work, i.e. akin to an inverse LLIN mapping. An interesting finding here was that observers did not expand the lightness of images as far as was possible in the destination gamut, which again supports the idea of an optimal level rather than maximization being preferred. Different levels of chroma expansion were also compared, and a 59% increase was preferred over both 33% and 97% expansions, which agrees well with Yendrikhovskij *et al.*'s (1998) findings.

Finally, Kim *et al.* (2004) described the following approach in their paper on gamut expansion for HDTV:

1. The mapping takes place in a *luminance-linear space*, i.e. *YWV*, where the first dimension is the Y tristimulus value, $W = -0.53X - 0.687Y + 0.643Z$ (i.e. a blueness–yellowness channel, since response to the high-frequency part of the spectrum (i.e. blue: Z) has the weighted sum of the two other parts of the spectrum subtracted from it (i.e. red plus green: $X + Y$)) and $V = 1.82X - 1.48Y + 0.23Z$ (i.e. a redness–greenness channel). The polar equivalents of *WV* are referred to as chroma and hue in this method, and we will label them chroma_{WV} and hue_{WV} to differentiate them from the hue and chroma attributes used in the rest of the chapter. The result is a space where changes along each of the axes (luminance, blueness–yellowness and redness–greenness) are linear with weighted spectral power rather than being an attempt at linearity with perceived differences (as is the case with CIELAB and CIECAM02, which are more commonly used in gamut mapping). There is, therefore, greater similarity here with Adobe's *BPC* (Adobe, 2006) than with other gamut mapping methods.
2. Three expansion strategies are then proposed where a source range is mapped linearly onto a destination range:
   (a) along lines from the origin (i.e. absolute black) through the source color (giving a more natural result) – *vector mapping*;
   (b) along lines of constant Y and hue_{WV} (giving a more vivid result and being similar to LLIN in structure) – *chroma mapping*; or
   (c) along lines at a specific angle θ between options 2a and 2b that varies with luminance and is computed as $\theta = (90° - \theta_0)Y + \theta_0$, where θ_0 is the minimum mapping angle (providing a compromise) – *adaptive mapping* (Figure 10.24).

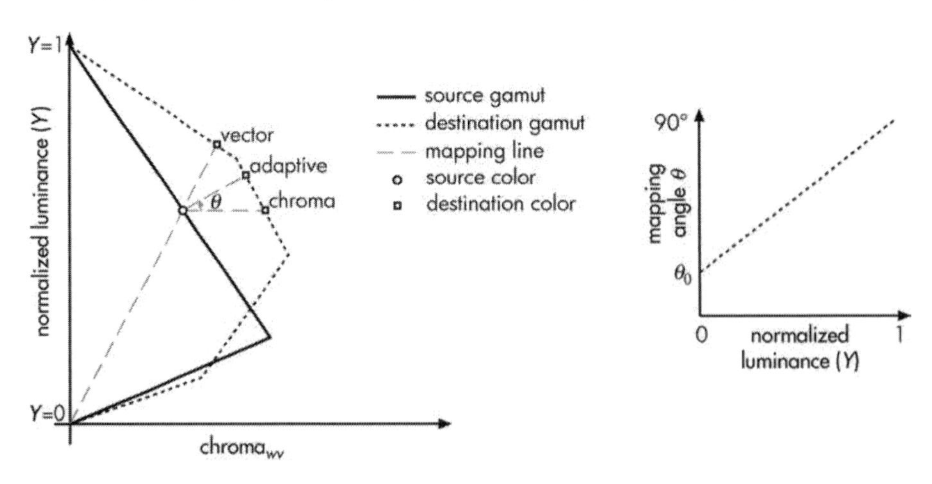

Figure 10.24 Kim *et al.*'s (2004) gamut expansion strategies.

10.2.2 Reduction algorithms applied to expansion

In his PhD thesis, Braun (1999: chapter 6) evaluated the gamut expansion performance of his gamut-compression algorithms (i.e. sigmoidal lightness compression and knee chroma compression towards the cusp) in a psychovisual experiment. When comparing alternative lightness expansion techniques, results showed that inverse-sigmoidal lightness expansion performed as well as or better than linear lightness expansion for the eight test images. This is likely due to it avoiding the darkening of images that results from linear lightness expansion. Alternative chroma expansions along lines from the lightness of the cusp on the lightness axis were also compared (linear versus knee-function expansion), and the effect of setting a cap on the maximum amount of chroma expansion was considered too. Here, setting the knee point at 50% versus 90% improved gamut expansion results, as it helped to maintain the smoothness of chromatic blends. This was found to be particularly important for the reproduction of skies. In terms of the amount of expansion, limiting it to 30% performed as well as doing expansion to the full destination gamut.

Finally, let us look at two MSc final projects from the *University of Derby's Color and Imaging Institute* that had gamut expansion as their subject.

First, Ling (2001) used the LLIN (Sara, 1984) and TOPO (MacDonald *et al.*, 2002) reduction algorithms and ran a pair comparison psychovisual experiment where the pleasantness of gamut extended images was evaluated. Nineteen test images were used, and the pleasantness of the originals and LLIN- and TOPO-transformed versions was judged. The results (Figure 10.25) show a clear preference for LLIN-expanded images over both original and TOPO-expanded images. LLIN gave the most pleasant results in 17 of the 19 cases, although the degree of improvement varied greatly – ranging from being dramatic for images 1 and 8 to being negligible for images 5 and 14.

Chen (2002) then compared LLIN, SLIN and an expansion algorithm that operated in display RGB for rendering printed images on a CRT monitor. For LLIN and SLIN the effect of expanding from the source image versus medium gamuts to either the entire destination gamut or only to halfway between source and destination was evaluated.

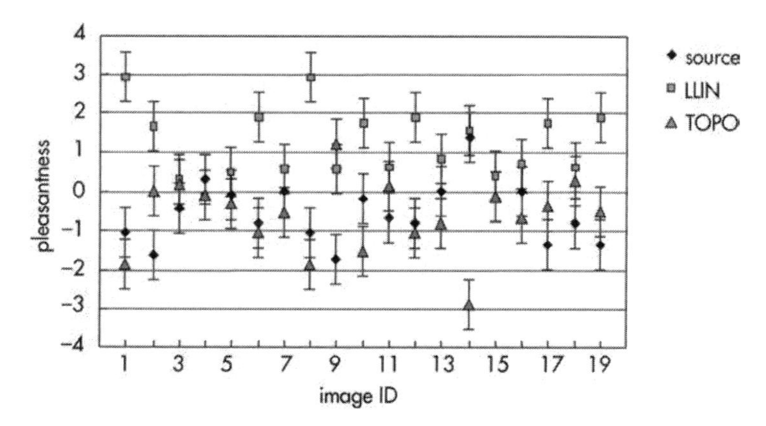

Figure 10.25 Pleasantness judgments and their 95% confidence intervals for source images and their expanded versions obtained using LLIN and TOPO.

What was found to work best overall again was LLIN using medium (rather than image) gamuts.

Finally, Hirokawa *et al.* (2007) also report good results for expanding colors along lines of constant lightness, albeit using a nonlinear function.

Compared with the more than 50 gamut reduction algorithms that have been proposed and studied to date, only a handful of gamut extension methods exist, and drawing any conclusions from them is pretty risky. Nonetheless, it can be said safely that any further work in this area needs to use the LLIN algorithm (Sara, 1984) or its more adaptive variant proposed by Kim *et al.* (2004) as a benchmark.

10.3 SPATIAL GAMUT REDUCTION

All the GMAs discussed so far operated purely in a color domain and their effect could be expressed as a set of vectors from source to destination colors in the gamut mapping color space. This means that they resulted in the same destination color every time a given source color was gamut mapped. While such an approach can be very successful, it does not directly address one of the important requirements of gamut reduction, namely the preservation of image detail. In particular, minimum color difference gamut clipping algorithms are notorious for mapping a whole range of source colors onto a single destination color. Any source image detail in that part of color space was lost. Even compression algorithms do not address detail preservation, even though they aim to do so indirectly by compressing source ranges onto destination ranges (resulting in bijection between source and destination). However, it is often the case that detail that was clearly perceptible in the source is mapped onto differences that are below the perceptibility threshold or that are lost as a consequence of encoding quantization. The main point here is not the inability of color-by-color reduction algorithms to guarantee detail preservation (because they can), but the cost at which they are able to deliver it in terms of loss of overall contrast.

To address the need for local detail preservation while maintaining as much global source contrast as possible, gamut mapping needs to take into account local neighborhoods in images. This is because many of the possible color differences in a source medium gamut do not need to be preserved since they do not occur in a given source image. The way in which a color needs to be mapped into a destination gamut does not have to be the same regardless of where it is in a source image – where a color is present in a large uniform area it can be mapped in one way (e.g. to the visually closest destination color), whereas if it is surrounded by pixels with similar colors with which it forms a texture or pattern then visual similarity needs to be balanced against detail preservation and a compromise can be sought.

Before looking at specific spatial GMAs, let us first consider the example shown in Figure 10.26. Here, a source grayscale image (center) is mapped to a much reduced lightness range and the top and bottom of the figure respectively show the result of doing this in a color-by-color versus a spatial way. The nonspatial example is already one where the most successful current method (sigmoidal mapping) is used, and the result exhibits an attempt at balancing contrast preservation with detail preservation. However, the constraint of mapping each source color onto the same destination color regardless of its source neighborhood is too restrictive, and the bottom of the figure shows what can be

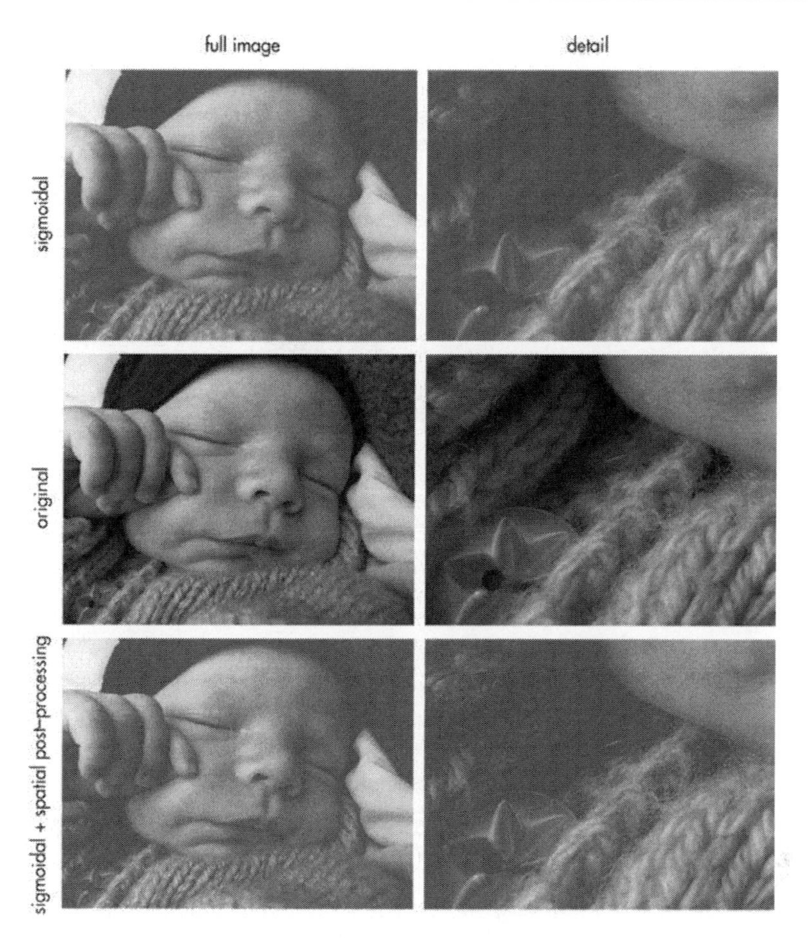

Figure 10.26 Example of spatial lightness mapping (bottom) compared with a nonspatial alternative (top).

achieved when that constraint is removed. Looking at the enlarged details in the right half of the figure shows how much more of the source detail is preserved in the spatially mapped destination while using the same destination lightness range as the color-by-color mapping. Taking a closer look at how source and destination lightnesses relate in the two cases in Figure 10.27 reveals how much variation there is in the destination lightnesses for each source lightness in the spatial case, while in the color-by-color case there is a one-to-one relationship. Note that the one-to-many relationship shown for the spatial case is specific to the image shown in Figure 10.26, and the same algorithm as was used here will result in a different one-to-many relationship for each image – dependent on the local relationships in each case.

Looking at the spatial GMAs proposed to date, three categories can be seen to emerge. The first are algorithms that apply color-by-color gamut mapping to low spatial frequencies and then add detail to the result, the second are algorithms that build on the Retinex model (Land, 1964) and the third category is based on image difference minimization.

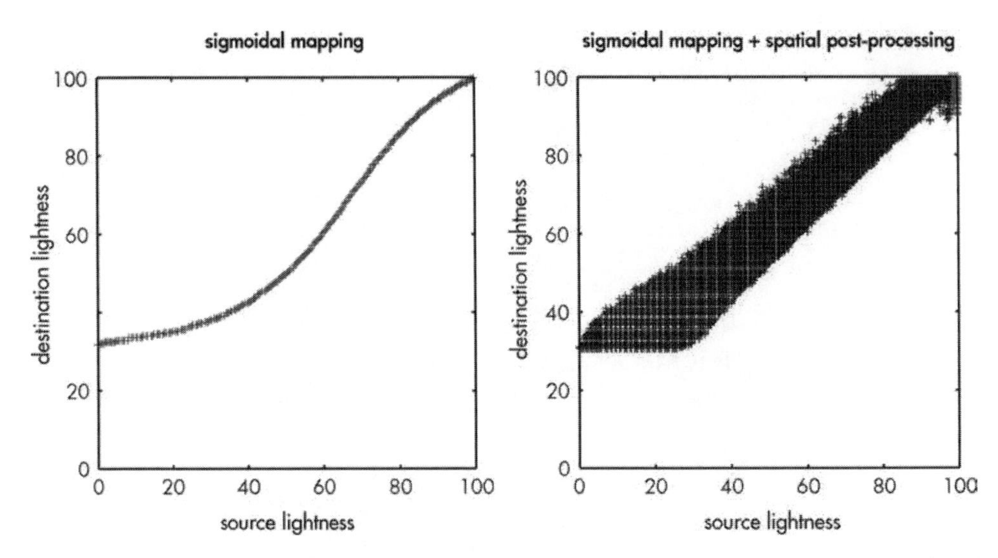

Figure 10.27 Source–destination lightness relationships for color-by-color (left) and spatial (right) lightness mapping of images in Figure 10.26.

10.3.1 Frequency-sequential methods

The first, pioneering spatial gamut mapping solution was proposed by Meyer and Barth (1989) and was followed by a gap of about 10 years before this type of approach was revisited. Their algorithm involved a spatially varying lightness mapping followed by piecewise linear chroma compression along lines of constant hue and (mapped) lightness. In other words, it had the structure of LLIN (Sara, 1984) and differed only in the lightness mapping, which used the following steps (Figure 10.28):

1. *Low-pass filtering* in the Fourier spatial frequency domain (Section 5.5.4) to remove spatial variation occurring at high frequencies, i.e. the detail, and preserving only representative lightnesses of image regions.
2. *Linearly compressing lightness range* of low-pass-filtered source image into destination lightness range.
3. *Adding high-pass component* of source image (i.e. the difference between the source image and its low-pass-filtered version – containing the high spatial frequency content) to its lightness compressed low-pass version.
4. *Clipping* to destination lightness range (since adding back high-pass component is likely to result in some OOG lightnesses).

The end result is a lightness mapping like that shown in Figure 10.26.

A related approach is the solution proposed by Balasubramanian *et al.* (2000; Bala *et al.*, 2001), where the key differences are that it is the detail from the difference between the source and color-by-color gamut mapped destination that is added to that destination, rather than the detail from the source directly, and that all the processing is in the spatial

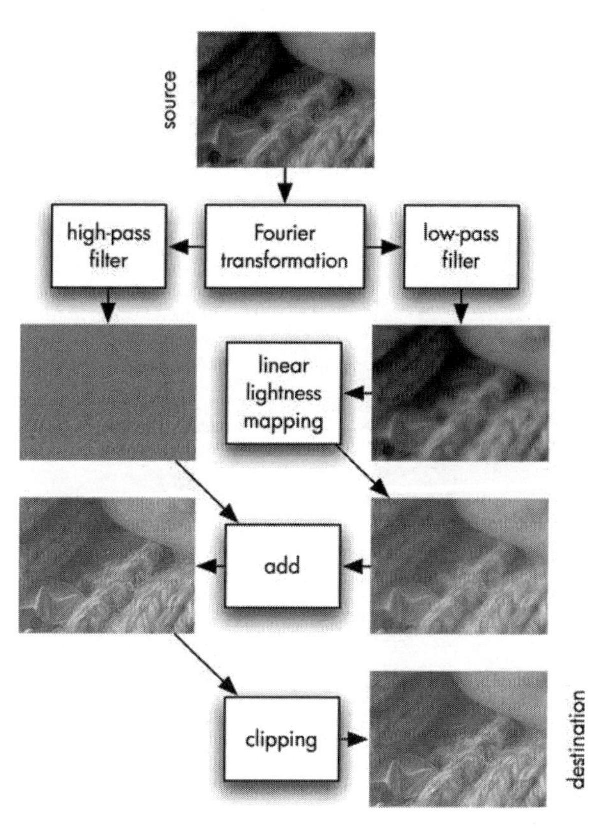

Figure 10.28 Stages of Meyer and Barth's spatial lightness mapping (illustrated on detail from image in Figure 10.26).

domain rather than the frequency domain (which can allow for faster execution) (Figure 10.29). The source image is first processed using a chroma-preserving gamut-clipping. Then, detail is extracted from the difference between the original and gamut-mapped luminance (or lightness) channels using a high-pass filtering method. The example filter given in reference is simply a difference from local mean (DLM) and (interestingly) the filter window size is content dependent, with window sizes of 3×3 and 15×15 pixels used for business graphics and photographic content respectively. The detail information obtained in this way is added to the gamut-mapped image, which is processed again using a luminance- (or lightness-)preserving gamut-clipping method. The details of the algorithm are as follows:

1. Optional initial lightness mapping (e.g. IPI; Section 10.1.1).
2. Chroma-preserving gamut clipping or minimum Euclidean distance gamut clipping in CIELAB.
3. Calculating luminance difference between source and gamut-clipped images as $\Delta Y = Y_{\mathrm{O}} - Y_{\mathrm{GC}}$.
4. Extracting the lost edges from the difference image using a high-pass filter. This can be computed as $\Delta Y' = k\{\Delta Y - \sum[\Delta Y(i,j)/N]$, where $\Delta Y(i,j)$ are the pixels in the

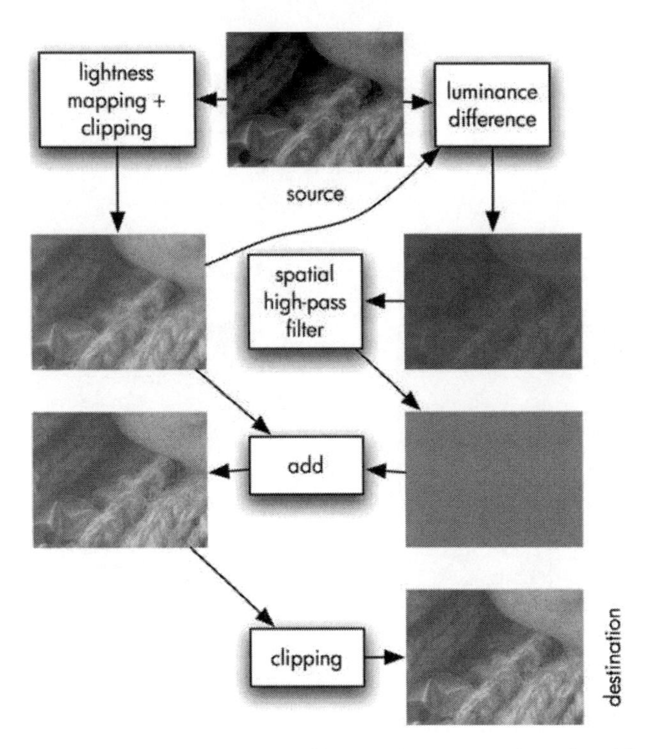

Figure 10.29 Stages of Balasubramanian *et al.*'s spatial mapping algorithm (illustrated on detail from image in Figure 10.26).

filtering window, N defines window size and k allows for control over how much detail to add back (in the original paper, $k - 1$).

To deal with different kinds of image content (i.e. photographic or synthetic), the authors propose an adaptive algorithm for deciding between the two filter sizes. They define $a_1 = \sum |\Delta Y(i,j)|$ as the local variance with the trial filter size of 5×5 and then analyze it as follows:

(a) if a_1 is not in the range of $[t_1,\ t_2]$, use filter size of 3×3;

(b) if a_1 is in the range of $[t_1,\ t_2]$, calculate $a_2 = \sum |\Delta Y(i,\ j)|$ with trial filter size of 15×15:

(i) if $a_2/a_1 > t_3$, use filter size of 3×3;

(ii) if $a_2/a_1 < t_3$, use filter size of 15×15.

Note that t_1, t_2 and t_3 are chosen in a trial-and-error way.

5. Add $\Delta Y'$ to the gamut-clipped Y' (i.e. the result of the first two steps) to compensate for the lost spatial information.
6. Apply point-wise luminance- (or lightness-)preserving gamut clipping or clipping towards the lightness of the cusp on the lightness axis with the intention of preserving the spatial detail added in step 5. This is necessary since step 5 is likely to give some OOG colors.

In the psychovisual evaluation performed by its authors, the use of the initial IGI lightness mapping performs best when preferred reproduction is wanted, whereas no initial lightness mapping is best for accurate reproduction. Color Plate 15 shows the application of this algorithm in CIELAB, and it is also worth noting that it has also performed well when evaluated by others (e.g. Morovič and Wang, 2003a; Bonnier *et al.*, 2006).

An alternative that combines elements from the two above spatial GMAs was proposed by Morovič and Wang (2003a). Here, the processing is in the spatial domain, whereas it is the difference between high- and low-frequency source content that gets added to color-by-color gamut-mapped destination images. What is new here is the use of all three color appearance dimensions (rather than only lightness/luminance) and of multiple levels of spatial frequency content, i.e. instead of a division only between high and low frequencies, here a series of frequency bands is used. The reason for doing this is to attempt a preservation not only of high-frequency detail, but also of lower frequency intra-image differences. The algorithm proceeds as follows (Figure 10.30):

1. *Multi-resolution decomposition* of the source image with a choice of filter size r (e.g. 3, 7, 13) and number of decomposition bands n (e.g. 2, 4, 6; note that at $n = 2$ the approach is most similar to previous ones). The same DLM filtering as proposed by Bala *et al.* (2001) is used to obtain low-pass versions of an image. Note that the image is filtered in all three orthogonal channels of a color appearance space (e.g. CIECAM).

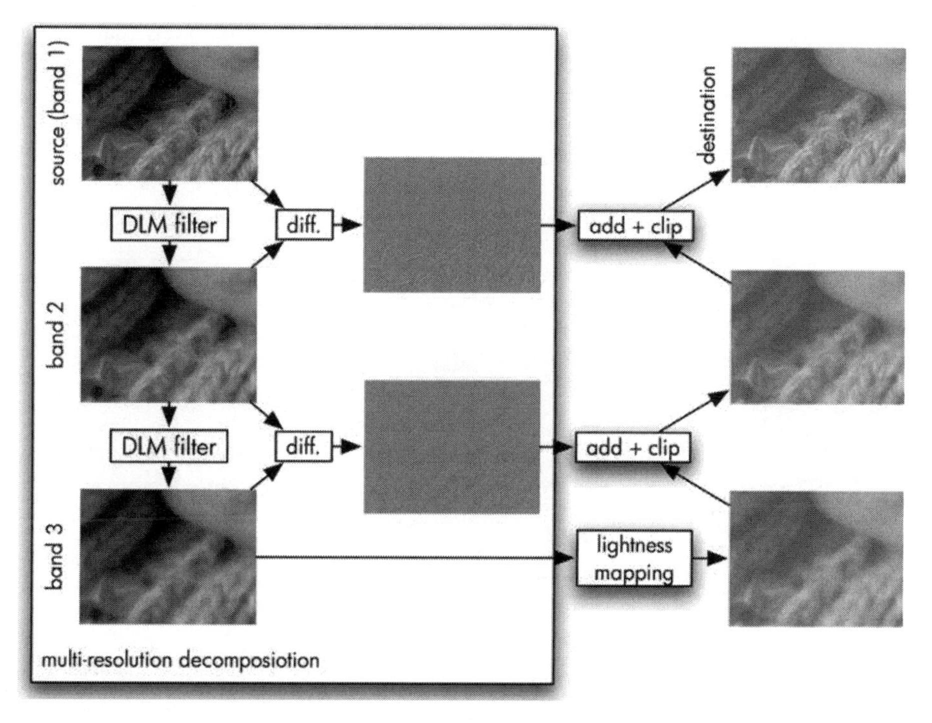

Figure 10.30 Stages of Morovič and Wang's spatial mapping algorithm (shown for $n = 3$ and $r = 5$ and illustrated on detail from image in Figure 10.26).

2. *Lightness compression* of the lowest spatial-frequency band either in a linear or sigmoidal way (Section 10.1.1). Note that this is optional and no explicit, initial compression is necessary with this method.
3. *Initial gamut mapping* using the hue-preserving minimum ΔE clipping method for the lowest spatial frequency band (band n). This step establishes approximate colors for destination pixels, which will be modified by subsequent steps.
4. *Addition of difference* between the current band i and next higher band $i-1$ from the source image decomposition to the gamut-mapped reproduction at band i. Here, the absolute difference can be added or it can first be linearly compressed according the ratio of the lightness ranges of the destination and source media.
5. *Subsequent gamut mapping*, e.g. using clipping towards the center of the lightness axis to preserve more spatial detail.
6. If $n > 2$ and $i > 2$, go to step 4; otherwise terminate.

When applied with a filter size of 9×9, four decomposition levels, the hue-preserving minimum ΔE as both the initial and subsequent GMAs and the linear compression of inter-band differences in CIELAB, the result shown in Color Plate 16 is obtained.

Zolliker and Simon (2006b) then proposed two approaches: a basic and an extended one. The basic approach is essentially the same as Bala *et al.*'s (2001) method, with the change that differences are considered in all three color dimensions (like Morovič and Wang (2003a)) and that a Gaussian filter is used instead of a DLM filter. The point here is that while DLM gives differences between a pixel and each of its neighborhood pixels the same weight, a Gaussian filter gives more weight to neighborhood pixels that are closer than those that are farther. In their extended approach, the authors make weights for computing the lower spatial frequency image not only dependent on spatial distance (like they did in their basic approach), but also on color difference, i.e. neighborhood pixels that are very different are given less weight than those that are more similar (Figure 10.31).

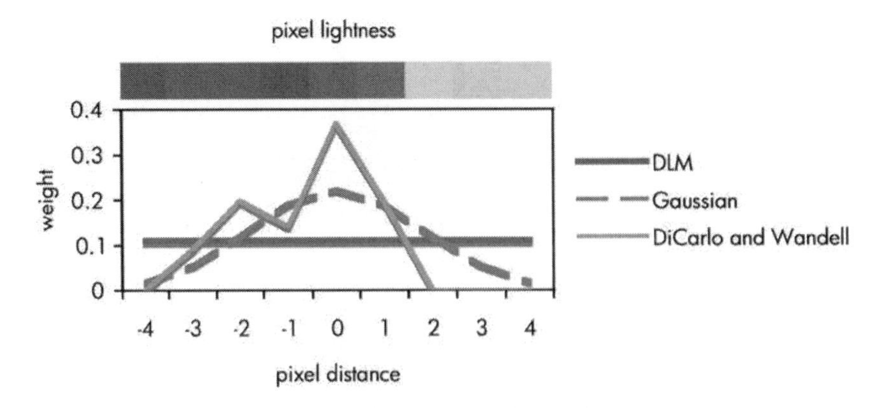

Figure 10.31 One-dimensional comparison of spatial filter weights: (a) DLM; (b) Gaussian; (c) DiCarlo and Wandell. Note, weights always add up to one and the DiCarlo and Wandell weights are for this particular set of pixels, whereas the others are for any nine-pixel neighborhood.

This type of filter is like the bilateral filter proposed by DiCarlo and Wandell (2000) in the context of high dynamic range imaging (HDRI), and its effect is a filtering that maintains more edge detail than a Gaussian filter does.

The use of bilateral filters (Tomasi and Manduchi, 1998), i.e. filters that weight both spatial and color distance simultaneously, is further carried forward by Bonnier *et al.* (2006, 2007), who describe two spatial algorithms. Both share the same starting point, which is a bilateral filtering of the source image that gives a low-pass image and (when subtracted from the source) also a high-pass image. The low-pass image is gamut mapped using a color-by-color algorithm (i.e. hue-preserving minimum color difference clipping) – all of which is like Zolliker and Simon's approach. The difference then comes in how the high-pass version of the source is added to the color-by-color gamut-mapped low-pass version. Here, two methods are proposed.

The first method, namely *spatial and color adaptive clipping* (*SCACLIP*), uses clipping and does it by selecting from among three alternative clipping directions (towards the minimum ΔE, the center or the lightness of the cusp). The choice is made in the following, novel way:

1. Add high-pass filtered version of the source image to gamut-mapped low-pass-filtered version (this is likely to move colors back OOG).
2. For each pixel, compute three alternative gamut clippings of it and its neighborhood (filtered using the bilateral filter) and compute how well each of them preserves the local neighborhood relationships ('energy') from step 1.
3. For each pixel, choose the clipping direction that preserves local relationships best.

The second method (SCACOMP) uses compression along lines towards the center of color space to an extent dependent on the magnitude of high-pass detail (i.e. a source color in the low-pass source image is compressed further inside the destination gamut if there is detail of greater magnitude around it than if its magnitude is lower).

Finally, Kolås and Farup (2007) have proposed a computationally efficient spatial GMA that consists of the following stages:

1. Compute SCLIP gamut-clipped version of source image.
2. Compute map of pixel-by-pixel difference in distance from center between source and SCLIP-mapped versions of image.
3. Filter result of stage 2 using an edge-preserving filter controlled by edge information from the source image. The values from stage 2 set the maximum, which therefore guarantees that the filtered values do not result in distances from the center representing OOG colors.
4. Apply result of stage 3 to source image (i.e. set distances from center for source image pixels to those resulting from stage 3).

The main benefit of this strategy is likely to be its speed, and it can be expected that for some combinations of source–destination gamuts (e.g. display to print) it is likely not to perform well due to its use of the mapping towards the center of the lightness axis. Such a mapping is likely to make excessive lightness changes when source and destination gamut shapes are significantly different.

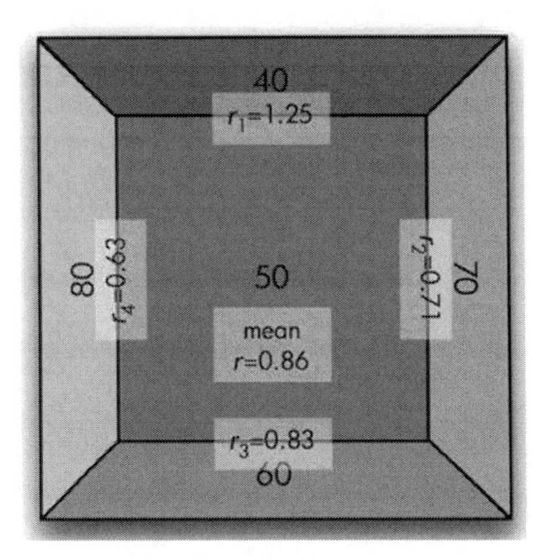

Figure 10.32 Retinex example (each surface's radiance shown at its center).

10.3.2 Retinex-based methods

An alternative approach to the ones discussed above is to apply Land's (1964) retinex model to the gamut mapping problem (McCann, 1999, 2002). The retinex model takes radiances of scene surfaces and for each surface predicts its perceived lightness by considering its relationships with neighboring surfaces. The reason for this is rooted in the makeup of the human visual system, where signals encode spatial relationships rather than absolute magnitudes (Section 2.4.2); a simplified view of the concepts behind retinex is illustrated in Figure 10.32. Here, the prediction of lightness for the central surface involves computing the radiance ratios between the surface's radiance X and each of its neighbors' radiances X' (i.e. $r_i = X/X'_i$) and finally averaging all the ratios a surface has with its neighbors. Note that this is a huge oversimplification; there are a host of alternative implementations of its idea that can involve considering ratios along paths in the scene (rather than only immediate neighbors), which is the original approach (Land, 1964), or it can be implemented in a multi-scale way (Frankle and McCann, 1983; Jobson *et al.*, 1996). Also, ratios are then clipped to be in a zero to some maximum (e.g. one) range (referred to as a 'reset').

The basic idea of the multi-scale retinex is next shown in Figure 10.33. Its first stage is the computation of a sequence of multiple representations of an image of decreasing scale (resolution). Within each scale, mean ratios between a pixel and its neighbors are computed and stored. The final ratios for the full-scale image are then obtained in the following way. First, the ratios in the lowest scale are multiplied with those of the scale above it, which conceptually involves scaling up the lower scale image to the next scale above it and then multiplying the ratios in the corresponding pixels. The result is a set of new ratios at the penultimate scale that now also take into account ratios at the lowest scale. The process is then repeated for all subsequent, higher scales until the highest scale is reached.

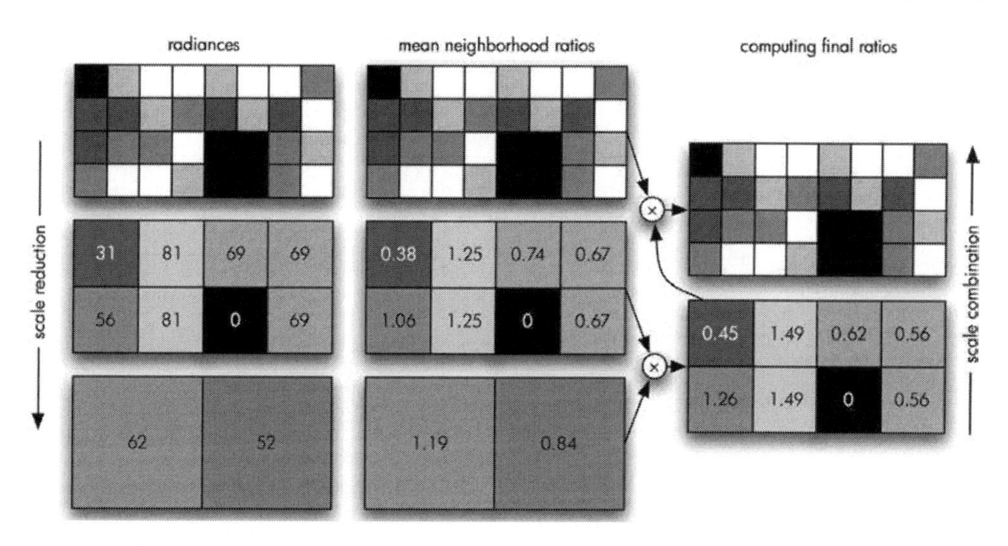

Figure 10.33 Multi-scale retinex stages.

The relevance of retinex to gamut mapping stems from its use by McCann (1999, 2002). Here, the starting point is a source image and an initial attempt at a destination image (we will call this the 'color-mapped' destination image), which is inside the destination gamut (how it is obtained is not specified, but it could be done using a color-by-color gamut reduction algorithm). Once the multi-scale decomposition of both images and the computation of ratios at each scale in each image are done, the final retinex-gamut-mapped image is computed by starting at the bottom of the source scale decomposition and using the ratios from the color-mapped image as maxima (i.e. 'reset' values). That is, the ratio used in the retinex-gamut-mapped image at each scale is the minimum of the ratios present in the source and color-mapped images. The aim here is to restrict ratios based on what they are like in an image known to be IG, while trying to keep them like they were in the source when possible.

While this is certainly an interesting and well-reasoned approach, it remains to be seen whether the kind of procedure proposed by McCann results in fully IG images (since it relies on indirect means) and how it compares with the methods in Section 10.3.1.

10.3.3 Image-difference minimization

Finally, a very elegant and long-term most promising solution to gamut mapping in general was proposed by Nakauchi *et al.* (1996, 1999). The beauty of their approach is in defining gamut mapping as an optimization problem of finding the destination image that is perceptually closest to a given source and has all of its pixels inside the destination gamut. The model of perceived image difference used by these authors involves band-pass filtering the Fourier-transformed CIELAB source and destination images and then predicting the images' difference by weighting differences in the various bands according to the human contrast sensitivity function. Specifically, the following difference metric PD

(D, S) is used between a source image S and a destination image D for a given spatial location (x, y) in the image pair:

$$\mathrm{PD}(D,S)_{(x,y)} = \sum_{X=L^*a^*b^*} \left\{ \sum_{i,j=-\omega}^{+\omega} [h^X_{(i,j)} (X_{S_{(x-i,y-j)}} - X_{D_{(x-i,y-j)}})] \right\}^2 \tag{10.21}$$

where $2\omega + 1$ is the filter size and $h^X (X \in \{L^*, a^*, b^*\})$ are filters that resemble human contrast sensitivity to lightness, redness–greenness and yellowness–blueness. The optimization algorithm then minimizes this difference metric while also constraining image pixels to be inside the destination gamut.

While being a very promising approach, its success depends greatly on the accuracy of the difference metric being minimized, and the one used in this paper has not been shown to work well. The challenge, therefore, is to have a robust metric of image difference; once that is available, Nakauchi *et al.*'s method will have great potential for providing the most similar gamut-restricted reproductions.

The idea of gamut mapping by image difference minimization is further developed by Kimmel *et al.* (2005), who propose an alternative to Nakauchi's PD metric that has the advantage of resulting in a unique destination image for convex destination gamuts. Kimmel *et al.*'s approach also shares some of the features of retinex-based methods.

Finally, Giesen *et al.* (2006b) proposed a color-by-color GMA whose parameters are optimized for a specific source image, and their method would also lend itself for use with spatial approaches.

10.3.4 Summary of spatial gamut reduction

Spatial gamut mapping has clearly got the greatest potential for delivering reproductions that are most like their originals, and it has been repeatedly shown to outperform color-by-color algorithms (this is the case in all the frequency sequential papers in Section 10.3.1, and it has also been shown specifically by Zolliker and Simon (2006b), who compared color-by-color and spatially modified versions of those algorithms with each other). Where highest subjective accuracy is sought, therefore, spatial algorithms are the way forward, and what stops them from being the only form of gamut mapping is their current infeasibility due to increased processing and resource requirements and there being some cases where image-dependent gamut mapping is not the most suitable solution (Section 10.1.2).

10.4 SPECTRAL GAMUT MAPPING

Having reviewed several categories of color GMAs, we will here take a look at gamut mapping in a spectral rather than a color domain. The context is that of multi-spectral color reproduction, where source content is captured in terms of spectral reflectance or power at each pixel and the destination device can also be addressed in terms of spectral rather than color properties. There is a substantial body of literature on multi-spectral color reproduction (e.g. Yamaguchi *et al.*, 2000; Berns *et al.*, 2003; Herzog and Hill, 2003; Rosen and Berns, 2005), but only a relatively small number of spectral gamut mapping papers have been published so far.

Instead of inputs being source color appearances and outputs modified appearances that are inside the destination's color gamut, in spectral reproduction the source and destination are in a spectral space (e.g. a 31-dimensional space for a 10 nm wavelength sampling between 400 and 700 nm) or in a lower dimensional space of linear bases in terms of which reflectances are represented. The reason for this latter case stems from most surfaces having spectral reflectances that are lower dimensional than the typical spectral sampling dimensions of measuring instruments. It has been found repeatedly that between three and eight dimensions are typically sufficient for representing the variation in measured spectra (Section 2.5; Krinov, 1947; Vrhel *et al.*, 1994).

To represent spectra in terms of a linear basis, that basis first needs to be obtained by analyzing measured spectra, e.g. using *characteristic vector analysis* (Maloney, 1986) or *principal component analysis* (PCA; Jolliffe, 2002). $R(\lambda)$, the reflectance at a given wavelength λ, can then be written as a weighted linear sum $R(\lambda) = B_1(\lambda)w_1 + B_2(\lambda)w_2 + \cdots + B_d(\lambda)w_d$, where d is the dimension of the linear model. This means that n-dimensional (e.g. $n = 31$) reflectances can be represented by d linear weights $[w_1, w_2, \ldots w_d]$ (d being much smaller than n) that uniquely define the reflectance within the known linear basis $[B_1(\lambda), B_2(\lambda), \ldots, B_d(\lambda)]$.

As a consequence, spectral gamut mapping typically takes place in spaces formed by weights for a set of bases (e.g. six) that can be used to represent natural spectra, and the first thing that is needed is a way of describing the spectral gamuts of a source and the destination that is to be used for reproducing it. Note that these spectral gamuts are polyhedra in the weight space of the chosen linear basis and, owing to their high-dimensional nature, it is convex hulls that are used for their determination (in spite of the issues that this is known to have in color space – Chapter 8).

Given the spectral (or linear basis weight) gamuts of a source and destination pair and the spectral reflectance (and its weights in the linear basis) of a source, the following approaches for gamut mapping it have been proposed to date.

The simplest approach clearly is to take a source spectrum and map it to the destination spectrum that is closest to it in the spectral space in terms of Euclidean distance (i.e. the spectral equivalent to simple minimum color difference clipping). Such *spectral distance minimizing clipping* is likely to result in objectionable differences, since spectral differences do not correlate well with human color response (e.g. imagine a pair of spectra that are very different at 400 nm and another that have the same difference at 500 nm – the first will be much more similar than the second, since the human visual system is much more sensitive at 500 nm than at 400 nm).

Having realized this limitation of pure spectral distance minimization, the earliest work on spectral gamut mapping by Chau and Cowan (1996), which predates any other publication on this subject by almost 10 years, had already looked at better alternatives. What they set out to do was to find the mapping for source spectra that are outside a given destination gamut that minimizes color difference under a chosen light source and also reduces spectral differences.

To do this, Chau and Cowan used a special set of bases instead of those obtained directly using the statistical analyses mentioned above. These bases (Wyszecki, 1953, 1958) are ones where the first three (called *fundamental* bases) account for the color response under a chosen illuminant and the remaining ones (called *metameric blacks*) only encode the metameric components under that light source. In other words, any changes to the weights in the metameric black bases result in no change to tristimulus values under the chosen light source. Given such a representation, Chau and Cowan's mapping consists first of matching the source's weights in the three fundamental bases (when possible – otherwise minimizing distance) and then taking the source's residual reflectance (i.e. the reflectance obtained by subtracting the source's representation only in the fundamental bases from the source's reflectance) and minimizing distance between it and its closest representation in the metameric black bases. In other words, the mapping consists first in minimizing color difference under a chosen illuminant and then minimizing the distance between the source's metameric component and the metameric components available in the destination gamut. Chau and Cowan also provide for a way of minimizing distance under more than a single light source.

Taking essentially the same idea of separating that part of spectra which account for color response from the remaining, metameric part is also at the basis of Derhak and Rosen's (2004) *LabPQR* space. Here, the first three dimensions are those of CIELAB (i.e. nonlinear transformations of the fundamental bases used by Chau and Cowan (1996)) and the remaining three dimensions *PQR* are metameric blacks. Given this space, the color gamut of a set of reflectances is represented by their CIELAB gamut and by a *PQR* gamut at each IG $L^*a^*b^*$ location. Gamut mapping in *LabPQR* (Rosen and Derhak, 2006) then involves two stages for each source reflectance:

1. Gamut mapping of the source reflectance's color under a chosen illuminant into the destination's color gamut under the same illuminant – here, any color GMA can be used (i.e. anything from Sections 10.1–10.3) and the result is destination $L^*a^*b^*$ values. Note that, even though the space is *LabPQR*, the gamut mapping of $L^*a^*b^*$ values does not have to be in CIE *LAB* itself; instead, those $L^*a^*b^*$ values could first be converted to another colorimetric space (e.g. CIECAM02, IPT), gamut mapped and then converted back to $L^*a^*b^*$.
2. Mapping of the source *PQR* values to the *PQR* gamut available in the destination at the destination *Lab* values. Here, the two most relevant alternatives are:
   (a) Closest *PQR* (i.e. minimizing Euclidean distance in *PQR*) – this is most similar to Chau and Cowan's (1996) method.
   (b) Closest scaled *PQR*, since *PQR* is obtained using PCA, where the first component *P* accounts for more variation in spectra than does the second, and so on; in this variant, the *PQR* space is scaled in each dimension in turn proportionally with the amount of variation that each accounts for. The result is differences that take into account the relative importances of the *PQR* dimensions.

The idea of treating the fundamental and metameric bases separately and differently certainly makes good intuitive sense. It will be interesting to see how the solutions proposed to date perform when evaluated by other groups than the ones who developed them and to see whether novel ways of mapping the metameric components will emerge.

For completeness' sake, and owing to the limited amount of spectral gamut mapping work, it is also worth mentioning three other studies. First, Bakke *et al.*'s (2005) proposal of clipping source spectra outside the destination gamut in an up to eight-dimensional

PCA-computed linear basis along lines towards the gamut's center. The gamut computation here was performed using the convex hull method (Section 8.2.1); this is essentially the use of the SCLIP idea (Section 9.5.3), but applying it in a spectral rather than a color space. What is not clear is whether this has any advantages over the other methods mentioned above.

Second, a special case of spectral gamut mapping has been developed by Bell and Alexander (2004) for use when rendering three-dimensional scenes whose surfaces have their spectral properties defined. Their approach consists in computing ranges of light and material parameters that will result in rendered images that are inside a given destination spectral gamut. In their own words, the motivation for their work is the 'growing disparity between the high dynamic range images produced by spectral renderers and the limitations of display gamuts and low-dimensional colour management standards.'

Also in this scene rendering context is the earlier *device-directed rendering* method by Glassner *et al.* (1995), which (although described in a color context) is also applicable in the spectral domain (and the authors themselves acknowledge this). The basic idea here is to adjust light and material parameters by first rendering a scene using the initially desired parameters and seeing whether it results in any OOG values. If it does, then a color-by-color gamut-mapped version of the rendered image is computed and for each surface this results in color differences between its rendering using the original parameters and corresponding IG colors. These differences are then used to compute adjusted parameters and the process is repeated until the whole scene is IG. The authors report that this method converges.

10.5 GAMUT MAPPING FOR SPECIAL APPLICATIONS

Finally, it is also important to at least mention the growing body of gamut mapping work developed for special, niche applications. The intention here is simply to give visibility to color and imaging contexts where the kinds of general-purpose algorithms discussed so far do not provide the desired result or to contexts where it might not be apparent that gamut mapping is considered directly.

Starting with the least esoteric, Braun *et al.* (1999b) developed specific color-by-color GMAs specifically for business graphics. The reason for these needing a special approach is because of their desired reproduction property being different and their reproductions needing 'differentiation among colors, smoothness of sweeps, and purity of the primaries.' The solution that worked best was one where source hues are rotated about halfway towards destination hues (which is similar to Johnson's (1992) hue shift) on top of a simple device mapping where print CMYs are obtained from display RGBs so that $C = 1 - R$, $M = 1 - G$ and $Y = 1 - B$ (for device color values normalized to [0, 1]).

While the GMAs discussed in this book are meant for source and destination media that give access to all hues and have significant hue and lightness combinations available at each hue, there are also printed media that use only two inks – *duotones*. The resulting gamuts are very different in shape from the gamuts of trichromatic (or multi-primary) systems, and gamut mapping to them presents special challenges. Here, Power *et al.* (1996) and Stollnitz (1998: 31–35) present a solution, which notes that the gamut of a two-ink system forms a bilinear surface in color space (i.e. it is lower dimensional than the

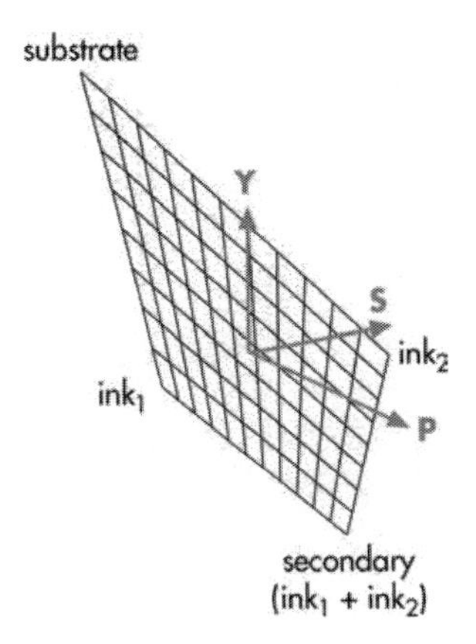

Figure 10.34 The *YSP* space used for mapping to duotone gamuts.

color space). Mapping to that bilinear surface then consists of first representing each source color in the following coordinate system defined by the two-ink system, where (Figure 10.34)

- the first dimension is that of luminance Y;
- the second dimension is orthogonal to Y and closest to the line joining the two inks' colors (labeled 'ink-spreading' S); and
- the third dimension is orthogonal to the YS plane (labeled 'projection' P).

To attempt the preservation of as many differences in the source, the following three-stage mapping takes place in the *YSP* space:

1. The Y range of the source is mapped onto the Y range of the duotone destination using linear, sigmoidal or some other method.
2. The source S range along each mapped Y is then mapped onto the destination range.
3. Finally, the P values of mapped source colors are projected onto the duotone gamut and since at each YS combination the duotone gamut has a single P value the P values of the source are simply set to that value.

Related in terms of the magnitude of gamut difference that needs to be overcome, but quite different in nature, is the work on transforming color images into ones where those with deficient color vision can better distinguish differences in them. Here, too, the destination gamut is like that of a duotone (or monotone), and a solution to the problem of mapping to it has been proposed by Rasche *et al.* (2005). Also aimed at those with deficient color vision is the work of Han (2004), who proposes a hardware-supported, real-time gamut mapping architecture. Instead of discussing how to gamut map from a full trichromatic source to a representation suitable for dichromats or monochromats, this paper instead focuses on optimizing the hardware implementation of any gamut mapping that can be stored in a three-dimensional LUT.

Continuing along the lines of efficient implementation is also Yang and Kwok's (2003) gamut clipping in the YIQ and LSH color spaces (which are luminance-linear and popular in image processing). The key feature of their work is to provide an efficient way of altering saturation S without having to do the full forward and inverse transformation between a colorimetric RGB and the LSH spaces. As a consequence, their implementation is 27% more efficient than a conventional one.

Specific solutions for gamut mapping in the following contexts have also been proposed:

- ray-traced prisms and rainbows (Musgrave, 1989);
- stitching together of several images (e.g. for panoramic views – Pham and Pringle, 1995);
- watermarking (Wu *et al.*, 2004), where a gamut mapping to a convex hull is used to make an error-diffusion operation bounded;
- restoration (Gaiani *et al.*, 2005), where luminance is tone mapped and gamut clipping is subsequently applied (e.g. towards the white point);
- recording CGI effects onto film using a laser recorder (Ishii, 2002) – here, a soft-clipping along lines towards the lightness of the cusp was reported to be successful.

Finally, gamut differences are also an issue in tiled wall displays, where a solution for getting consistent color between displays is to adjust the transfer functions of each display to match them to the central tendency of the entire display set – even though this method is referred to as 'gamut matching' (Wallace *et al.*, 2003) it is a form of calibration rather than gamut mapping as such (i.e. calibrating a set of devices also makes their gamuts similar).

10.6 SUMMARY

To conclude this survey of >90 gamut mapping publications, reviewed in the present and previous chapters, a look will be taken at some of their statistics. The intention here is to shed some light on both the dominant themes in this area in the last 30 years and to highlight some of the variety too.

Looking at how many gamut mapping papers were published per year (Figure 10.35), for example, shows a clear peak around the year 2000, followed by a noticeable dip in activity

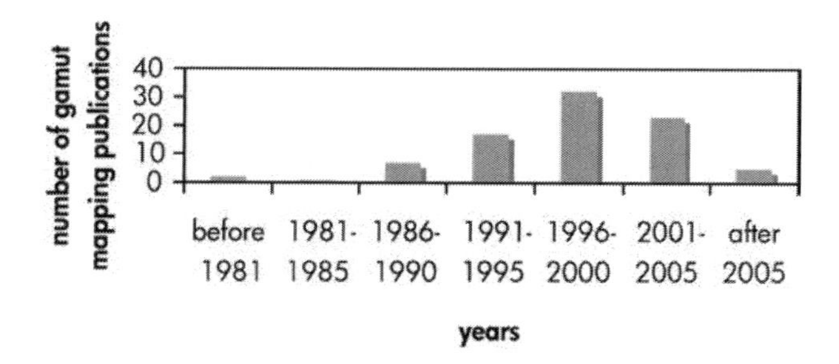

Figure 10.35 Number of gamut mapping publications over time.

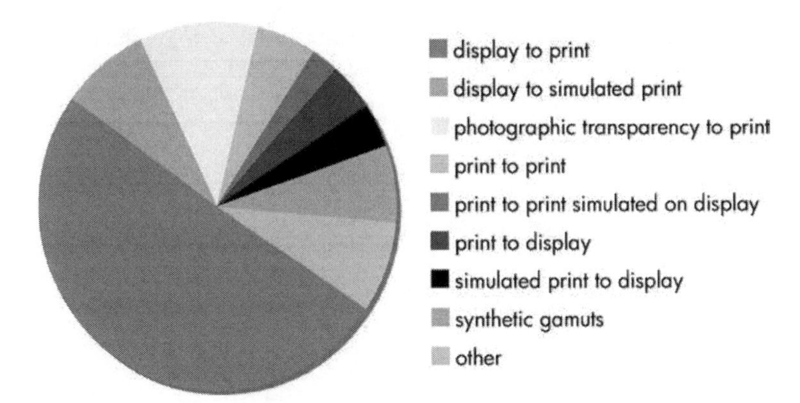

Figure 10.36 Source and destination gamut combinations (simulation stands for simulation on a display).

in this area. The reason for this could have been disillusionment with the diversity and contradictory nature of results from individual studies and the lack of a robust way forward in evaluating alternative solutions (see Chapter 12).

There is also a distinct focus on gamut mapping where only at most the display and print media are involved – 75% of all reported studies belong to this category – and the most frequently used setup by far is one where a displayed original is reproduced in print (Figure 10.36). The reduction versus expansion balance is also heavily in favor of the former, with 88% of studies concerned with how to fit a larger gamut into a smaller gamut.

In an attempt to provide an overview of the gamut mapping work published to date, the present chapter first looked at color-by-color gamut reduction techniques followed by a brief look at gamut expansion. Spatial reduction algorithms were then followed by methods that map between gamuts in the spectral rather than color domain; finally, a brief glance at solutions for niche contexts was provided. Where there is consensus in the literature, an attempt was made to identify it, but what was ominously lacking in this chapter were statements about one approach performing better than others (with some scarce exceptions). The reason for this lies in the tremendous difficulty of evaluating GMAs in a way that allows for comparison between individual evaluation studies, and this will in fact be the subject of Chapter 12. First, though, let us close the loop by addressing the way in which GMAs fit into the various color management approaches described in Chapter 4.

11

Gamut Mapping Algorithms and Color Management Systems

The aim of this brief chapter will be to link the survey of GMAs presented in Chapters 9 and 10 to the color management systems reviewed in Chapter 4 by focusing on what GMAs are applicable and/or most suited to each case. Closed-loop, sRGB, ICC and WCS color management approaches will be looked at in turn and references will be made to relevant work on either integrating or analyzing gamut mapping in color management. Note that it is highly recommended that this chapter be read after Chapters 4, 9 and 10, as it makes extensive references to them and builds on their contents.

11.1 GAMUT MAPPING ALGORITHMS FOR CLOSED-LOOP COLOR MANAGEMENT

In closed-loop color management (Section 4.3) an LUT is provided between source and destination device color space coordinates and gamut mapping can be either implicit (when that LUT is set up manually) or explicit (when it is computed using a color reproduction workflow). In the explicit case the color management architecture only tends to limit the solution to color-by-color algorithms that map between a source and a destination medium gamut and any GMA of this category (e.g. see Chapter 9 and Sections 10.1 and 10.2) can be used. In fact, not even this limitation is inherent in closed-loop systems, and any of the gamut mapping approaches described in Chapters 9 and 10 could be used. This degree of freedom in how to do things is a true advantage of closed systems.

11.2 GAMUT MAPPING ALGORITHMS FOR sRGB COLOR MANAGEMENT

With sRGB color management (Section 4.4) things are not too different in terms of freedom, since the mapping to and from sRGB is unconstrained. Therefore, it is possible for devices to use any of the color-by-color or spatial gamut mapping algorithms from Chapters 9 and 10 when they either map from their device color space to sRGB or vice versa. The only limitation this architecture imposes is that everything needs to pass through the sRGB space (and gamut) in the middle. What it implies most certainly is the inability to use this approach for spectral reproductions (due to the reduction to three dimensions in the interchange space), and it also means that any given source to destination mapping is not controlled by a single system.

Therefore, the dilemma of how much color and or image enhancement to perform arises, since any one device does not know what another device will do. For example, if both a digital camera and a printer apply a chroma boost, then the result will be unpleasantly colorful and the same would also be true of 'enhancing' any attribute that has an optimum preferred level (e.g. contrast).

Since the sRGB architecture does not allow for choosing different desired reproduction properties, it needs to use a single gamut mapping approach to satisfy all types of content and applications. As a consequence, it is unlikely that gamut clipping would be used here; much more likely options are algorithms that both compress and expand gamuts (i.e. Sections 10.1–10.3). Of these, then, it is algorithms that behave differently in different parts of color space that are prime candidates, as they allow for one type of behavior for near-neutral colors and memory colors and another for the gamut's extremes. Therefore, approaches like Spaulding *et al.*'s (1995) *UltraColor* or Zeng's (2006) *spring–primary* are good examples of frameworks that allow for this kind of adaptability, and devices also have the option of supplementing this type of algorithm with spatial gamut mapping post-processing (e.g. as suggested by Zolliker and Simon (2006b)).

11.3 GAMUT MAPPING ALGORITHMS FOR ICC COLOR MANAGEMENT

When considering gamut mapping for the ICC color management framework (Section 4.5) it is important to distinguish among its rendering intents (i.e. desired reproduction properties) and between the specification up to version 2 and the current version 4. Note again that the PCS's use of CIELAB or CIEXYZ does not mean that gamut mapping needs to take place in one of these spaces, but only that its result needs to be encoded.

For the *media-relative* or *ICC-absolute colorimetric rendering intents*, any color-by-color clipping algorithm (Chapter 9) is applicable and what is most advisable here is the use of a weighted ΔE clipping with weights that result in OOG colors being mapped to the most similar looking destination colors (e.g. $[kJ,\ kC,\ kH] = [1,\ 2.6,\ 1.3]$ in CIECAM02 (Morovič *et al.*, 2007)). The only limiting factor here is that gamut mapping needs to be provided for a grid in the PCS (i.e. either *XYZ* or *LAB*), which affects the accuracy of the mapping. Having such a quantized representation also means that it is challenging to address the actual gamut boundary of a device or medium (Bhachech *et al.*, 2006). This is

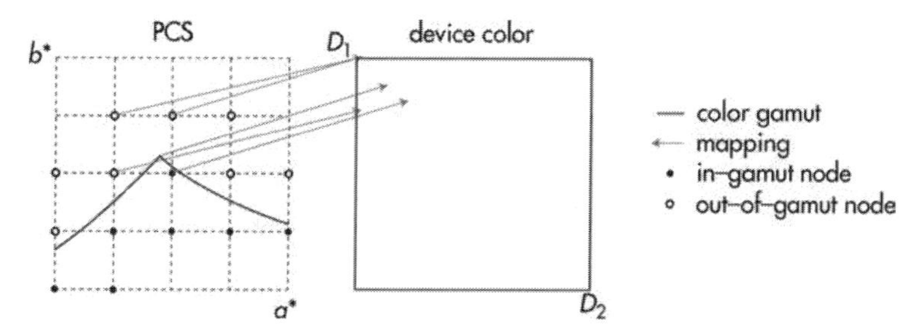

Figure 11.1 Gamut boundary colors get interpolated to device color values inside the color gamut.

because, unless gamut surface colors fall on grid points (and all of them never can), the interpolation process will map the gamut surface into the interior of the device color space and, therefore, of the gamut. Take, for example, the yellow primary of a destination device's gamut. If it happens to land on a grid node, then a source color that matches it can be reproduced as it. However, if that yellow lands between grid points, then it will have two types of neighboring node: some that are IG, containing device color values that correspond to them, and some that are OOG, containing device color values corresponding to gamut surface colors. Interpolating between such neighboring nodes will assign the yellow primary (which, by definition, is on the gamut surface) device color values that will result in IG colors (Figure 11.1). For a solution to this limitation, see Bhachech *et al.*'s (2006) extrapolation algorithm.

The *perceptual rendering intent*, then, is likely to be best served by a gamut compression rather than clipping algorithm, as it is more important here to balance the preservation of detail with that of color than to focus on color alone. Algorithms that work for the sRGB workflow are good candidates, as well as others like SGCK (CIE, 2004d) that have some track record for giving good reproductions of photographic originals. Being compression algorithms, they need to know both the source and the destination gamut, and here we have to consider the ICC v. 2 and v. 4 specifications separately.

In v. 2 there is no source gamut when mapping from the PCS to device color and no destination gamut for mapping from device color to the PCS. This, in turn, is a serious limitation, as gamut mapping algorithms are simply presented with the entire PCS encoding oblong in *LAB* or *XYZ* as either a source or destination. This effectively means that there is no useful gamut at one end of the attempted compression. Note that this lack of color gamut defined in the PCS is not an issue at all for the colorimetric rendering intents, since colorimetric reproduction requires no simultaneous knowledge of both source and destination – the source simply needs to map accurately from device color to colorimetry and the destination needs to take whatever colorimetry it is asked to reproduce and either match it or give the closest reproduction it has available in its gamut (for whichever interpretation if 'closest' is wanted). Rather than giving up on compression, what is typically done in practice is to choose a gamut for the PCS to which or from which to map and to provide clipping from the remainder of the PCS to that gamut. Typically, it is sRGB or Adobe RGB that is chosen, and if the source happens to be in that gamut, then good results can be achieved. The problem arises when the assumption of PCS gamut made at one end (e.g. assuming that the source is sRGB in an output profile) does not match the state at the other (e.g. having an Adobe RGB source profile).

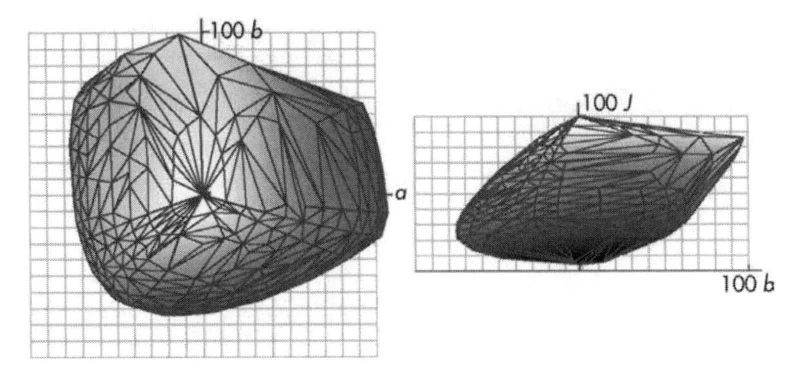

Figure 11.2 The reference medium gamut in CIECAM02.

To address this issue in the v. 4 specification, the ICC have defined a gamut in the PCS for use with the perceptual rendering intent. This gamut is called the *reference medium gamut* (RMG; Figure 11.2), is derived from measurements of prints made using various technologies and has been standardized by the ISO (2006). Having this single, explicit gamut defined in the PCS results in a situation like that of sRGB workflows, where color transformations pass through a specific, intermediate color gamut and the ambiguity of v. 2 is removed. For more on the details and implications of the RMG, see Holm *et al.* (2006).

The saturation rendering intent is the least frequently used one in the ICC context, and the most suitable GMAs for are ones like those presented by Braun *et al.* (1999b).

Finally, it is also worth noting the work of Büring *et al.* (2000), who have explored the concept of *inverse gamut mapping*, where the mappings from device color space to the PCS would mirror (i.e. be better invertible with regard to) the mappings encoded in the PCS to device color LUTs of ICC profiles. Shimizu *et al.* (1998) then describe an approach that adapts GMAs defined in continuous color space (i.e. all of those in Chapters 9 and 10) for implementation in quantized LUTs to get the most out of the available encoding resolution.

Note also that any implementation of gamut mapping in an LUT (as is the case here in the ICC approach, as well as in many instances of the closed-loop and sRGB approaches) will mean that the result will differ from what it would have been had the algorithm been applied directly to source colors. The consequences, though, can be both negative (i.e. the mapping is not exactly what was intended in the algorithm and some color's – mainly primaries and other extremes on the gamut surface – may not be mapped as desired) and positive (as LUT implementation can effectively result in smoothing of transitions).

11.4 GAMUT MAPPING ALGORITHMS FOR WCS COLOR MANAGEMENT

The Microsoft WCS (Section 4.6) allows for the implementation of any of the color-by-color algorithms in Chapters 9 and 10, since it provides the gamut mapping stage with information about both source and destination gamuts. By default, WCS comes with three *baseline* gamut mapping algorithms: *minimum color difference* (*MinCD*), *BasicPhoto* and *Hue Mapping* implemented in CIECAM02 (Microsoft, 2007).

Here, *MinCD* is a clipping algorithm that uses a weighted ΔE minimization whereby the weights for ΔJ on the one hand and ΔC and ΔH on the other hand are different and a function of C. Here, the weight for ΔJ is small for low chromas and then increases until it reaches unity at a predefined threshold as follows:

$$w_J = k_2 - k_1(C - C_{\max})^n \tag{11.1}$$

where $k_2 = 1,\ k_1 = 0.75/(C_{\max})^n,\ C_{\max} = 100,\ n = 2,\ C$ is the source color's chroma and $C_{\max}$ is the maximum chroma in the source (Bourgoin, 2007). Like in the ICC framework, there are absolute and relative variants of this algorithm.

The *BasicPhoto* algorithm is meant for preferred color reproduction and has been derived from SGCK (CIE, 2004d), which starts with a chroma-dependent, sigmoidal lightness compression followed by a knee mapping along lines towards the lightness of the cusp on the lightness axis with a knee at 90% of the range. *BasicPhoto* differs from SGCK by using knee mapping along lines of constant lightness after the initial lightness mapping. Also, the black points used by *BasicPhoto* are the darkest colors in the two gamuts, rather than the darkest neutrals as SGCK specifies.

Finally, the *Hue Mapping* algorithm is targeted at the applications for which the ICC specifies the saturation rendering intent. The first stage here is a hue change that moves source hues depending on the hues of the source and destination primaries and secondaries, like in the case of Johnson's (1992) CARISMA algorithm. Linear lightness mapping is applied next (Equation (10.1)), followed by a 'shear' mapping that is like the TRIA algorithm (Morovič, 1998). Since the resulting intermediate mapped colors are still likely to be OOG, further modification is applied where lightnesses are moved towards the lightness of the destination cusp's lightness and any remaining differences are finally clipped to the destination gamut (Microsoft, 2007).

What is most attractive in WCS from the point of view of gamut mapping is that it allows for the provision of gamut mapping solutions by third parties in the form of '*gamut map models*'.

11.5 TYPES OF COLOR MANAGEMENT CONTEXT

Looking at the above color management architectures, we can identify three types of context in which gamut mapping is applied (Figure 11.3):

1. *Gamut mapping from unknown source to known destination gamut*, e.g. ICC v. 2 perceptual rendering intent or any ICC colorimetric rendering intent BToA tags. Here, a gamut mapping needs to be provided from an encoding of color space (e.g. the ICC PCS oblong) to a known destination gamut.
2. *Gamut mapping to/from intermediate gamut*, e.g. sRGB, ICC v. 4 perceptual rendering intent. Here, the mapping between a source and a destination is the result of combining two gamut mapping operations: from source to intermediate and from intermediate to destination.
3. *Gamut mapping from source to destination gamut*, e.g. closed-loop color management, WCS. Here, both source and destination are known at the time of mapping between the two.

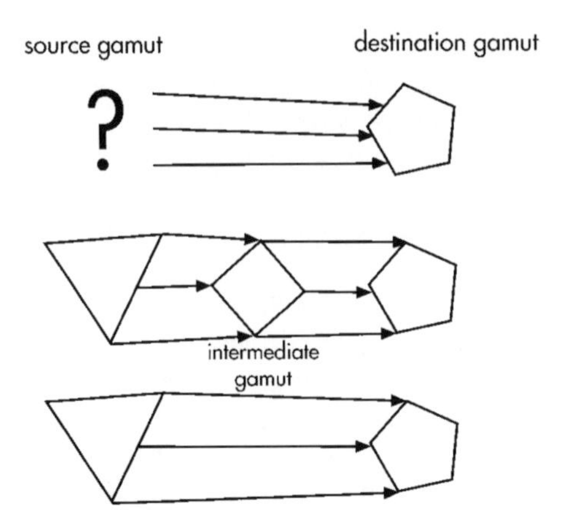

Figure 11.3 Alternative gamut mapping contexts in color management. Top: unknown to known; middle: via intermediate; bottom: between known gamuts.

11.6 SPATIAL AND SPECTRAL GAMUT MAPPING

Given the above discussion of gamut mapping in color management systems, it is worth noting that the majority of color management is done on a color-by-color basis using LUTs.

When spatial gamut mapping is desired in these contexts, it is most directly applicable to the closed-loop and sRGB cases, since here it is up to the system or device provider to do whatever they like. In the ICC and WCS contexts, spatial gamut mapping is not supported by the frameworks themselves, but it is possible to use the color-by-color mappings they allow for in a separate application that then provides an extension to spatial processing.

The situation is more complex for spectral gamut mapping, which cannot be provided in the sRGB, ICC or WCS frameworks and would only directly fit into the closed-loop case (due to its inherent flexibility and lack of specific definition). Here, it is worth noting the efforts to define an interchange space for spectral color reproduction in the form of *LabPQR* (Derhak and Rosen, 2004), which could be used instead of the ICC PCS and allow for spectral data interchange in six dimensions.

11.7 CONCLUSIONS

An attempt was made here to discuss the relationship of alternative color management approaches and the GMAs discussed in Chapters 9 and 10. It was shown that there are currently three alternatives in terms of what gamuts an algorithm can map between when implemented in color management. Finally, some thought was given to spatial and spectral gamut mapping.

12

Evaluating Gamut Mapping Algorithms

Having a vast variety of proposed gamut mapping solutions (Chapters 9 and 10) is certainly a good starting point, but what is needed to go beyond such heterogeneity is some way of judging whether the latest algorithm outperforms its predecessors.

To illustrate this, let us consider an example from another area of color science, i.e. that of color difference ΔE equations. Here there is a body of psychophysical data consisting of measurements of pairs of stimuli (in terms of spectral power, reflectance or tristimulus values) and corresponding color difference judgments made for them by groups of observers. Existing ΔE formulae are then tested against this data set and their performance can be quantified by comparing their predictions with observer judgments. Any new ΔE proposal can then be easily compared against previous solutions in terms of its predictive powers (Figure 12.1). There is, therefore, a sense of progress in color difference prediction in that today's ΔE formulae more accurately predict color difference judgments than those of 30 years ago.

Turning to gamut mapping, we are presented with a dramatically different view, as there is no data set against which new algorithms can be tested and neither is there a sense of progress, as it cannot be said in general that the latest GMAs perform better than those proposed 20 years ago (although there are some specific exceptions that will be addressed later). All that individual evaluation experiments tend to say is how the algorithms that were simultaneously compared performed relative to each other (Figure 12.2).

In this chapter we will start by examining why gamut mapping lacks a reference data set, followed by an overview of how GMAs have been evaluated in the past, the limited conclusions that can be drawn from existing evaluations and finally the CIE's *Guidelines for the Evaluation of Gamut Mapping Algorithms* (CIE, 2004d). In the second half of the chapter we will look at two significant examples of how specific properties of GMAs can be studied directly. This will consist of overviews of Sun's (2002) systematic study of the influence of source image characteristics on the performance of GMAs and Zolliker and Simon's (2006a) evaluation of GMA continuity.

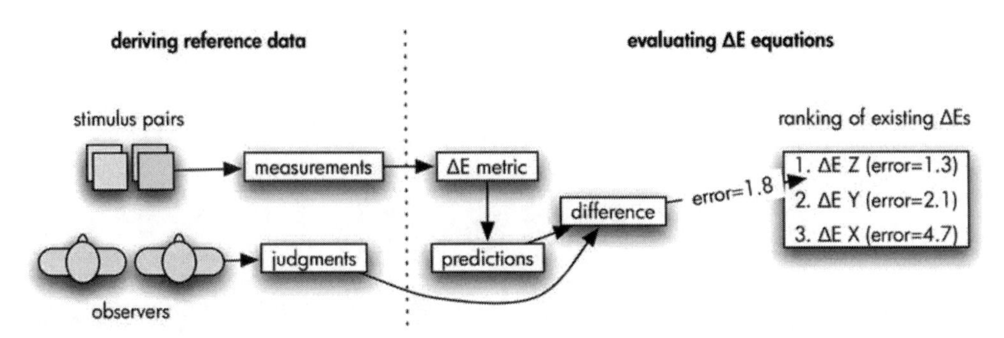

Figure 12.1 Evaluating color difference ΔE equations.

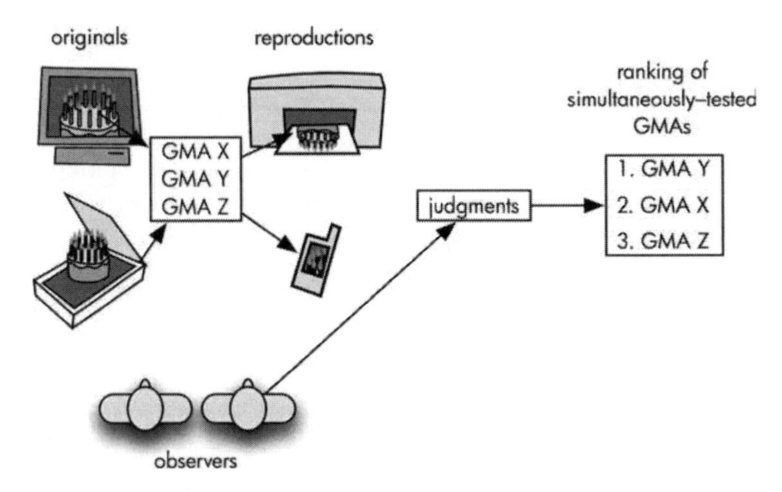

Figure 12.2 Evaluating GMAs.

12.1 WHY IS THERE NO REFERENCE DATA SET IN GAMUT MAPPING ALGORITHM EVALUATION?

Looking at the color difference example, we can identify the following aspects of its evaluation:

1. *Manageable population of inputs.* The population consists of all possible pairs from among all possible uniformly colored stimuli presented on top of each possible stimulus in turn and having all possible spatial arrangements of the two stimuli against their background (e.g. stimulus pair touching, having an only just perceptible gap, having a large gap). Combinatorially, this is a vast population, but it is one where there are a small number of dimensions: background, first stimulus, second stimulus and distance between stimuli.

2. *Easily quantifiable inputs.* Stimulus pair and background can be expressed in tristimulus (or spectral) terms and the gap between stimuli can be measured too.
3. *Relatively observer-neutral judgment.* The experience of color difference magnitude is relatively unaffected by the past experiences and professional needs of an observer (i.e. photographers and estate agents are likely to judge a similar difference ratio given two pairs of stimuli – they are likely to differ in the significance of those differences though, but that is another matter).
4. *Relatively few evaluation techniques that can be interrelated.* Even though there are alternatives for how to perform color difference evaluation, they relate well to each other and, given results using different approaches, it is possible to compare them at least approximately.
5. *Easy evaluation of ΔE equation's predictive powers.* Given existing inputs, a ΔE equation can predict the differences between them and these can, in turn, be compared with difference magnitudes judged by observers. There is, therefore, no need for costly and time-consuming new psychophysical work when a new ΔE is to be tested.

All existing ΔE equations, therefore, can be easily ranked in terms of their predictive powers and the evaluation of new proposals can be readily automated. Note that this is clearly somewhat of a simplification and idealization of how ΔE equations are evaluated, but the general features are believed to be correct and serve the purpose of contrast with the *status quo* of gamut mapping evaluation.

The case of gamut mapping evaluation is very different from the color difference one in all of the aspects listed above:

1. *Population of inputs is significantly greater and not well characterized.* Here, the inputs consist of:
   (a) *Any possible source image.* This alone is a dramatic source of undersampling in any particular evaluation experiment, and it is currently not even possible to represent the population of this aspect systematically (unlike the color difference case, where one can at least identify the gamut of all possible colors).
   (b) *Any possible source–destination medium combination.* Given any single source image, its reproduction will differ on different destination media (seen under their viewing conditions), and how it is mapped will also be affected by the source medium (and its viewing conditions) in which it is present.
2. *Any possible choice of desired reproduction property.* Note also that this is an open-ended list, as new reproduction properties can be defined as and when necessary.
3. *Quantification of inputs involves more resources.* Since inputs are images and medium gamuts, they cannot be represented as simply as uniform color stimuli.
4. *Judgments are observer dependent.* The strength of this claim varies depending on the choice of desired reproduction property, and while observer dependence is more like the color difference case for accurate reproduction, it differs strongly when preferred reproduction is the aim. Here, the preferences of a fine artist are likely to be very different from those of a forensic scientist, and having judgments by one will say very little about what the other would have said.
5. *Huge variety in evaluation setups and the use of ratio-scale data.* As will be shown later in this chapter and as was already hinted at in Chapter 3, there is tremendous variety in

GMA evaluation. This, in turn, makes the 'scores' that a GMA receives in one evaluation essentially incomparable with the 'score' another GMA receives when evaluated differently. What also adds to the difficulty is that many GMAs are evaluated only in relative terms, i.e. whether they are better or worse than other GMAs against which they were compared and by how much. This can result in two sets of GMAs having identical relative scores with one set consisting of well-performing algorithms while the other set's GMAs perform badly.

6. *Past data cannot be used to evaluate new algorithms.* Since past evaluations are about the extent to which a reproduction made using a past GMA exhibited a chosen reproduction property, they cannot be compared directly with a new GMA. Given a new GMA, a past source image will be mapped to a new destination one and it is necessary again to perform a psychovisual evaluation of the extent to which it exhibits the chosen reproduction property. The reason for this is that GMAs are not predictive (i.e. resulting in a prediction of what an observer will say), but prescriptive (i.e. specifying what should be done to source colors given medium gamuts).

A consequence of this nature of GMA evaluation is a substantial body of literature reporting how various small sets of GMAs (typically around five at a time) compare with each other when applied typically to the reproduction of around five images for a single pair of source and destination media. Since each of these sets of results is obtained for such different conditions, it is not possible to judge from them whether GMA X from experiment A performs better than GMA Y from experiment B. To do that it would be necessary to set up a psychovisual experiment C where GMAs X and Y are evaluated simultaneously and under the same conditions. This, however, is very time and resource intensive and is a major hindrance on the way towards improved gamut mapping results.

A reasonable objection to this picture might be to say that it should be possible to *predict computationally* the extent to which a reproduction exhibits a chosen reproduction property especially if the property is visual similarity. Given colorimetric pixel-by-pixel data about source and gamut-mapped destination images plus information about viewing conditions, it should be possible to predict observer responses. This is certainly a very attractive idea and one that will hopefully become reality in the future. However, at present there is only a catalogue of failures of various degrees that just highlight the difficulty (e.g. Morovič and Sun, 2003; Bando *et al.*, 2005).

While it is unlikely that gamut mapping will ever be evaluated like ΔE equations are, Section 12.4 will present the CIE's *Guidelines for the Evaluation of Gamut Mapping Algorithms* (CIE, 2004d) that aim at having the results of individual evaluation experiments at least interrelatable. First, though, the degree of variety in gamut mapping evaluation will be made explicit by surveying the literature published to date.

12.2 CHARACTERISTICS OF PUBLISHED GAMUT MAPPING ALGORITHM EVALUATION

To drive home the point about how varied GMA evaluation is, this section will present some statistics about the approaches described in the >90 publications where the GMAs discussed in Chapters 9 and 10 were presented and evaluated.

There is a great variety in how many GMAs were compared with each other in the various reported cases and it is important to note that it is often only a single GMA that is

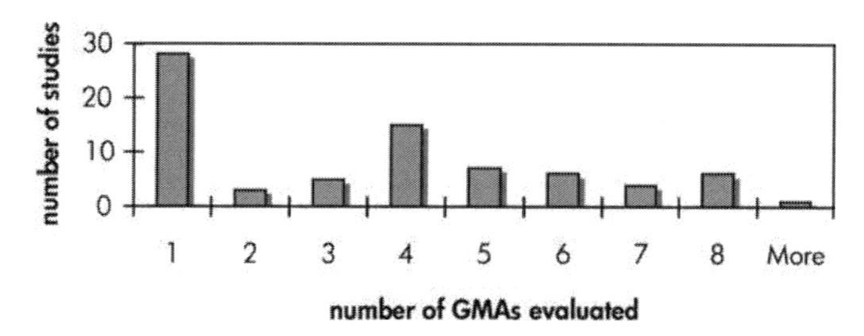

Figure 12.3 Number of GMAs evaluated simultaneously in published studies.

proposed and evaluated in isolation, without comparison with alternatives. When alternatives are considered, there is a great deal of variety in how many are used in any one study (Figure 12.3). On the other hand, there is an overwhelming focus (84%) on judging reproductions in terms of *accuracy* (defined in a variety of ways) versus preference.

Looking at the methods of evaluation in these studies, we can see that 59% do use rigorous psychovisual procedures, 17% report observations from informal visual evaluation and the remaining studies stop at simply proposing a new GMA without providing an indication of its performance.

Of the cases where psychovisual methods are used, pair comparison is by far the most popular (83%), with category judgment (10%) and ranking (7%) making up the rest. An important consequence is that the vast majority of existing psychovisual data about gamut mapping performance only gives information about how algorithms performed relative to each other within a single study. This, in turn, makes interrelating results from separate studies impossible in general.

The final aspects to note are the numbers of observers and test images used (Figure 12.4), where it can be seen that the typical study involves only between one and

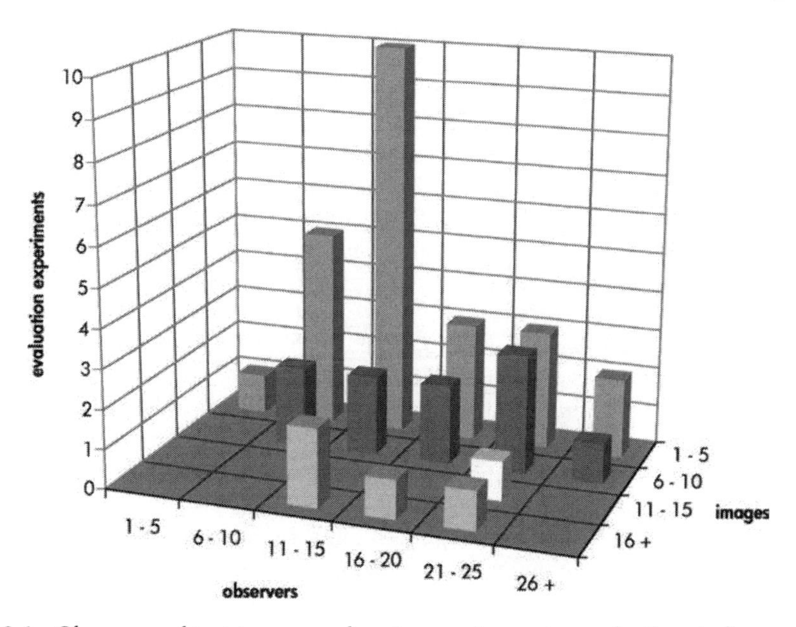

Figure 12.4 Observer and test image numbers in gamut mapping evaluation studies.

five test images. There are only a handful of cases where 16 or more test images were processed using GMAs and visually evaluated.

The risks of using such small numbers of test images were shown to be such as to make the results of the evaluation more strongly dependent on the choice of test images than of GMAs. Given a set of 15 test images used for evaluating six algorithms, six subsets of five images could be chosen to have each of the algorithms in turn in the top performing group (Morovič and Wang, 2003b). In the years following this study, the number of test images used in gamut mapping evaluation has increased and there are examples of using 15 (Bonnier *et al.*, 2006), 20 (Dugay *et al.*, 2007) and even 250 (Zolliker *et al.*, 2005), which is likely to lead to more robust results.

12.3 WHAT CAN BE SAID ON THE BASIS OF EXISTING EVALUATION RESULTS?

As has been suggested in Chapters 9 and 10, it is possible to identify some trends when looking at GMAs, and their evaluation and a summary of these will be provided next.

In the case of *gamut clipping to the visually most similar destination color,* evaluation is much more like that of color difference prediction and there is greatest consensus here. The existing literature exhibits remarkable convergence and it is universally reported that the preservation of chroma can be traded of more than that of lightness and hue by a factor of about two. The latest study here reports significantly better results when performing the weighted color difference minimization (Equation (9.1)) in CIECAM02 than in CIELAB and recommends dividing weights of [kJ, kC, kH] = [1, 2.6, 1.3] (Morovič *et al.*, 2007), which agrees well with the previous work (Katoh and Ito, 1996; Ebner and Fairchild, 1997; Wei and Sun, 1997; Wei *et al.*, 1997).

This level of agreement is only found in three other cases:

1. *color-by-color gamut expansion,* where lightness expansion followed by chroma expansion along lines of constant hue and expanded lightness is repeatedly reported to perform well;
2. *spectral gamut mapping,* where the use of representing reflectances using linear bases where the first three ('fundamental bases') account for the color signal under chosen illumination and the remaining ones are 'metameric blacks' is very popular; and
3. *spatial gamut mapping,* where algorithms using spatial operations are universally reported to outperform their color-by-color counterparts.

However, at least the first two are likely to be due to their relative immaturity and small number of published studies available to date.

Concerning color-by-color gamut compression algorithms (which have been studied most so far), a list of trends was given in Section 10.1.4 and consisted of the following: making gamut mapping more adaptive to source and destination gamuts, sacrificing some detail preservation to preserve more contrast, maintaining a core region of the source gamut where no change is made, mapping nonlinearly along mapping paths, mapping from source image rather than medium gamut and categorizing colors. Note, though, that these are merely trends rather than instances of convergence and independent verification.

What is most notable as regards the work published on color-by-color gamut compression is instead the lack of convergence, which is believed to be partly due to the great variety of how individual GMA evaluations were conducted and the resulting impossibility to interrelate their results.

12.4 CIE GUIDELINES

In a response to the above situation, the CIE has recommended a specific way of GMA evaluation that is geared towards making the results of individual studies interrelatable. Most important, the CIE's (2004d) *Guidelines for the Evaluation of Gamut Mapping Algorithms* specify how to report aspects of GMA evaluation and make the use of a specific, single test image and a pair of GMAs obligatory.

The role of the obligatory 'Ski' test image – available at http://www.colour.org/tc8–03/ (Color Plate 1) – and the obligatory pair of GMAs (both to be performed in CIELAB):

1. HPMINDE – hue preserving minimum ΔE and
2. SGCK – chroma-dependent sigmoidal lightness mapping followed by knee scaling towards the cusp

results in an anchor that allows interrelating experimental results. If two experiments, where whatever GMAs are tested on whatever images, also include this image and GMA pair, then there is a basis for interrelating them by taking the scores for the anchor pair to be the same. For example, in Figure 12.5 we see the results of two pair comparison

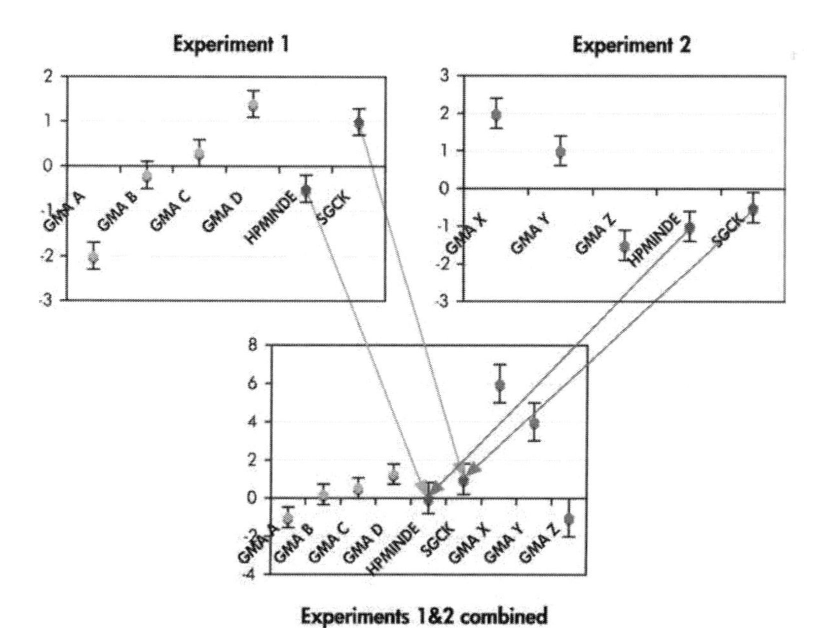

Figure 12.5 Interrelating experimental evaluation results.

experiments evaluating different sets of GMAs. Without the anchor pair and application of the test GMAs to the 'Ski image,' the GMA scores cannot be interrelated at all, but with the presence of HPMINDE and SGCK in both experiments the scores for the other algorithms can be placed on the same scale. Note that this is strictly only possible when the exact same experimental conditions are used (especially the same source and destination media), but the process yields at least approximate results even when that is not so.

Finally, it is important to point out that the question of how to interrelate experiments given the presence of an anchor pair is still only loosely defined and needs to be tested and further refined. Also, it is worthwhile emphasizing that the HPMINDE and SGCK algorithms are used in the guidelines exclusively to provide an *anchor* for bringing the results of multiple experiments closer together and that they in no way represent a recommendation by the CIE as baseline or default GMAs.

In addition to the image and GMA pair, the guidelines specify how to report details of the following aspects of GMA evaluation:

1. *Test images.* Apart from the obligatory 'Ski' image, the guidelines specify that at least three additional test images must be used and that they need to be made available to the CIE in electronic format in terms of the colorimetric data used as input to the evaluated GMAs.
2. *Media.* Mappings between reflective print, transparencies, displays and virtual media (e.g. wide gamut color spaces) are foreseen, and only media that have acceptable uniformity, repeatability and viewing geometry independence characteristics and that can be successfully characterized are recommended for use in the experiments.
3. *Viewing conditions* are specified in detail in the guidelines in terms of illumination, viewing environment, border and surround and specific instructions for reflective, transmitting and self-luminous media. Three alternative viewing conditions are specified for comparing images between print and displays, ranging from an absolute colorimetric match between the medium white points to each medium being viewed in a way that is standard for it in isolation.
4. *Measurement.* In general, measurements should be taken under conditions that are as similar as possible to those in the experiment where the media will be used. To comply with the *Guidelines*, details of the instruments and measurement procedures used need to be given and recommended procedures are provided for various media.
5. *Gamut boundaries.* The method used for computing the gamut boundaries of source and destination color reproduction media needs to be reported. Gamut boundaries themselves need to be reported in CIELAB in terms of lightness versus chroma plots at the primary and secondary hue angles of the destination medium and the projection of the gamuts onto the a^*b^* plane also needs to be shown. CIELAB needs to be used for this reporting regardless of the color space in which the gamut mapping is carried out.
6. *GMAs.* Source code for HPMINDE and SGCK is provided in 'C' (http://www.colour.org/tc8-03/), and the details of other GMAs evaluated in an experiment must be reported.
7. *Mapping color space.* Here, all that is obligatory is to report what color space was used, including values of its parameters. The guidelines also recommend the CIECAM97s

space, following Moroney's (2000b) guidelines for setting its parameters, and IPT (Ebner and Fairchild, 1998b) as the hue-corrected color space of choice.

8. *Experimental method.* What needs to be reported here are: training given to observers, script (or task description), choice of observers, reliability and repeatability of observers, viewing technique, user interface (if any) and method of data analysis used. At least 15 observers and the use of one of the psychovisual methods of pair comparison, category judgment or ranking are obligatory.

Finally, the guidelines also provide advice on how they can be applied to three typical workflows: the ROMM wide gamut color space (Spaulding *et al.*, 2000) to print, CRT to print and transparency to print. As there are a large number of instructions that need to be adhered to, a checklist is also included to help determine whether a given experiment complies with the guidelines.

Since their publication in 2004 the guidelines have been very widely adopted, so that essentially all published GMA evaluation conducted since then adheres to them either in full or at least in its key aspects. Some examples are the following studies: Fukasawa *et al.* (2005), reporting the results of nine printer manufacturers evaluating GMAs for office printing; Zolliker *et al.* (2005), using 250 test images to evaluate GMAs; Bonnier *et al.* (2006), comparing spatial GMAs.

With this potential for consistency in evaluation in place there is a growing possibility of arriving at a set of algorithms that could serve as baseline solutions for gamut mapping. What is meant by 'baseline' here is the potential to deliver *acceptable* solutions, since it is unlikely that a small set of algorithms could be defined that would be *optimal* for all combinations of source, destination, desired reproduction property and application. It is much more likely that, even if baseline algorithms existed, there would still be specialized solutions that would outperform them in any particular case. This might not make the prospect of baseline algorithms sound very attractive, but it should be borne in mind that, today, the choice of GMA is a very complex and trial-and-error intensive task. Having at least a reasonable starting point, which a baseline would provide, would allow both for a broader use of explicit gamut mapping, where color reproduction is less critical, and for more time to be spent on optimization, where maximizing color reproduction performance is essential. A baseline algorithm would also allow for the provision of similar color reproduction across different devices, which would be of value in, for example, multimedia advertising.

12.5 STUDYING SPECIFIC PROPERTIES OF GAMUT MAPPING ALGORITHMS

Instead of taking a set of GMAs and trying to determine how well they deliver the chosen desired reproduction property, another way of evaluating gamut mapping is to quantify either how specific factors affect it or to what extent its output has chosen properties. In this section we will look at one example each: first, a study that tried to understand what it is about images that makes the success of GMAs image-dependent and, second, a study that quantified the continuity of GMA transformations.

12.5.1 Image dependence

A key limitation of current GMAs is that their performance is dependent both on the magnitude and nature of source–destination gamut differences and on what images are being reproduced. Even given a single source–destination gamut pair and an algorithm that is best in class for some images, it may fail badly for others. This image dependence has long been known in gamut mapping research, and every single evaluation experiment published to date reports differences between how GMAs perform for different images. Taking the per-image GMA scores from published work where those data are provided, computing coefficients of determination r^2 between each image's GMA scores and those of all other images in turn and reporting the mean of these coefficients for each experiment results in the picture in Table 12.1. In the vast majority of cases the results for individual images have very weak correlation (with a few exceptions that are due to either just similar variants of a GMA being compared or due to very similar images being used), with the mean coefficient of determination for the experiments shown here being $r^2 = 0.40$.

The question then arises of what it is about images that makes GMAs perform differently for them, and there is a clear tendency in many studies to attribute this to 'image type,' e.g. whether the image is an outdoor landscape, a portrait, or a business graphic. Such hypotheses, however, were arrived at on the basis of seeing how GMAs performed for images that differed in terms of many of their attributes – only one of which was their 'image type.'

To address this lack of understanding, Sun (2002) describes a series of experiments in which a number of increasingly more complex image characteristics were tested in terms of their effect on GMA performance. The key to Sun's experiments is their use of 'artificial' test image sets, where all images are identical in terms of a chosen property (i.e. gamut, lightness or chroma histogram or full three-dimensional color histogram) in each such set, and their comparison to a set of regular images, differing in all characteristics. Given such image sets it is possible to see whether GMA performance differences are reduced when differences in a chosen characteristic are removed. For example, does removing image

Table 12.1 Inter-image GMA score correlation in 21 psychovisual experiments.

Reference	Data source	Mean r^2
MacDonald *et al.* (1995)	Appendix	0.10
Wei *et al.* (1997)	Tables 2 and 5	0.51, 0.33
Morovič (1998)	Appendices D, E, F (glossy & plain paper)	0.47, 0.67, 0.43, 0.16
Braun (1999)	Appendices G (CIELAB/H&B/E&F), P (Xpress/MajestiK), S and W	0.46, 0.41, 0.57, 0.91, 0.61, 0.61, 0.25
Braun *et al.* (1999b)	Figure 5	0.08
Katoh and Ito (1999)	Figure 7	0.64
MacDonald *et al.* (2001)	Figure 7	0.28
Motomura (2000)	Figure 7	0.21
Newman and Pirrotta (2000)	Tables 2–4	0.23
Chen *et al.* (2001)	Table 8	0.35
Zeng (2006)	Figure 10	0.16

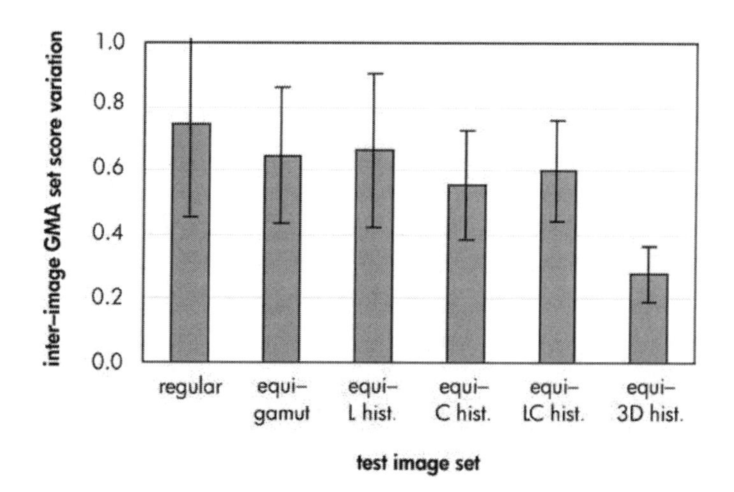

Figure 12.6 Effect of test image set properties on inter-image GMA performance differences.

lightness histogram differences (Color Plate 2), while still leaving other differences (including image type), make GMAs perform more similarly for the individual images than when gamut differences are present?

Following the evaluation of the effect of gamut, lightness histogram, chroma histogram, lightness and chroma histogram and finally three-dimensional color histogram, Sun found that there was a significant reduction in inter-image differences for GMA performance only when the whole set of test images shared the same full, three-dimensional color histogram (Figure 12.6). Image sets with the same three-dimensional histogram had about two-thirds less variation than regular test image sets, which suggests that gamut mapping decisions could be successfully made if the color histograms of source images were considered (Sun and Zheng, 2004). These findings, of color statistics playing a dominant role in image reproduction also agree with Montag and Kasahara's (2001) work on the role of color in overall image quality and with Sun and Morovič's (2002) experimental investigation of what image aspects observers consider when they compare a reproduction with its original. The results also point to one-third of variation not being due to color-statistical properties of images, but instead being consequences of spatial and cognitive features, which also need to be dealt with to achieve fully automatic color reproduction.

12.5.2 Continuity

While some aspects of the gamut mapping transformation, such as whether an algorithm is going to preserve a source color's hue predictor in the chosen mapping space, are explicit in an algorithm's design, others are implicit. One such aspect is the continuity of the transformation, i.e. whether small changes in the source also result in small changes in the destination or whether, in the discontinuous case, small source changes in some cases

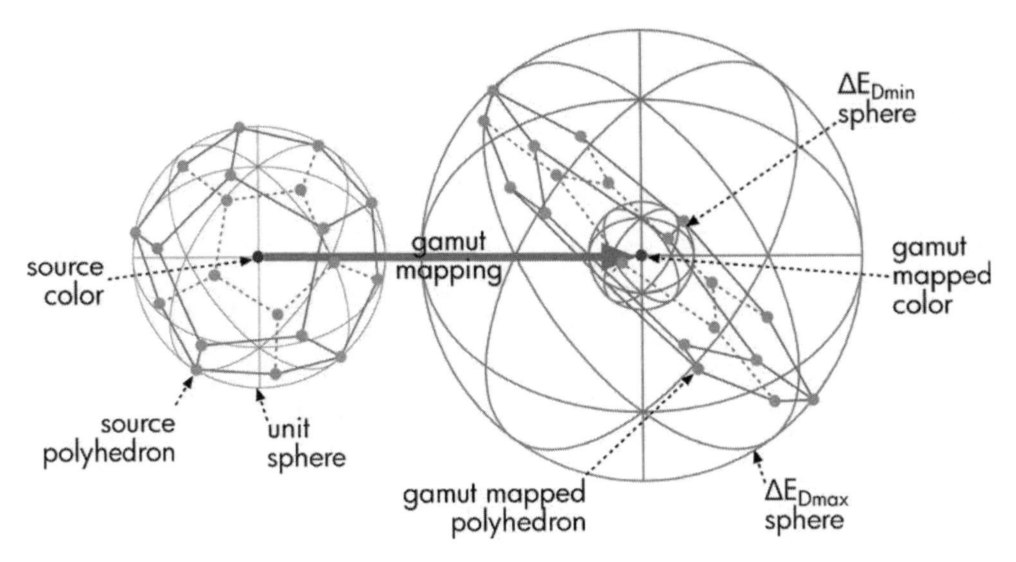

Figure 12.7 Example of gamut mapping a source color and its unit sphere inscribed polyhedron.

result in large destination changes after the mapping is applied. Zolliker and Simon (2006a) have proposed the following procedure for quantifying it:

1. Randomly pick a color from the source gamut such that a unit sphere with it at the center is also entirely in gamut.
2. Approximate the unit sphere by a regular polyhedron.
3. Map the chosen color (from step 1) and the vertices of its enclosing polyhedron (from step 2) using the GMA whose continuity is to be tested.
4. Compute the distances between the mapped chosen color and mapped polyhedron vertices and find their minimum $\Delta E_{D\text{min}}$ and maximum $\Delta E_{D\text{max}}$. Also compute the mapped polyhedron's volume (plus the radius of a sphere with the same volume ΔE_{Vol}) and its surface (plus the radius of the sphere with the same surface ΔE_{Srf}) (Figure 12.7).

Repeating this computation for a number of samples then yields frequency histograms for all of the distances computed in step 4 where '$\Delta E_{D\text{max}}$ [values] much larger than one are an indication for discontinuities[, ...] clipping to the surface results in $\Delta E_{\text{Vol}} = 0$ and $\Delta E_{D\text{min}} = 0$, clipping to an edge additionally yields $\Delta E_{\text{Srf}} = 0$, and for clipping to a corner we even find $\Delta E_{D\text{max}} = 0$.' Alternative algorithms can then be compared in terms of such histograms, and the frequencies of having zeros in any of them express the likelihood and nature of the resulting clipping.

12.6 SUMMARY

The majority of GMA evaluation published to date shows great heterogeneity in the methods used and does not provide a reliable basis for identifying solutions with the greatest potential for success. To address this lack of interrelatability of individual studies, the CIE published guidelines for evaluating GMAs that has the potential for providing a solid basis to identifying baseline solutions to the gamut mismatch problem. In addition to GMA evaluation methodology, this chapter also presented work done on identifying what it is about images that makes an algorithm succeed or fail for it, which has found that it is predominantly an image's full color histogram that is the key here. Finally, a method for studying the continuity of gamut mapping transformation was also presented.

13
Conclusions

The intention of the preceding chapters was to convey a sense of the current understanding of how differences in color ranges are dealt with when attempting the reproduction of visual content. Since gamut mapping is a very active area of academic research and industrial engineering, the focus was more on exposing the landscape at whose heart gamut mapping operates, rather than to provide a cookbook with recipes for currently popular flavors that may become last-season soon.

Chapter 2's whirlwind tour of color science attempted to position color firmly in itself as the result of highly complex, time-variant interactions between a viewer (with their individual physiological and neurological makeup and sum total of past visual experiences), objects in their environment (with their manifold color formation mechanisms) and illumination and viewing conditions. This was done to dispel any simplistic notions of color as a physical property (and consequently as something that can be measured directly and is the property solely of objects). Such a notion would be highly counterproductive in the context of color reproduction and had to be tackled right from the start.

Color reproduction was then analyzed conceptually in terms of its constituent parts – color information, media, viewing conditions, transformation and desired reproduction property – in Chapter 3 and the question was posed of what the desired results of color reproduction can be. The reason for their detailed treatment is that the choice of how a reproduction should relate to its original is then the yardstick by which an entire color reproduction effort is judged. Leaving it implicit is simply reckless and can lead to a fluctuation of the criteria by which color reproduction is judged in practice. This, in turn, seriously undermines any attempts that one may make to optimize color reproduction performance. Two alternatives of how a given color reproduction's quality can be judged, i.e. a measurement-based one and a psychovisual one, were then discussed at some length.

Chapter 4 zoomed in on the transformation part of color reproduction, presented both a theoretical and alternative practical views and emphasized that the end points of any color reproduction process are color experiences elicited by physical stimuli resulting from imaging devices controlled using digital data, rather than such digital data itself.

Having set the scene, gamut mapping was introduced in Chapter 5, with an emphasis on sketching its features in rough strokes that were then refined in the remainder of the book.

The intention here was to define the alternative terminologies used by the CIE versus ISO, to shed light on some misconceptions and then to sketch out the basic components of GMAs and the factors that affect their performance.

Issues related to the use of color-predictive spaces for the implementation of gamut mapping were reviewed in Chapter 6, with an emphasis on spelling out the consequences of imperfect predictions, surveying some of the most popular color appearance spaces in which GMAs are implemented and mentioning the role of nonappearance spaces in this context. Chapter 7 then introduced some very basic geometrical concepts that are necessary for the comprehension of GMAs.

The multiple challenges of computing color gamut boundaries were the subject of Chapter 8, where emphasis was placed on exploring the ambiguities of sampling density versus gamut concavity, on introducing the main types of gamut computation algorithms and finally to provide examples of some of the most pertinent color gamuts.

A case study of gamut mapping to colors with the smallest color difference was the background against which full detail was provided in Chapter 9 on how a single, specific source color travels through a color reproduction transformation. The intention here was to apply the theory of the preceding chapters and make it more concrete. Other gamut clipping strategies were also introduced here.

Chapter 10 finally delivered the 'meat' of this book, by attempting to provide a survey of all GMAs published to date. This involved covering color-by-color reduction as well as expansion algorithms, spatial GMAs, spectral GMAs and finally a number of solutions for highly specialized applications (e.g. restoration, duotone printing) were presented.

The fit between the GMAs presented in this book and color management systems was briefly touched upon in Chapter 11, and Chapter 12 concluded the treatment of gamut mapping by looking at how GMAs (as opposed to single reproductions) can be evaluated. Here, the key points were that a lot of past evaluation is so heterogeneous as to be not interrelatable, that the CIE have published *Guidelines for the Evaluation of Gamut Mapping Algorithms* (CIE, 2004d) to address the issue and that specific aspects of gamut mapping – the image dependence of GMA performance and the continuity of GMA transforms – have also been studied directly.

All that remains to be done at this point is to give some obligatory thought to the next steps in how color gamut differences may be dealt with in the future. First, though, it is worth calling to mind J. K. Galbraith's dictum: 'We have 2 classes of forecasters: Those who don't know and those who don't know they don't know,' to which I can only say that I am sure it will always be true.

With that caveat, I will venture, though, to suggest that the following are likely to be areas of focus in the coming years:

1. Further development of gamut expansion due to emergence of wider gamut imaging devices and the existence of vast amounts of past, narrow gamut content.
2. Greater focus on spectral gamut mapping, going hand in hand with the greater proliferation of multi-primary imaging devices that allow for spectral reproduction.
3. Increased use of spatial gamut mapping, enabled by increases in computational power available to color management.
4. The possible definition of a standard, baseline gamut mapping solution for accurate color reproduction.

5. The continued development and commercialization of multiple, proprietary and competing solutions to gamut mapping for preferred color reproduction.

On the more wishful side I would hope for a *'paradigm shift'* (Kuhn, 1962) in gamut mapping, i.e. a change in the basic assumptions, methods, accepted constraints and essentially the implicit worldview of current gamut mapping research, rather than the *normal science* style of incremental, but relatively minor advances that we have been witnessing over the last few years. This, however, is infinitely more easily said than done, and while the answer to the question of what such a new paradigm might be eludes me, I would like to conclude this book by sketching out a gamut mapping utopia:

1. Accurate gamut mapping is a standardized commodity where, by default, the reproduction that looks most similar to a given original is provided. Greatest similarity here is a statistical concept (i.e. most similar is what would be chosen as most similar by a significant proportion of the population). The domain for accurate gamut mapping can be either color or spectral, and users have the choice to adjust accurate gamut mapping by expressing their preference for how to trade off among similarity in terms of individual colors, detail, contrast, smoothness, etc. Each of these attributes is predicted perfectly (it is a utopia after all).
2. There is a standard, baseline preferred reproduction algorithm that can be used when predictability or consistency among reproductions made using different imaging technologies is needed.
3. Competing, proprietary solutions for optimal preferred color gamut mapping abound and allow users to make choices in terms of their color preferences. A 'language' for expressing preference has been developed and is part of the collective unconscious (in the same way in which preferences for how to have a steak cooked can be expressed).

This view, therefore, is clearly one where the continued development of gamut mapping remains an activity even in a utopian setting, where its role becomes to provide a means for satisfying the perpetually changing visual preferences of individuals and societies.

References

Adobe (2006) *Adobe Systems' Implementation of Black Point Compensation*, Adobe Systems, retrieved 19 October 2006 from http://www.color.org/AdobeBPC.pdf.

Anton H. A. (2005) *Elementary Linear Algebra*, John Wiley and Sons.

Bakke A. M., Farup I., Hardeberg J. Y. (2005) Multispectral Gamut Mapping and Visualization – a First Attempt, *Proceedings of SPIE*, vol. 5667, pp. 193–200.

Bakke A. M., Hardeberg J. Y., Farup I. (2006) Evaluation of Gamut Boundary Descriptors, *IS&T/SID 14th Color Imaging Conference*, Scottsdale, AZ, pp. 50–55.

Bala R. (2003) Device characterization, *Digital Color Imaging Handbook*, Sharma G. (*ed.*), CRC Press, pp. 269–384.

Balasubramanian R., Dalal E. (1997) A method for quantifying the color gamut of an output device, *Proceedings of SPIE*, vol. 3018, pp. 110–116.

Balasubramanian R., de Queiroz R., Eschbach R., Wu W. (2000) Gamut Mapping To Preserve Spatial Luminance Variations, *IS&T/SID 10th Color Imaging Conference*, Scottsdale, AZ, pp. 122–128.

Bala R., DeQueiroz R., Eschbach R., Wu W. (2001) Gamut mapping to preserve spatial luminance variations. *Journal of Imaging Science and Technology*, vol. 45., no. 5, pp. 436–443.

Bando E., Hardeberg J. Y., Connah D. (2005) Can gamut mapping quality be predicted by colour image difference formulae?, *Proceedings of SPIE*, vol. 5666, pp. 180–191.

Barber C. B., Dobkin D. P., Huhdanpaa H. T. (1996) The Quickhull Algorithm for Convex Hulls, *ACM Transactions on Mathematical Software*, vol. 22, pp. 469–483.

Bartleson C. J. (1984) *Optical Radiation Measurement. Vol. 5 – Visual Measurements*, Bartleson C. J. and Grum F. (eds.), Academic Press Inc.

Bartleson C. J., Breneman E. J. (1967) Brightness Perception in Complex Fields, *Journal of the Optical Society of America*, vol. 57, no. 7, pp. 953–957.

Bell I. E., Alexander S. K. (2004) A Spectral Gamut–Mapping Environment with Rendering Parameter Feedback, *Eurographics 2004*.

Berlin B., Kay P. (1969) *Basic Color Terms*, University of California Press, Berkeley and Los Angeles, CA.

Berns R. S. (1992) Color WYSIWYG: A Combination of Device Colorimetric Characterization and Appearance Modeling, *SID International Symposium – Digest of Technical Papers*, Society for Information Display, Playa del Rey, CA, pp. 549 – 552.

Berns R. S., Motta R. J., Gorzynski M. E. (1993a) CRT Colorimetry. Part I: Theory and Practice, *Color Research and Application*, vol. 18, no. 5, pp. 299–314.

Berns R. S., Gorzynski M. E., Motta R. J. (1993b) CRT Colorimetry. Part II: Metrology, *Color Research and Application*, vol. 18, no. 5, pp. 315–325.

Berns R. S., Taplin L. A., Imai F. H., Day E. A., Day D. C. (2003) Spectral Imaging of Matisse's Pot of Geraniums: A Case Study, *IS&T/SID 11th Color Imaging Conference*, Scottsdale, AZ, pp. 149–153.

Bhachech M., Shaw M., DiCarlo J. (2006) Improved Color Table Inversion Near The Gamut Boundary, *IS&T/SID 14th Color Imaging Conference*, Scottsdale, AZ, pp. 44–49.

Bonnier N., Schmitt F., Brettel H., Berche S. (2006) Evaluation of Spatial Gamut Mapping Algorithms, *IS&T/SID 14th Color Imaging Conference*, Scottsdale, AZ, pp. 56–61.

Bonnier N., Schmitt F., Hull M., Leynadier C. (2007) Spatial and Color Adaptive Gamut Mapping: A Mathematical Framework and Two New Algorithms, *IS&T/SID 15th Color Imaging Conference*, Albuquerque, NM, pp. 267–272.

Boust C., Bretel H., Viénot F., Alquié G. (2005) Color Enhancement of Digital Images by Experts and Preference Judgments by Observers, *Journal of Imaging Science and Technology*, vol. 50, no. 1, pp. 1–11.

Bradley R. A., Terry M. E. (1952) Rank analysis of incomplete block designs I: The method of paired comparisons, *Biometrika*, vol. 39, pp. 324–45.

Bradley R. A. (1984) Paired comparisons: Some basic procedures and examples, *Handbook of Statistics*, Krishnaiah P. R. and Sen P. K. (*eds.*), Elsevier Science Publishers, vol. 4, pp. 299–326.

Braun G. J. (1999) *A Paradigm for Color Gamut Mapping of Pictorial Images*, Ph.D. Thesis, Rochester Institute of Technology, Rochester, NY.

Braun G. J., Fairchild M. D. (1997) Techniques for Gamut Surface Definition and Visualization, *IS&T/SID 5th Color Imaging Conference*, Scottsdale, AZ, pp. 147–152.

Braun G. J. and Fairchild M. D. (1999) Image Lightness Rescaling Using Sigmoidal Contrast Enhancement Functions, *Journal of Electronic Imaging*, **8**/4:380–393.

Braun G. J., Fairchild M. D., Ebner F. (1998) Color Gamut Mapping in a Hue–Linearized CIELAB Color Space, *IS&T/SID 6th Color Imaging Conference*, Scottsdale, AZ, pp. 163–168.

Braun G. J., Spaulding K. (2002) Method for Evaluating the Color Gamut and Quantization Characteristics of Output–Referred Extended–Gamut Color Encodings, *IS&T/SID 10th Color Imaging Conference*, Scottsdale, AZ, pp. 99–105.

Braun K. M., Balasubramanian R., Eschbach R. (1999a) Development and Evaluation of Six Gamut Mapping Algorithms for Pictorial Images, *IS&T/SID 7th Color Imaging Conference*, Scottsdale, AZ, pp. 144–148.

Braun K. M., Balasubramanian R., Harrington S. J. (1999b) Gamut–Mapping Techniques For Business Graphics, *IS&T/SID 7th Color Imaging Conference*, Scottsdale, AZ, pp. 149–154.

Brill M. H. (1992) Tradeoffs in VDU Monitor Calibration and in Color Correction, *TAGA Proceedings*, pp. 903–916.

Brill M. H. (2006) Irregularity in CIECAM02 and its Avoidance, *Color Research and Application*, vol. 31, no. 2, pp. 142–145.

Buckley R. R. (1978) *Reproducing pictures with non–reproducible colors*, MSc. thesis, Massachusetts Institute of Technology, Cambridge, MA, February 1978.

Bourgoin M. (2007) Gamut Mapping In WCS, Part 2: the MinCD baseline gamut mapping model, retrieved 9 July 2007 from http://blogs.msdn.com/color_blog/archive/2006/03/15/552198.aspx.

Büring H., Herzog P. (2002) Gamut Mapping Along Curved Lines, *Color Image Science – Exploiting Digital Media*, MacDonald L.W., Luo M. R. (eds.), John Wiley and Sons, pp. 343–353.

Büring H., Herzog P., Jung E. (2001) Investigating Inverse Gamut Mapping of Current Color Management Tools, *IS&T/SID 9th Color Imaging Conference*, Scottsdale, AZ, pp. 263–266.

Byrne A., Hilbert D. R. (1997) *Readings on Color, Vol. I: The Philosophy of Color*, MIT Press, Cambridge, MA.

Carroll J., Neitz J., Neitz M. (2002) Estimates of L:M cone ratio from ERG flicker photometry and genetics, *Journal of Vision*, vol. 2, no. 8, pp. 531–542, retrieved 18 January 2006 from http://journalofvision.org/2/8/1/.

Cazals F., Giesen J., Pauly M., Zomorodian A. (2006) The conformal alpha shape filtration, *The Visual Computer*, vol. 22, pp. 531–540.

Chau W. W. K and Cowan W. B. (1996) Gamut Mapping Based on the Fundamental Components of Reflective Image Specifications, *Proceedings of 4th IS&T/SID Color Imaging Conference*, pp. 67–70.

Chen H. S., Kotera H. (2000) Three–dimensional gamut mapping method based on the concept of image–dependence, *Proceedings of IS&T's NIP16 Conference*, pp. 783–786.

Chen H. S., Omamiuda M., Kotera H. (1999) Gamut Mapping Method Adaptive to Hue–Divided Color Distributions, *Proceedings of 7th IS&T/SID Color Imaging Conference*, pp. 289–294.

Chen H. S., Omamiuda M., Kotera H. (2001) Gamma–Compression Gamut Mapping Method Based on the Concept of Image–to–Device, *Journal of Imaging Science and Technology*, vol. 45, no. 2, pp. 141–151.

Chen X. (2002) *Investigation of Gamut Extension Algorithms*, MSc Thesis, Colour & Imaging Institute, University of Derby.

Cholewo T. J., Love S. (1999) Gamut Boundary Determination Using Alpha–Shapes, *Proceedings of 7th IS&T/SID Color Imaging Conference*, pp. 200–204.

CIE (1995) *CIE 116:2001: Industrial colour–difference evaluation*, CIE Central Bureau, Vienna.

CIE (1996) *CIE 122:1996: The Relationship between Digital and Colorimetric Data for Computer–Controlled CRT Displays*, CIE Central Bureau, Vienna.

CIE (2001) *CIE 142:2001: Improvement to industrial colour difference evaluation*, CIE Central Bureau, Vienna.

CIE (2004a) *CIE 15:2004: Colorimetry*, 3rd Edition, CIE Central Bureau, Vienna.

CIE (2004b) *CIE 159:2004: A colour appearance model for colour management systems: CIECAM02*, CIE Central Bureau, Vienna.

CIE (2004c) *CIE 162:2004: Chromatic adaptation under mixed illumination condition when comparing softcopy and hardcopy images*, CIE Central Bureau, Vienna.

CIE (2004d) *CIE 156:2004: Guidelines for the evaluation of gamut mapping algorithms*, CIE Central Bureau, Vienna.

Clark J. A. (1977) A Method of Scaling with Incomplete Pair–Comparison Data, *Educational and Psychological Measurement*, vol. 37, no. 3, pp. 603–611.

Cui C. (2001) Gamut Mapping With Enhanced Chromaticness, *IS&T/SID 9th Color Imaging Conference*, Scottsdale, AZ, pp. 257–262.

Cui G., Luo M. R., Rigg B. (2005) Colour difference evaluation under illuminant A, *Proceedings of 10th Congress of the International Colour Association – AIC Colour 05*, Granada, Spain, pp. 575–578.

Dartnall H. J. A., Bowmaker J. K., Mollon J. D. (1983) Human visual pigments: microspectrophotometric results from the eyes of seven persons. *Proceedings of the Royal Society of London, B 220*, pp. 115–130.

de Berg M., van Kreveld M., Overmars M., Schwarzkopf O. (2000) *Computational Geometry*, Springer.

de Fez M. D., Luque J. M., Capilla P., Pérez–Carpinell J., Díez M. A. (1998) Colour memory matching analysed using different representation spaces, *Journal of Optics*, IOP Publishing Ltd., England, vol. 29, pp. 287–297.

Delaunay B. (1934) Sur la sphère vide, *Izvestia Akademii Nauk SSSR, Otdelenie Matematicheskikh i Estestvennykh Nauk*, vol. 7, pp. 793–800.

Derhak M. W., Rosen M. R. (2004) Spectral Colorimetry Using LabPQR – An Interim Connection Space, *IS&T/SID 12th Color Imaging Conference*, Scottsdale, AZ, pp. 246–250.

De Valois R. L., Abramov I., Jacobs G. H. (1966) Analysis of Response Patterns of LGN Cells, *Journal of the Optical Society of America*, vol. 56, pp. 966–977.

De Valois K. K., De Valois R. L. (2002) Human Visual System – Spatial Visual Processing, *Encyclopedia of Imaging Science and Technology*, John Wiley and Sons, retrieved 24 February 2006 from DOI: 10.1002/0471443395.img032.

DiCarlo J. M., Wandell B. A. (2000) Rendering high dynamic range images. *Proceedings of SPIE*, vol. 3965, pp. 392–401.

Dugay F., Ivar F., Hardeberg J. (2007) Perceptual Evaluation of Gamut Mapping Algorithms, *Color Research and Application*, submitted for publication.

Ebner F., Fairchild M. D. (1997) Gamut Mapping from Below: Finding the Minimum Perceptual Distances for Colors Outside the Gamut Volume, *Color Research and Application*, vol. 22, pp. 402–413.

Ebner F., Fairchild M. D. (1998a) Finding constant hue surfaces in color space, *SPIE Proceedings*, vol. 3300, pp. 107–117.

Ebner F., Fairchild M. D. (1998b) Development and Testing of a Color Space (IPT) with Improved Hue Uniformity, *IS&T/SID 6th Color Imaging Conference*, Scottsdale, AZ, pp. 8–13.

Eco U. (2003) *Mouse or Rat? Translation as Negotiation*, Phoenix Press, London.

Edelsbrunner H., Mücke E. P. (1994) Three–Dimensional Alpha Shapes, *ACM Transactions on Graphics*, vol. 13, no. 1, pp. 43–72.

Engeldrum P. G. (1986) Computing Color Gamuts of Ink–jet Printing Systems, *SID Proceedings*, vol. 27, pp. 25–30.

Engeldrum P. G. (2000) *Psychometric Scaling: A Toolkit for Imaging Systems Development*, Imcotek Press, Winchester, MA.

Evans R. M. (1943) Visual Processes and Color Photography, *Journal of the Optical Society of America*, vol. 33, no. 11, pp. 579–614.

Fairchild M. D. (1993) Chromatic adaptation in hard copy/soft copy comparisons, *SPIE Proceedings*, vol. 1912, pp. 47–61.

Fairchild M. D. (2005) *Color Appearance Models*, John Wiley & Sons, Chichester, UK.

Fairchild M. D., Johnson G. M. (2004) iCAM framework for image appearance, differences, and quality, *Journal of Electronic Imaging*, vol. 13, no. 1, pp. 126–138.

Fairchild M. D., Johnson G. M. (2005) On the salience of novel stimuli: Adaptation and image noise, *IS&T/SID 13th Color Imaging Conference*, Scottsdale, AZ, pp. 333–338.

Farnand S. P. (1995) The effect of image contents on color difference perceptibility. *Proceedings of the 4th IS&T/SID Color Imaging Conference*, Scottsdale, AZ, pp. 101–104.

Farup I., Hardeberg J., Amsrud M. (2004) Enhancing the SGCK Colour Gamut Mapping Algorithm, *Proceedings of CGIV: 2nd European Conference on Colour Graphics, Imaging and Vision*, pp. 520–524.

Fedorovskaya E., de Ridder H., Blommaert F. J. J. (1997) Chroma variations and perceived quality of color images of natural scenes, *Color Research and Application*, vol. 22, no. 2, pp. 96–110.

Fernandez S., Fairchild M. D. (2002) Observer preferences and cultural differences in color reproduction of scenic images, *IS&T/SID 10th Color Imaging Conference*, Scottsdale, AZ, pp. 66–72.

Field G. G. (1997) *Color and Its Reproduction*, Second Edition, Graphic Arts Technical Foundation, Sewickley, PA.

Finlayson G. D., Hordley S. D. (2000) Improving gamut mapping color constancy, *IEEE Transactions on Image Processing*, vol. 9, no. 10, pp. 1774–1783.

Forsyth D. (1990) A novel algorithm for color constancy, International Journal of Computer Vision, vol. 5, no. 1, pp. 5–36.

Frankle J., McCann J. (1983) Method and apparatus for lightness imaging, *US Patent #4384336.*

Fraser B. (2004) *Real World Color Management*, Peachpit Press.

Frisken S. F., Perry R. N., Rockwood A. P., Jones T. R. (2000) Adaptively Sampled Distance Fields: A General Representation of Shape for Computer Graphics, *SIGGRAPH*, ACM, pp. 249–254.

Fukasawa K., Ito A., Qunigoh M, Nakaya F., Shibuya T., Shimada H., Yaguchi H. (2005) Suitable Printer Color Reproduction for Office Environment, *Proceedings of the 13th IS&T/SID Color Imaging Conference*, Scottsdale, AZ, pp. 185–188.

Gaiani M., Micoli L.L., Russo M. (2005) The Monuments Restoration Yard: a Virtualization Method and the Case of Study of Sala delle Cariatidi in Palazzo Reale, Milan, *Proceedings of the ISPRS Working Group V/4 Workshop 3D–ARCH 2005: "Virtual Reconstruction and Visualization of Complex Architectures"*, Mestre–Venice, Italy, 22–24 August 2005.

Gentile R. S., Walowit E., Allebach J. P. (1990) A comparison of techniques for color gamut mismatch compensation, *Journal of Imaging Technology*, vol. 16, pp. 176–181.

Getty Images (2004) *Color Management: a guide*, Getty Images, pp. 7, retrieved on 27 September 2006 from http://creative.gettyimages.com/en–us/marketing/services/Color_management.pdf

Gescheider G. A. (1985) *Psychophysics, Method, Theory, and Application*, Second Edition, Lawrence Erlbaum Associates.

Giesen J., Schuberth E., Simon K., Zolliker P. (2005) Toward image–dependent gamut mapping: fast and accurate gamut boundary determination, *Proceedings of SPIE*, vol. 5667, pp. 201–210.

Giesen J., Schuberth E., Simon K., Zolliker P. (2006) A Kernel Approach to Gamut Boundary Determination, *Proceedings of the 14th European Signal Processing Conference (EUSIPCO).*

Giesen J., Schuberth E., Simon K., Zolliker P., Zweifel O. (2006) Image–Dependent Gamut Mapping as Optimization Problem, *IEEE Transactions on Image Processing*, vol. 16, no. 10, pp. 2401–2410.

Giorgianni E. J., Madden T. E. (1998) *Digital Color Management: Encoding Solutions*, Addison Wesley.

Glassner A. S., Fishkin K. P., Marimont D. H., Stone M. C. (1995) Device–Directed Rendering, *ACM Transactions on Graphics*, vol. 14, no. 3, pp. 58–76.

Gonzalez R. C., Woods R. E. (1993) *Digital Imaging Processing*, Addison Wesley.

Gordon J., Holub R., Poe R. (1987) On the Rendition of Unprintable Colors, *TAGA Proceedings*, pp. 186–195.

Gose E., Johnsonbaugh R., Jost S. (1996) *Pattern Recognition and Image Analysis*, Prentice Hall Inc., pp. 211.

Granger E. M. (1995) Gamut Mapping for Hard Copy Using the ATD Color Space, *SPIE Proceedings*, vol. 2414, pp. 27–35.

Gray H. (1918) *Anatomy of the Human Body*, (ed.) Lewis W. H., 20th ed., Lea & Febiger, Philadelphia.

Green P. J. (2001) A test target for defining media gamut boundaries, *Proceedings of SPIE*, vol. 4300, pp. 105–113.

Green P. J., Luo M. R. (2002) Gamut Mapping For High Quality Print Reproduction, *Color Image Science – Exploiting Digital Media*, MacDonald L.W., Luo M. R. (eds.), John Wiley and Sons, pp. 319–341.

Guan S. S., Luo M. R. (1999) Investigation of parametric effects using small colour–differences, *Color Research and Application*, vol. 24, pp. 331–343.

Guild J. (1931) The colorimetric properties of the spectrum. *Philosophical Transactions of the Royal Society of London*, vol. A230, pp. 149–187.

Guth S. L. (1989) Unified Model for Human Color Perception and Visual Adaptation, *SPIE Proceedings*, vol. 1077, pp. 370.

Han D. (2004) Real–Time Color Gamut Mapping Architecture and Implementation for Color–Blind People, *Computer Human Interaction – 6th Asia Pacific Conference (APCHI 2004)*, Rotorua, New Zealand, 29 June – 2 July 2004, Springer Verlag, pp. 133–142.

Han K., Kang B. H., Cho M. S. (1999) A Simultaneous Gamut Compression With Extended Line Of Cusps, *Proceedings of IEEE Region 10 Conference – TENCON 99 'Multimedia Technology for Asia–Pacific Information Infrastructure'*, pp. 927–929.

Handley J. (2001) Comparative Analysis of Bradley–Terry and Thurstone–Mosteller Paired Comparison Models for Image Quality Assessment. *IS&T PICS 2001 Conference Proceedings*, Montréal, Canada, pp. 108–112.

Handley J. C., Babock J. S., Pelz J. B. (2004) Experimental congruence of interval scale production from paired comparisons and ranking for image evaluation, *SPIE Proceedings*, vol. 5294, pp. 211–221.

Haneishi H., Miyata K., Yaguchi H., Miyake Y. (1993) A New Method for Color Correction in Hardcopy from CRT Images, *Journal of Imaging Technology*, vol. 37, no. 1, pp. 30–36.

Hansen T., Olkkonen M., Walter S., Gegenfurtner K. R. (2006) Memory modulates color appearance, *Nature Neuroscience*, vol. 9, pp. 1367–1368.

Heckaman R. L., Fairchild M. D. (2006a) Effect of DLP projector white channel on perceptual gamut, *Journal of the SID*, vol. 14, no. 9, pp. 755–761.

Heckaman R. L., Fairchild M. D. (2006b) Expanding Display Color Gamut Beyond the Spectrum Locus, *Color Research and Application*, vol. 31, no. 6, pp. 475–482.

Heidelberger Druckmaschinen AG (2003) Expert Guide – Color Management, 08/2003, *Heidelberger Druckmaschinen AG*, retrieved on 5 June 2007 from HTTP://www.heidelberg.com/wwwbinaries/bin/files/dotcom/en/products/prinect/color_management_eng.pdf.

Herzog P. G. (1998) Further Development of the Analytical Color Gamut Representations, *SPIE Proceedings*, vol. 3300, pp. 118–128.

Herzog P. G., Hill B. (2003) Multispectral Imaging and its Applications in the Textile Industry and Related Fields, *PICS: The Digital Photography Conference*, Rochester, NY, pp. 258–263.

Herzog P. G., Büring H. (1999) Optimizing Gamut Mapping: Lightness and Hue Adjustments, *IS&T/SID 7th Color Imaging Conference*, Scottsdale, AZ, pp. 160–166.

Herzog P. G., Büring H. (2000) Optimizing Gamut Mapping: Lightness and Hue Adjustments, Journal of Imaging Science and Technology, vol. 44, no. 4, pp. 334–342.

Herzog P. G. and Müller M. (1997) Gamut Mapping Using an Analytical Color Gamut Representation, *SPIE Proceedings*, vol. 3018, pp. 117–128.

Hill A. R. (1997) How we see colour, *Colour Physics for Industry*, Second Edition, McDonald R. (ed.), Society of Dyers and Colourists, Bradford, UK.

Hirokawa T., Inui M., Morioka T., Azuma Y. (2007) A Psychophysical Evaluation of a Gamut Expansion Algorithm Based on Chroma Mapping II: Expansion within Object Color Data Bases, NIP23, Anchorage, AK, pp. 175–179.

Hofer H., Carroll J., Neitz J., Neitz M., Williams D. R. (2005) Organization of the Human Trichromatic Cone Mosaic, *Journal of Neuroscience*, vol. 25, no. 42, pp. 9669–9679.

Holm, J. (2005) Some considerations in the development of color rendering and gamut mapping algorithms, *SPIE Proceedings*, vol. 5678, pp. 137–143.

Holm J. (2006) Capture Color Analysis Gamuts, *IS&T/SID 14th Color Imaging Conference*, Scottsdale, AZ, pp. 108–113.

Holm J., Tastl I., Johnson T. (2006) Definition & Use of the ISO 12640–3 Reference Color Gamut, *IS&T/SID 14th Color Imaging Conference*, Scottsdale, AZ, pp. 62–68.

Hoshino T. (1991) A Preferred Color Reproduction Method for the HDTV Digital Still Image System, *IS&T Proceedings: Symposium on Electronic Photography*, pp. 27–32.

Hoshino T. (1994) *Color Estimation Method for Expanding a Color Image for Reproduction in a Different Color Gamut*, U.S. Patent 5,317,426.

Hoshino T., Berns R. S. (1993) Color Gamut Mapping Techniques for Color Hard Copy Images, *SPIE Proceedings*, vol. 1909, pp. 152–164.

Hubel D. H. (1989) *Eye, Brain and Vision*, Scientific American Library, New York.

Hubel P. M. (1999) The perception of color at dawn and dusk, *IS&T/SID 7th Color Imaging Conference*, Scottsdale, AZ, pp. 48–51.

Hung P. C., Berns R. S. (1995) Determination of Constant Hue Loci for a CRT Gamut and Their Predictions Using Color Appearance Spaces, *Color Research and Application*, vol. 20, pp. 285–295.

Hunt R. W. G. (1970) Objectives in colour reproduction. *Journal of Photographic Science*, vol. 18, pp. 205–215.

Hunt R. W. G. (1995a) *Measuring Colour*, Second Edition, Ellis Horwood Ltd.

Hunt R. W. G. (1995b) *The Reproduction of Colour in Photography, Printing & Television*, Fifth Edition, Fountain Press, Kingston–upon–Thames, UK.

Hunt R. W. G. (2001) Saturation, Superfluous or Superior?, *IS&T/SID 9th Color Imaging Conference*, Scottsdale, AZ, pp. 147–152.

Hunt R. W. G., Pitt I. T., Winter L. M. (1974) Preferred reproduction of blue sky, green grass and Caucasian skin in colour photography, *Journal of Photographic Science*, vol. 22, pp. 144–150.

Hunter D. R. (2004) MM Algorithms For Generalized Bradley–Terry Models, *The Annals of Statistics*, vol. 32, no. 1, pp. 384–406.

Hunter D. R. (2006) *MATLAB code for Bradley–Terry models*, retrieved 4 May 2006 from http://www.stat.psu.edu/~dhunter/code/btmatlab/.

Hunter D. R., Lange K. (2004) A Tutorial on MM Algorithms, *The American Statistician, vol.* 58, pp. 30–37.

ICC (1994) *InterColor Profile Format*, International Color Consortium, retrieved 9 March 2006 from http://www.color.org/icc30.pdf.

ICC (2001) *Specification ICC.1:2001–04 File Format for Color Profiles*, International Color Consortium, retrieved 9 March 2006 from http://www.color.org/ICC_Minor_Revision_for_Web.pdf.

ICC (2004) *Specification ICC.1:2004–10 (Profile version 4.2.0.0) Image technology colour management — Architecture, profile format, and data structure*, International Color Consortium, retrieved 9 March 2006 from http://www.color.org/ICC1V42.pdf.

ICC (2006) *White Paper #9: Common Color Management Workflows & Rendering Intent Usage*, retrieved 10 March 2006 from http://www.color.org/ICC_white_paper_9_workflow.pdf.

IEC (1999) *IEC 61966–2–1: Multimedia systems and equipment – Colour measurement and management – Part 2–1: Colour management – Default RGB colour space – sRGB*, IEC, Switzerland.

Inui M. (1993) Fast Algorithm for Computing Color Gamuts, *Color Research and Application*, vol. 18, pp. 341–348.

Inui M., Hirokawa T., Azuma Y., Tajima J. (2004) Color Gamut of SOCS and its Comparison to Pointer's Gamut, *IS&T's NIP20: 2004 International Conference on Digital Printing Technologies*, Salt Lake City, Utah, pp. 410–415.

Ishii A. (2002) Color space conversion for the laser film recorder using 3–D LUT, *SMPTE Journal*, vol. 111, no. 11, pp. 525–532.

ISO (2000) *ISO 3664:2000. Viewing conditions – Prints, transparencies and substrates for graphic arts technology and photography*, ISO.

ISO (2003) *ISO/TR 16066:2003. Graphic technology – Standard object colour spectra database for colour reproduction evaluation (SOCS)*, ISO.

ISO (2004) *ISO 22028–1:2004, Photography and graphic technology — Extended colour encodings for digital image storage, manipulation and interchange — Part 1: Architecture and requirements*, ISO.

ISO (2004b) *12647–2:2004, Graphic technology — Process control for the production of half–tone colour separations, proof and production prints — Part 2: Offset lithographic processes*, ISO.

ISO (2006) *ISO 12640–3. Graphic technology – Prepress digital data exchange – Part 3: CIELAB standard colour image data (CIELAB/SCID)*, Draft Standard, ISO.

Ito M., Katoh N. (1995) Gamut Compression for Computer Generated Images, *Extended Abstracts of SPSTJ 70th Anniversary Symposium on Fine Imaging*, pp. 85–88 (in Japanese).

Ito M. and Katoh N. (1999) Three–Dimensional Gamut Mapping Using Various Color Difference Formulæ and Color Spaces, *Proceedings of SPIE*, vol. 3648, pp. 83–95.

Jarvis R. A. (1973) On the identification of the convex hull of a finite set of points in the plane, *Information Processing Letters*, vol. 2, pp. 18–21.

Jobson D. J., Rahman Z., Woodell G. A. (1996) Retinex Image Processing: Improved Fidelity To Direct Visual Observation, *Proceedings of the 4th IS&T/SID Color Imaging Conference*, Scottsdale, AZ, pp. 124–126.

Johnson A. J. (1979) Perceptual Requirements of Digital Picture Processing, Paper presented at IARAIGAI symposium and printed in part in *Printing World*, 6 February 1980.

Johnson A. J. (1992) *Colour Appearance Research for Interactive System Management and Application – CARISMA, Work Package 2 – Device Characterisation*, Report WP2–19 Colour Gamut Compression (For a description of the CARISMA algorithm see also (Morovič and Luo, 1999, pp. 271–274)).

Johnson G. M., Fairchild M. D. (2006) Image Appearance Modelling, in *Colorimetry: Understanding The CIE System. CIE Colorimetry 1931 – 2006*, Schanda J. (ed.), CIE, Vienna.

Jolliffe I. T. (2002) *Principal Component Analysis*, Springer–Verlag New York Inc.

Kang B. H., Cho M. S., Morovič J., Luo M. R. (2001) Gamut Extension Development Based on Observer Experimental Data, *IS&T/SID 9th Color Imaging Conference*, pp. 158–162.

Kang B. H., Cho M. S., Morovič J. and Luo M. R. (2002) Gamut Compression Algorithm Development using Observer Experimental Data, *Colour Image Science: Exploiting Digital Media*, MacDonald L. W. and Luo M. R. (eds.), John Wiley & Sons, pp. 259–290.

Kasson J. M., Nin S. I., Plouffe W., Hafner J. L. (1995) Performing Color Space Conversions with Three–Dimensional Linear Interpolation, *Journal of Electronic Imaging*, vol. 4, no. 3, pp. 226–250.

Katoh N. (1994) Practical method for appearance match between soft copy and hard copy, *Proceedings of SPIE*, vol. 2170, pp. 170–181.

Katoh N., Ito M. (1996) Gamut Mapping for Computer Generated Images (II), *IS&T/SID* 4th *Color Imaging Conference*, Scottsdale, AZ, pp. 126–129.

Katoh N., Ito M. (1999) Applying Non–Linear Compression to the Three–Dimensional Gamut Mapping, *IS&T/SID* 7th *Color Imaging Conference*, Scottsdale, AZ, pp. 155–159.

Katoh N., Ito M., Ohno S. (1999) Three–dimensional Gamut Mapping Using Various Color Difference Formulae and Color Spaces, *Journal of Electronic Imaging*, vol. 8, no. 4, pp. 365–379.

Katoh N., Nakabayashi K., Ito M., and Ohno S. (1998) Effect of Ambient Light on Color Appearance of Softcopy Images: Mixed Chromatic Adaptation for Self–luminous Displays, *Journal of Electronic Imaging*, vol. 7, pp. 794–806.

Kim M. C., Shin Y. C., Song Y. R., Lee S. J., Kim I. D. (2004) Wide Gamut Multi–Primary Display for HDTV, *CGIV 2004 – Second European Conference on Color in Graphics, Imaging and Vision*, Aachen, Germany, April 2004, pp. 248–253.

Kim S. D., Lee C. H., Kim K. M., Lee C. S., Ha Y. H. (1998) Image–dependent gamut mapping using a variable anchor point, *SPIE Proceedings*, SPIE, Bellingham, WA, vol. 3300, pp. 129–137.

Kimmel R., Shaked D., Elad M., Sobel I. (2005) Space–dependent color gamut mapping: a variational approach, *IEEE Transactions on Image Processing*, vol. 14, pp. 796 –803.

Kipphan H. (2001) *Handbook of Print Media. Technologies and Production Methods*, Kipphan H. (*ed.*), Springer.

Koh K. W., Tastl I., Nielsen M., Berfanger D. M., Zeng H., Holm J. (2003) Issues Encountered in Creating a Version 4 ICC sRGB Profile, *IS&T/SID 11th Color Imaging Conference*, Scottsdale, AZ, pp. 232–237.

Kolås Ø., Farup I. (2007) Efficient hue–preserving and edge–preserving spatial color gamut mapping, *IS&T/SID 15th Color Imaging Conference*, Albuquerque, NM, pp. 207–212.

Kolb H., Fernandez E., Nelson R. (2006) *Webvision: The Organization of the Retina and Visual System*, retrieved 25 January 2006 from http://webvision.med.utah.edu/index.html

Kotera H., Saito R. (2003) Compact description of 3–D image gamut by r–image method, *Journal of Electronic Imaging*, vol. 12, no. 4, pp. 660–668.

Kress W., Stevens M. (1994) Derivation of 3–Dimensional Gamut Descriptors for Graphic Arts Output Devices, *TAGA Proceedings*, pp. 199–214.

Krinov E. L. (1947) Spectral Reflectance Properties of Natural Formations, *Proceedings of the Academy of Sciences of the USSR.*

Kubelka P., Munk F. (1931) Ein Beitrag zur Optik der Farbanstriche, *Zeitschrift für technische Physik*, vol. 12, pp. 593–601.

Kuehni R. G. (2003) *Color Space and Its Divisions: Color Order from Antiquity to the Present*, John Wiley & Sons, Chichester, UK.

Kuehni R.G. (2005) Focal color variability and unique hue stimulus variability, *Journal of Cognition and Culture*, vol. 5, pp. 409–426.

Kuhn T. S. (1962) *The Structure of Scientific Revolutions*, University of Chicago Press.

Kuo C., Zeise E., Lai D. (2006) Robust CIECAM02 Implementation and Numerical Experiment within an ICC Workflow, *IS&T/SID 14th Color Imaging Conference*, Scottsdale, AZ, pp. 215–219.

Laihanen P. (1987) Colour Reproduction Theory based on the Principles of Colour Science, *IARAIGAI Conference Proceedings – Advances in Printing Science and Technology*, vol. 19, pp. 1–36.

Lammens J., Morovič J., Nielsen M. and Zeng H. (2005) Adaptive re–rendering of CMYK separations for digital printing, *AIC '05*, Granada, Spain, 1211–1214.

Land E. H. (1964) The Retinex, *American Scientist*, vol. 52, pp. 247–264.

Lavendel L. (2003) Introduction to Computer Graphics, *The Guild Handbook of Scientific Illustration*, Hodges E. R. S. (ed.), John Wiley and Sons.

Lee C. S., Lee C. H. and Ha Y. H. (2000) Parametric Gamut Mapping Algorithms Using Variable Anchor Points, *Journal of Imaging Science and Technology*, vol. 44, no. 1, pp. 68–73.

Li C., Luo M. R., Hunt R. W. G. (1999) The CIECAM97s2 Model, *Proceedings of the 6th IS&T/SID Color Imaging Conference*, Scottsdale AZ, pp. 262–263.

Lin H., Luo M. R., MacDonald L. W., Tarrant A. W. S. (2001) A Cross–Cultural Colour–Naming Study: Part II—Using a Constrained Method, *Color Research and Application*, John Wiley & Sons, Chichester, UK, vol. 23, no. 3, pp. 193–208.

Ling Y. (2001) *Investigation of a Gamut Extension Algorithm*, MSc Thesis, Colour & Imaging Institute, University of Derby.

Livingstone M., Hubel D. (1988) Segregation of form, color, movement, and depth: anatomy, physiology, and perception. *Science*, vol. 240, no. 4853, pp. 740–49.

Lo M. C. (1995) *The LLAB Model for Quantifying Colour Appearance*, Ph.D. Thesis, University of Loughborough.

Luo M. R. (1998) Colour science, in *Colour Image Processing Handbook*, Sangwine S. J., Horne R. E. N. (*eds.*), Chapman & Hall, London, pp. 26–66.

Luo M. R., Cui G., Rigg B. (2001) The development of the CIE 2000 colour difference formula, *Color Research and Application*, vol. 26, pp. 340–350.

Luo M. R., Li C. J., Hunt R. W. G., Rigg B., Smith K. J. (2003) The CMC 2002 Colour Inconstancy Index: CMCCON02, *Coloration Technology*, vol. 119, no. 280–285.

Luo M. R. and Rigg B. (1987) BFD(l:c) Colour Difference Formula, Part 1 and Part 2, *Journal of the Society of Dyers and Colourists*, pp. 126–132.

MacAdam D. L. (1935) Maximum visual efficiency of colored materials, *Journal of the Optical Society of America*, vol. 25, pp. 36.

MacAdam D.L. (1942) Visual sensitivities to color differences in daylight, *Journal of the Optical Society of America*, vol. 32, pp. 247–274.

MacDonald L. W., Luo M. R., Scrivener S. A. R. (1990) Factors Affecting the Appearance of Coloured Images on a Video Display Monitor, *Journal of Photographic Science*, vol. 38, pp. 177–186.

MacDonald L. W., Morovič J. (1995) Assessing the Effects of Gamut Compression in the Re–production of Fine Art Paintings, *Proceedings of 3rd IS&T/SID Color Imaging Conference*, pp. 194–200.

MacDonald L. W., Morovič J. and Saunders D. (1995) Evaluation of Colour Fidelity for Reproductions of Fine Art Paintings, *Museum Management and Curatorship*, vol. 14., no. 3, pp. 253–281.

MacDonald L. W., Morovič J., Xiao K. (2001) A Topographic Gamut Compression Algorithm, *Journal of Imaging Science and Technology*, vol. 46, no. 3, pp. 228–236.

MacDonald L. W., Morovič J., Xiao K. (2002) Evaluating Gamut Mapping Algorithms. In *Colour Image Science: Exploiting Digital Media*, MacDonald L. W. and Luo M. R. (eds.), John Wiley & Sons, pp. 291–318.

Mahy M. (1997) Calculation of Color Gamuts Based on the Neugebauer Model, *Color Research and Application*, vol. 22, pp. 365–374.

Mahy M. (2002) Colour gamut determination, *Colour Engineering: Achieving Device Independent Colour*, Green P. and MacDonald L. W. (eds.), John Wiley & Sons, pp. 297–318.

Maloney L. T. (1986) Evaluation of Linear Models of Surface Spectral Reflectance with Small Numbers of Parameters, *Journal of the Optical Society of America A*, vol. 3, pp. 1673–1683.

Maltz M. S. (2005) Tetrahedralizing of Point Sets Using Expanding Spheres, *IEEE Transactions on Visualization and Computer Graphics*, vol. 11, no. 1, pp. 102–109.

Marcu G. (1998) Gamut Mapping in Munsell Constant Hue Sections, *Proceedings of the 6th IS&T/SID Color Imaging Conference*, Scottsdale AZ, pp. 159–162.

Marcu G., Abe S. (1996) Gamut Mapping for Color Simulation on CRT Devices, *SPIE Proceedings*, vol. 2658, pp. 308–315.

Martin P. R. (1998) Colour processing in the primate retina: recent progress, *The Journal of Physiology*, vol. 513, no. 3, pp. 631–638.

Marr D. (1982) *Vision: A Computational Investigation into the Human Representation and Processing of Visual Information*. W. H. Freeman, San Francisco.

Maund B. (2002) Color, *The Stanford Encyclopedia of Philosophy (Fall 2002 Edition)*, Edward N. Zalta (*ed.*), retrieved 1 December 2005 from http://plato.stanford.edu/archives/fall2002/entries/color/.

McCann J. J. (1999) Color Gamut Measurements and Mapping: The Role of Color Spaces, *SPIE Proceedings*, vol. 3648, pp. 68–82.

McCann J. J., Rizzi A. (2007) Camera and visual veiling glare in HDR images, *Journal of the SID*, vol. 19, no. 9, pp. 721–730.

MCSL (2006) *Munsell Renotation Data*, Munsell Color Science Laboratory, retrieved 16 August 2006 from http://www.cis.rit.edu/mcsl/online/munsell_data/real.dat

Meyer J., Barth B. (1989) Color Gamut Matching for Hard Copy, *SID 89 Digest*, pp. 86–89.

Microsoft (2006) *Windows Image Color Management*, retrieved 20 April 2006 from http://www.microsoft.com/whdc/device/display/color/default.mspx.

Microsoft (2007) *WCS Gamut Map Model Profile Schema and Algorithms*, retrieved 4 July 2007 from http://msdn2.microsoft.com/en–us/library/ms536899.aspx.

Miller S. (2000) Noncoherent sources, in *Electro–Optics Handbook*, 2nd Edition, Waynant R. W., Ediger M. N. (*eds.*), McGraw–Hill, pp. 2.1–2.36.

Montag E. D., Fairchild M. D. (1997) Psychophysical Evaluation of Gamut Mapping Techniques Using Simple Rendered Images and Artificial Gamut Boundaries, *IEEE Transactions on Image Processing*, vol. 6, pp. 977–989.

Montag E. D., Kasahara H. (2001) Multidimensional Analysis Reveals Importance of Color for Image Quality, *IS&T/SID 9th Color Imaging Conference*, Scottsdale, AZ, pp. 17–21.

Moroney N. (1998) A comparison of CIELAB and CIECAM97s, *IS&T/SID 6th Color Imaging Conference*, Scottsdale, AZ, pp. 17–21.

Moroney N. (2000) Assessing hue constancy using gradients, *Proceedings of SPIE*, vol. 3963, pp. 294–300.

Moroney N. (2000b) Usage Guidelines for CIECAM97s, *IS&T PICS 2000 Conference Proceedings*, Portland, OR, pp. 164–168.

Moroney N. (2003) Unconstrained web–based color naming experiment, *Proceedings of SPIE*, vol. 5008, pp. 36–46.

Morovič J. (1998) *To Develop a Universal Gamut Mapping Algorithm*, Ph.D. Thesis, condensed format edition, University of Derby.

Morovič J. (1999) Colour Reproduction – Past, Present and Future, *Libro de Actas – V Congreso Nacional de Color (Proceedings of the 5th National Congress on Colour)*, 9th–11th June 1999, Terrassa, Spain, pp. 9–15.

Morovič J. (2003) Gamut Mapping, *Digital Color Imaging Handbook*, Sharma G. (*ed.*), CRC Press, 639–685.

Morovič J. (2007) Fast Computation of Multi–Primary Color Gamuts, *15th IS&T/SID Color Imaging Conference*, Albuquerque, NM, pp. 228–232.

Morovič J., Arnabat J., Vilar J. (2007) Visually Closest Cross–Gamut Matches Between Surface Colors, *Proceedings of the 15th IS&T/SID Color Imaging Conference*, Albuquerque, NM, pp. 273–277.

Morovič J., Lammens J. M. (2006) Colour Management, in *Colorimetry: Understanding The CIE System. CIE Colorimetry 1931 – 2006*, Schanda J. (*ed.*), CIE, Vienna.

Morovič J., Luo M. R. (1997) Gamut Mapping Algorithms Based on Psychophysical Experiment, *Proceedings of the 5th IS&T/SID Color Imaging Conference*, pp. 44–49.

Morovič J., Luo M. R. (1999) Developing Algorithms for Universal Colour Gamut Mapping, *Colour Engineering: Vision and Technology*, MacDonald L. W., Luo M. R. (*eds.*), John Wiley & Sons, Chichester, England, pp. 253–282.

Morovič J., Luo M. R. (2000) Calculating Medium and Image Gamut Boundaries for Gamut Mapping, *Color Research and Application*, vol. 25, pp. 394–401.

Morovič J., Sun P. L. (2000) The Influence of Image Gamuts on Cross–Media Colour Image Reproduction, *IS&T/SID 8th Color Imaging Conference*, pp. 324–329.

Morovič J., Sun P. L. (2000b) How different are Colour Gamuts in Cross–Media Colour Reproduction?, *Colour Image Science Conference*, Derby, 169–182.

Morovič J., Sun P. L. (2003) Predicting Image Differences In Colour Reproduction From Their Colorimetric Correlates, *Journal of Imaging Science and Technology*, vol. 47, pp. 509–516.

Morovič J., Sun P. L. and Morovič P. (2001) The gamuts of input and output colour reproduction media, *Proceedings of SPIE*, vol. 4300, pp. 114–125.

Morovič J., Morovič P. M. (2003) Determining Colour Gamuts of Digital Cameras and Scanners, *Color Research and Application*, vol. 28, no. 1, pp. 59–68.

Morovič J., Morovič P. M. (2005) Can Highly Chromatic Stimuli Have A Low Color Inconstancy Index?, *IS&T/SID 13th Color Imaging Conference*, Scottsdale, AZ, pp. 321–325.

Morovič J. and Wang Y. (2003) A Multi–Resolution, Full–Colour Spatial Gamut Mapping Algorithm, *IS&T/SID 11th Color Imaging Conference*, Scottsdale, AZ, pp. 282–287.

Morovič J. and Wang Y. (2003b) Influence Of Test Image Choice On Experimental Results, *IS&T/SID 11th Color Imaging Conference*, Scottsdale, AZ, pp. 143–148.

Morovič P. M., Morovič J. (2006) Spectral discrimination of color input devices, *IS&T/SID 14th Color Imaging Conference*, pp. 102–107.

Mortimer A. (1991) *Colour reproduction in the printing industry.* Pira International, Leatherhead, UK.

Motomura H. (1999) Categorical color mapping for gamut mapping: II. Using block average image, *Proceedings of SPIE*, vol. 3648, pp. 108–119.

Motomura H. (2000) Gamut Mapping using Color–Categorical Weighting Method, *IS&T/SID 8th Color Imaging Conference*, Scottsdale, AZ, pp. 318–323.

Motomura H. (2001) Categorical Color Mapping Using Color–Categorical Weighting Method — Part I: Theory, *Journal of Imaging Science and Technology*, vol. 45, p. 117–129.

Motomura H. (2002) Analysis of gamut mapping algorithms from the viewpoint of color name matching, *Journal of the Society for Information Display*, vol. 10, pp. 247–254.

Munsell A. H. (1941) *A Color Notation*, 9th Edition, Munsell Color Company, Baltimore.

Murch G. M., Taylor J. M. (1989) Color in Computer Graphics: Manipulating and Matching Color, *Eurographics Seminar: Advances in Computer Graphics V*, Springer Verlag, pp. 41–47.

Musgrave F. K. (1989) Prisms and Rainbows: A Dispersion Model for Computer Graphics, *Graphics Interface '89*, pp. 227–234.

Nakauchi S., Imamura M., Usui S. (1996) Color Gamut Mapping by Optimizing Percep–tual Image Quality, *Proc. 4th IS&T/SID Color Imaging Conf.*, IS&T, Springfield, VA, pp. 63–67.

Nakauchi S., Hatanaka S., Usui S. (1999) Color gamut mapping based on a perceptual image difference measure, *Color Research and Application*, vol. 24, pp. 280–291.

Nassau K. (2001) *The Physics and Chemistry of Color: The Fifteen Causes of Color*, Second Edition, John Wiley and Sons Ltd., Chichester, UK.

Neitz J., Carroll J., Yamauchi Y., Neitz M., Williams D.R. (2002) Color perception is mediated by a plastic neural mechanism that remains adjustable in adults, *Neuron*, vol. 35, pp. 783–792.

Nemcsics A. (1987) Color Space of the Coloroid Color System, *Color Research and Application*, vol. 12, pp. 135–146.

Neugebauer H. E. J. (1937) Die theoretischen Grundlagen des Mehrfarbenbuchdrucks, *Zeitschrift für wissenschaftliche Photographie*, vol. 36, no. 4, pp. 73–89.

Neumann L., Neumann A. (2004) Gamut Clipping and Mapping based on Coloroid System, *Proceedings of Second European Conference on Colour in Graphics, Imaging and Vision*, pp. 548–555.

Newman T. and Pirrotta E. (2000) The Darker Side of Colour Appearance Models and Gamut Mapping, *Proceedings of Colour Image Science 2000 Conference*, University of Derby, UK, pp. 215–223.

Nezamabadi M., Berns R. S. (2005) The Effect of Image Size on the Color Appearance of Image Reproductions, *IS&T/SID 13th Color Imaging Conference*, Scottsdale, AZ, pp. 79–84.

O'Rourke J. (2000) *Computational Geometry in C*, Second Edition, Cambridge University Press.

Ohta N., Robertson A. (2005) *Colorimetry: Fundamentals and Applications*, John Wiley and Sons, Chichester, UK.

Pankow D. (2005) *Tempting the Palette – A Survey of Color Printing Processes*, Second Edition, RIT Cary Graphic Arts Press, Rochester, NY, USA.

Pariser E. G. (1991) An Investigation of Color Gamut Reduction Techniques, *IS&T's 2nd Symposium on Electronic Publishing*, pp. 105–107.

Pellegri P., Schettini R. (2003) Gamut boundary determination for a colour printer using the Face Triangulation Method, *Proceedings of SPIE*, vol. 5008, pp. 92–100.

Pham B. and Pringle G. (1995) Color Correction for an Image Sequence, *IEEE Computer Graphics and Applications*, vol. 15, no. 3, pp. 38–42.

Pointer M. R. (1980) The gamut of real surface colors, *Color Research and Application* vol. 5, no. 3, 145–155.

Power J. L., West B. S., Stollnitz E. J., Salesin D. H. (1996) Reproducing color images as duotones. *Proceedings of SIGGRAPH 96*, ACM, New York, pp. 237–248.

Preparata F. P., Shamos M. I. (1985) *Computational Geometry: An Introduction*, Springer.

Pujol J., Martínez–Verdú F., Capilla P. (2003) Estimation of the Device Gamut of a Digital Camera in Raw Performance Using Optimal Color–Stimuli, *PICS*, pp. 530–535.

Rabbitz R. (1994) Fast collision detections of moving convex polyhedra, *Graphics Gems IV*, Heckbert P. (ed.), Academic Press, pp. 83–109.

Rasche K., Geist R., Westall J. (2005) Reproduction of Color Images for Monochromats and Dichromats, *IEEE Computer Graphics and Applications*, pp. 2–10.

Reinhard E., Ward G., Pattanaik S., Debevec P. (2005) *High Dynamic Range Imaging*, Morgan Kaufmann, San Francisco, CA, USA.

Rhodes P. A. (2002) Colour Notation Systems, *Colour Engineering – Achieving Device Independent Colour*, (eds.) Green P. and MacDonald L. W., John Wiley & Sons, Chichester, UK, pp. 105–126.

Richter K. (1980) Cube–root color spaces and chromatic adaptation, *Color Research and Application*, vol. 5, pp. 25–43.

Rizzi A., Pezzetti M., McCann J. J. (2007) Glare–limited Appearances in HDR Images, *IS&T/SID 15th Color Imaging Conference*, Albuquerque, NM, pp. 293–298.

Rosen M. R., Berns R. S. (2005) *Spectral Reproduction Research for Museums at the Munsell Color Science Laboratory, Proceedings of NIP21*, pp. 73–77.

Rosen M. R., Derhak M. W. (2006) Spectral Gamuts and Spectral Gamut Mapping, *Proceedings of SPIE*, vol. 6062, pp. 60620K.

Rosenblueth A., Wiener N. (1945) The role of models in science, *Philosophy of Science*, vol. 12, pp. 316–321.

Ruetz B. (1994) *Color Printing Method and Apparatus Using an Out–of–Gamut Color Table*, Canon Information Systems, U.S. Patent 5,299,291.

Saito R., Kotera H. (2000) Extraction of Image Gamut Surface and Calculation of it Volume, *IS&T/SID 8th Color Imaging Conference*, Scottsdale, AZ, pp. 330–334.

Sano C., Song T., Luo M. R. (2003) Colour Differences for Complex Images, *IS&T/SID 11th Color Imaging Conference*, Scottsdale, AZ, pp. 121–126.

Sara J. J. (1984) *The Automated Reproduction of Pictures with Nonreproducible Colors*, Ph.D. Thesis, Massachusetts Institute of Technology.

Schläpfer K. (1994) *Color Gamut Compression – Correlations Between Calculated and Measured Values*, IFRA Project Report, EMPA, 8 August 1994.

Schrödinger E. (1920) Theorie der Pigmente von größter Leuchtkraft, *Annalen der Physik* (Paris), vol. 62, pp. 603–622.

Scott J. (2006) Point in triangle test, retrieved 13 September 2006 from http://www.blackpawn.com/texts/pointinpoly/default.html.

Seetzen H., Heidrich W., Stuerzlinger W., Ward G., Whitehead L., Trentacoste M., Ghosh A., Vorozcovs A. (2004) High Dynamic Range Display Systems, *SIGGRAPH 2004, ACM Transactions on Graphics*, vol. 23, no. 3, pp. 760–768.

Sharma A. (2003) *Understanding Color Management*, Delmar.

Sharma G. (2003) Color fundamentals for digital imaging, in *Digital Color Imaging Handbook*, Sharma G. (ed.), CRC Press, Boca Raton, Fl., USA, pp. 51–90.

Shaw M. (2006) Gamut estimation using 2D surface splines, *Proceedings of SPIE*, vol. 6058, ref. 605807.

Shimizu M., Semba S., Suzuki S., Murashita K. (1998) Gamut Mapping Algorithm Suitable for Implementation to Device Profiles, *IS&T/SID 6th Color Imaging Conference*, Scottsdale, AZ, pp. 330–334.

Silvestrini N., Fischer E. P. (2005) Colour order systems in art and science, retrieved 13 January 2006 from http://www.colorsystem.com/index.htm.

Sinclair R. S. (1997) Light, light sources and light interactions, in *Colour Physics for Industry*, Second Edition, McDonald R. (ed.), Society of Dyers and Colourists, Bradford, UK.

Spaulding K. E., Ellson R. N., Sullivan J. R. (1995) UltraColor: A New Gamut Mapping Strategy, *SPIE Proceedings*, vol. 2414, pp. 61–68.

Spaulding K. E., Woolfe G. J. and Giorgianni E. J. (2000) Reference Input/Output Medium Metric RGB Color Encodings (RIMM/ROMM RGB), *IS&T PICS 2000 Conference Proceedings*, Portland, OR, pp. 155–163.

Stevens S.S. (1946) On the theory of scales of measurement, *Science*, vol. 103, pp. 677–680.

Stokes M. (1991) *Colorimetric Tolerances of Digital Images*, MSc Thesis, RIT.

Stollnitz E. J. (1998) *Reproducing Color Images with Custom Inks*, Ph.D. Thesis, University of Washington.

Stone M. C., Cowan W. B., Beatty J. C. (1988) Color Gamut Mapping and the Printing of Digital Color Images, *ACM Transactions on Graphics*, vol. 7, pp. 249–292.

Stone M. C., Wallace W. E. (1991) Gamut Mapping Computer Generated Imagery, *Graphics Interface '91*, pp. 32–39.

Sun P. L. (2002) *Influence Of Image Characteristics On Colour Gamut Mapping For Accurate Reproduction*, Ph.D. Thesis, University of Derby.

Sun P. L., Morovič J. (2002) What Differences Do Observers See In Colour Image Reproduction Experiments? *Color in Graphics Imaging and Vision (CGIV)*, pp. 181–186.

Sun P. L., Zheng Z. W. (2004) Selecting appropriate gamut mapping algorithms based on a combination of image statistics, *Proceedings of the SPIE*, vol. 5667, pp. 211–219.

Suzuki S., Shimizu M., Semba S. (1999) High–accuracy Color Reproduction (Color Management Systems), *FUJITSU Science and Technology Journal*, vol. 35, pp. 240–247.

TAMU (2006) *Statistics 30X Class Notes*, Texas A&M University, retrieved on 19 May 2006 from http://www.stat.tamu.edu/stat30x/notes/node150.html.

Taplin L. A., Johnson G. M. (2004) When Good Hues Go Bad, *CGIV 2004 – Second European Conference on Color in Graphics, Imaging and Vision*, Aachen, Germany, April 2004, pp. 348–352.

Tastl I., Moroney N., Bhachech M., Holm J. (2005) ICC Color Management and CIECAM02, *IS&T/SID 13th Color Imaging Conference*, Scottsdale, AZ, pp. 217–223.

Taylor J. M., Murch G. M. and McManus (1989) Tektronix HVC: A Uniform Perceptual Color System, *SID Digest of Technical Papers*, SID, San Jose, CA

Thompson E. (1995) *Color Vision*, Routledge, London, pp. 245.

Thurstone L. L. (1927) A Law of Comparative Judgment, *Psychological Review*, vol. 34, pp. 273–286.

Tomasi C., Manduchi R. (1998) Bilateral Filtering for Gray and Color Images, Proceedings of the 1998 IEEE International Conference on Computer Vision (ICCV), Bombay, India, pp. 839–847.

Torgerson W. S. (1958) *Theory and Methods of Scaling*, Wiley, New York, pp. 159–204.

UGRA (1995) *UGRA GAMCOM Version 1.1 – Program for the Color Gamut Compression and for the comparison of calculated and measured values*, UGRA, St. Gallen, 17 July 1995.

Uroz J., Morovič J., Luo M. R. (2002) Perceptibility Thresholds of Colour Differences in Large Printed Images. *Colour Image Science: Exploiting Digital Media*, MacDonald L. W. and Luo M. R. (*eds.*), John Wiley & Sons, 49–73.

van Gestel M., Draaisma H. (2001) Color Gamut Mapping along Electrostatic Field Lines, *Proceedings of NIP17: International Conference on Digital Printing Technologies*, Fort Lauderdale, Fl., pp. 801–803.

Viggiano J. A. S., Moroney N. M. (1995) Color Reproduction Algorithms and Intent, *IS&T/SID 3rd Color Imaging Conference*, Scottsdale, AZ, pp. 152–154.

Viggiano J. A. S. and Wang J. (1992) A Comparison of Algorithms for Mapping Color Between Media of Differing Luminance Ranges, *TAGA Proceedings*, vol. 2, pp. 959–974.

Vos J. (1984) Disability glare – a state of the art report, *CIE Journal*, vol. 3, no. 2, pp. 39–53.

Vrhel M. J., Gershon R., Iwan L. S. (1994) Measurement and analysis of object reflectance spectra, *Color Research and Application*, vol. 19, no. 1, pp. 4–9.

Wandell B. A. (1995) *Foundations of Vision*, Sinauer Associates, Sunderland, MA.

Wei R. Y. C., Shyu M. J., Sun P. L. (1997) A New Gamut Mapping Approach Involving Lightness, Chroma and Hue Adjustment, *TAGA Proceedings*, TAGA, Rochester, NY, pp. 685–702.

Wei R. Y. C., Sun P. L. (1997) A Psychophysical Comparison of Tone Reproduction Models Under Different Color Appearance Models, *Journal of the Chinese Association of Graphics Science & Technology*, Taiwan (R. O. C.), pp. 13–27.

Weisstein E. W. (2006a) *Kolmogorov–Smirnov Test*, MathWorld – A Wolfram Web Resource, retrieved 27 March 2006 from http://mathworld.wolfram.com/Kolmogorov–SmirnovTest.html.

Weisstein E. W. (2006b) *Normal Distribution*, MathWorld – A Wolfram Web Resource, retrieved 3 May 2006 from http://mathworld.wolfram.com/NormalDistribution.html.

Weisstein E. W. (2006c) *Correlation Coefficient*, MathWorld – A Wolfram Web Resource, retrieved 3 May 2006 from http://mathworld.wolfram.com/CorrelationCoefficient.html.

Wen C. H., Lee J. J., Liao Y. C. (2001) Adaptive Quartile Sigmoid Function Operator for Color Image Contrast Enhancement, *IS&T/SID 9th Color Imaging Conference*, Scottsdale, AZ, pp. 280–285.

Wen S. (2006) Display gamut comparison with number of discernible colors, *Journal of Electronic Imaging*, vol. 15, no. 4, 043001.

Westphal J. (1987) *Color: A Philosophical Introduction*, First Edition, Blackwell Publisher, Oxford.

Wikipedia (2005a) History of art, *Wikipedia*, retrieved 23 November 2005 from http://en.wikipedia.org/wiki/Art_history.

Wikipedia (2005b) Fresco, *Wikipedia*, retrieved 17 November 2005 from http://en.wikipedia.org/wiki/Fresco.

Wikipedia (2006a) Electromagnetic radiation, *Wikipedia*, retrieved 27 January 2006 from http://en.wikipedia.org/wiki/Electromagnetic_radiation.

Wikipedia (2006b) Psychometrics, *Wikipedia*, retrieved 20 April 2006 from http://en.wikipedia.org/wiki/Psychometrics.

Wikipedia (2006c) Psychophysics, *Wikipedia*, retrieved 20 April 2006 from http://en.wikipedia.org/wiki/Psychophysics.

Wikipedia (2006d) Coefficient of determination, *Wikipedia*, retrieved 11 May 2006 from http://en.wikipedia.org/wiki/Coefficient_of_determination.

Wikipedia (2006e) Delaunay triangulation, *Wikipedia*, retrieved 6 November 2006 from http://en.wikipedia.org/wiki/Delaunay_triangulation.

Wikipedia (2007a) Mahalanobis distance, *Wikipedia*, retrieved 4 June 2007 from http://en.wikipedia.org/wiki/Mahalanobis_distance.

Wild C. J., Seber G. A. F. (1999) *Chance Encounters: A First Course in Data Analysis and Inference*, John Wiley and Sons.

Wolski M., Allebach J. P. and Bouman C. A. (1994) Gamut Mapping. Squeezing the Most out of Your Color System, *Proceedings of the 2nd IS&T/SID Color Imaging Conference*, pp. 89–92.

Wright W. D. (1928) A re–determination of the trichromatic coefficients of the spectral colours, *Transactions of the Optical Society*, vol. 30, pp. 141–164.

Wu C. W., Thompson G. R., Stanich M. J. (2004) Digital watermarking and steganography via overlays of halftone images, *Proceedings of the SPIE*, vol. 5561, pp. 152–163.

Wyszecki G. (1953) Valenzmetrische Untersuchung des Zusammenhanges zwischen normaler und anormaler Trichromasie, *Die Farbe*, vol. 2, pp. 39–52.

Wyszecki G. (1958) Evaluation of Metameric Colors, *Journal of the Optical Society of America*, vol. 48, pp. 451–454.

Wyszecki G., Stiles W. S. (2000) *Color Science – Concepts and Methods, Quantitative Data and Formulae*, Second Edition, John Wiley and Sons Ltd., Chichester, UK.

Xiao K., Li C., Luo M. R., Taylor C. (2004) Colour Appearance for Dissimilar Size, CGIV 2004, pp. 65–69.

Yamaguchi M., Murakami Y., Uchiyama T., Ohsawa K., Ohyama N. (2000) Natural Vision: Visual Telecommunication based on Multispectral Technology, *IDW '00*, pp. 1115–1118.

Yang C. C., Kwok S. H. (2003) Efficient gamut clipping for color image processing using LHS and YIQ, *Journal of Optical Engineering*, vol. 42, no. 3, pp. 701–711.

Yendrikhovskij S. N. (1998) *Color Reproduction and the Naturalness Constraint*, Ph.D. thesis, Technische Universiteit Eindhoven, The Netherlands, ISBN 90–386–0719–9.

Yendrikhovskij S. N. (1999) Image Quality: Between Science and Fiction, PICS 1999: Image Processing, Image Quality, Image Capture, Systems Conference, Savannah, Georgia, April 1999, pp. 173–178.

Yendrikhovskij S. N., Blommaert F. J. J., de Ridder H. (1998) Optimizing Color Reproduction of Natural Images, *IS&T/SID 6th Color Imaging Conference*, Scottsdale, AZ, pp. 140–145.

Yule J. A. C. (1967) *Principles of Color Reproduction – Applied to Photomechanical Preproduction, Color Photography, and the Ink, Paper and Other Related Industries*, John Wiley and Sons, Inc., New York.

Zeki S. (1993) *A Vision of the Brain*, Blackwell Scientific Publications, Oxford, UK.

Zeng H. (2001) Gamut Mapping in a Composite Color Space, *NIP17: International Conference on Digital Printing Technologies*, Fort Lauderdale, FL, pp. 797–800.

Zeng H. (2006) Spring–Primary Mapping: Combining Primary Adjustment and Gamut Mapping for Pictorials and Business Graphics, *IS&T/SID 14th Color Imaging Conference*, Scottsdale, AZ, pp. 240–245.

Zhang X., Wandell B. A. (1996) A spatial extension to CIELAB for digital color image reproduction, *Society for Information Display Symposium Technical Digest*, vol. 27, pp. 731–734.

Zhao X., (2007) Representing an N–Dimensional Device Color Gamut, *15th IS&T/SID Color Imaging Conference*, Albuquerque NM, pp. 278–282.

Zolliker P., Dätwyler M., Simon K. (2005) Gamut Mapping for Small Destination Gamuts, *AIC Colour 05 – 10th Congress of the International Colour Association*, pp. 345–348.

Zolliker P., Simon K. (2006a) Continuity of gamut mapping algorithms, *Journal of Electronic Imaging*, vol. 15, no. 1.

Zolliker P., Simon K. (2006b) Adding Local Contrast to Global Gamut Mapping Algorithms, *CGIV 2006 – Third European Conference on Color in Graphics, Imaging and Vision*, Leeds, UK, June 2006, pp. 257–261.

Zong C. (1999) *Sphere Packings*, Springer Verlag.

Index

a, 113
*a**, 28
Adaptation, 29
  chromatic, 30
  cognitive mechanisms of, 31
  complete, 30
  elastic versus plastic, 31
  incomplete, 31
  local, 33–4
  prediction of, 31–2
  spatial, 32
  time to complete, 30
  to intensity, 30
Adaptively sampled distance field, 150
Adobe RGB, 84
Alpha hull, *see* Alpha shape
Alpha shape, 148–150
  conformal filtration of, 150
Angular subtens, 33
Appearance predictors, 112
  cross-contamination of, 112
  linearity of, 113
  uniformity of, 113
ATD, 118

b, 113
*b**, 28
BasicPhoto, 245
Bipolar cells, 20
Bit-depth, 76
Black point compensation (BPC), 86, 119–120, 222
Bradley-Terry model, 66–7
Brightness, 14

*C**, 28
CAM *see* Color appearance, model
Camera, 38, 165
Candelas per square meter, 26
CARISMA, 203–4, 206–7, 213, 214
Category judgment, 62
  data analysis of, 67–69
CCLIP, 187
Centre-surround organization, 20–1
Characteristic vector analysis, 235
Chroma, 14
CIE Guidelines for the Evaluation of Gamut Mapping Algorithms, 253–5
CIECAM02, 32, 116–8
  applied to ICC PCS, 119
CIECAM97s, 116–8
CIELAB, 28, 116–8
  non-uniformity of, 28, 116
CIELUV, 28, 116–8
CII *see* Color, inconstancy index
Circumcircle, 131
cJab, 157
Clipping *see* Gamut, clipping
  effect on in-gamut colors, 181–2
  spectral distance minimizing, 235
  uncontrolled, 138–9
CLLIN, 203, 211
CMF *see* Color, matching function
Color, 12
  appearance attributes, 13
  constancy, 32
  distinguishable number of, 157–8
  ecosystem, 16–7
  ecosystem interactions, 29–35
  enhancement, 79–80
  experiences, 11–13, 22
  experiences - variation of, 29
  ideal prototypes of, 80
  inconstancy, 31
  inconstancy index (CII), 31
  information context, 46

Color (*Continued*)
information type, 46
matching function (CMF), 25
measurement, 25
memory, 35
naming, 13
nature of, 11–12
order system, 13
trichromatic matching of, 25
Color appearance, 13–16
absolute, 14, 114
attributes, 13–16
effect of background on, 33
effect of size on, 35
experienced, 112
Hunt model of, 118
model, 32, 79
model's psychophysical basis, 118
predicted, 112
prediction, 28, 111, 172
preservation of, 111
relative to viewing conditions, 14, 114–5
Color difference, 27
acceptability, 58
between images, 57
dividing weights for, 184
effect of spatial arrangement on, 29
equation, 27
multiplying weights for, 184
perceptibility threshold, 58
preferences, 29
units, 27
versus visual similarity, 29, 183–4
Color management, 39–40
closed-loop, 81–83
context of gamut mapping in, 245–6
sRGB, 83–85
ICC, 85–8
Color order system, 13
Munsell, 116
Color reproduction, 45–6
additive, 35–6
conceptual stages of, 76–81
device, 37–9, 46
medium, 37–9, 46, 91
purpose of, 46
subtractive, 36–7
Color space, 13, 21, 28, 111
choice for gamut mapping, 121
comparison, 116–8
effect on parameter optimization, 114
gamut representation in, 118
uniform, 28
Colorant, 37
differences necessitating gamut mapping, 108–9
Colorfulness, 14
loss due to gamut mapping, 180
optimization, 80
Colorimetric Quality, 47
Colorimetry, 25–9, 54
Coloroid, 118
Complex flow, 150
Cones, 19
absorbance, 19
ratios, 19
Contrast, 32
loss due to gamut mapping, 180
optimization, 80
sensitivity function (CSF), 33
simultaneous, 32
threshold, 33
Convex combination, 144
Convex hull, 144–7
limitations, 145
modified, 146–7
Coordinates, 123
Covariance matrix, 218
Crispening, 32
CSF *see* Contrast, sensitivity function
Cubic color space units, 157
Cusp, 187
CUSP, 206
CUSP2CUSP, 211, 213

dCMYK, 74
Decomposition
into spatial frequency bands, 103
multi-resolution, 229
ΔE *see* Color difference
ΔE_{2000}, 29
ΔE_{76}, 28
ΔE^*_{ab}, 28
ΔH, 113
Density, 120
colorimetric, 120
Desired color reproduction property, 4–5, 47
dimensions, 47
effect of choice on gamut mapping, 105–6

Destination, 46, 175, 182–3
Detail loss due to gamut mapping, 180–1
Device *see* Color reproduction, device
  calibration, 79
  characterization, 77–9
Device color space, 3, 38, 73–6, 172–4
  colorimetric meaning of, 75
  virtual, 75
Device-directed rendering, 237
Difference from local mean (DLM), 227
Dimensions, 123
Display, 38, 133, 163–4
Distribution
  bimodal, 63
  normal, 56
  unimodal, 63
Domain
  frequency, 102
  spatial, 102
dRGB, 74
Duplication, 47
Dynamic range, 33
  of natural scenes, 30, 93
  simultaneous, 93
Dynamic response, 30

Ekphrasis, 59
Electromagnetic
  radiation *see* Electromagnetic, spectrum"
  spectrum, 19, 21–2
Evans consistency principle, 202–3
Expanding spheres tetrahedralization, 150
Extrapolation, 243
Eye, 18–20

Field
  of view, 19
  receptive, 21
Filter
  bilateral, 231
  edge-preserving, 230–1
  Gaussian, 230–1
  high-pass, 226, 227
  low-pass, 226
  size, 228
Flexible sequential LGB, 153–4
Fourier
  spectrum, 102
  transform, 102
Fundamental bases, 236

Gamulyt, 143
Gamut, 92, 133
  capture color analysis, 140
  clipping, 97, 100
  compression, 97
  computation, 106
  concavity, 137–9
  core, 104, 205
  destination, 103–104, 205–6
  differences, 3–4, 106–7, 175–6
  discreteness versus continuity of, 135–7
  display versus printer, 134
  effect of sampling on, 137–9
  effect of viewing conditions on, 133–4
  exceeding spectrum locus, 161
  expansion, 220–4
  extension *see* Gamut, expansion
  of camera, 133–4
  of well behaved devices, 141
  Pantone, 162
  Perceptual, 119
  Physical, 141
  Pointer, 161
  recommended sampling of, 139
  reduction *see* Gamut mapping
  source, 103–4, 205
  volume, 157
Gamut boundary, 143
  descriptor (GBD), 143
  descriptor evaluation, 155–6
  generic descriptor, 143
  intersection, 158–9
  medium-specific descriptor, 143
  smoothing, 159–160, 206
Gamut mapping, 80–1, 91
  accurate, 92
  affecting factors, 105–7
  aims, 94–95
  algorithm (GMA) *see* Gamut mapping
  application statistics, 240
  as art, 94
  as science, 95
  baseline, 255
  building blocks, 97–105
  color constancy, 93
  color-by-color, 194–220
  context, 95–6
  continuity, 257–8
  for business graphics, 237
  for closed-loop color management, 241

Gamut mapping (*Continued*)
for color management, 241–6
for deficient color vision, 238
for duotones, 237–8
for ICC color management, 242–4
for niche applications, 237–9
for sRGB color management, 242
for Windows Color System, 244–5
future focus of, 262
history of, 5–7
image dependence of performance of, 256–7
importance, 89–90
inverse, 244
ISO, 92
need for, 107–9
optimization of implementation, 10
optimized for LUT implementation, 244
publication statistics, 239–240
spatial, 102–3, 224–234, 246
spectral, 234–7, 246
types, 96–97
utopia, 263
Gamut mapping evaluation, 247–259
experiment statistics, 250–2
inter-relating results, 253–4
measurement-based, 53–8
number of GMAs, 250–1
number of observers, 251
number of test images, 251–2
psychovisual, 58–70
reference data, 248–250
trends in results, 252–3
Ganglion cells, 20
GBD *see* Gamut boundary, descriptor
GCA, 209
GCUSP, 198, 212
Gift-wrapping, 144–5
GMA *see* Gamut mapping

h^*_{ab}, 28
High dynamic range imaging (HDRI), 10, 93
HPMINDE, 188, 253–4
Hue, 13
constancy *see* Hue, linearity
linearity, 28, 116
shift *see* Mapping, of hue
uniformity *see* Hue, linearity
Hue Mapping, 245

ICC, 39, 85
color management *see* Color management, ICC
device profile, 40, 85, 172
rendering intents, 49–50, 86–7
version 2, 243
version 4, 244
Illuminant, 22
A, 22
D50, 23
D65, 23
F11, 23
TL84, 23
Image, 91
causes of different appearances, 2–3
content, 107–8
difference metric, 233–4
difference minimization, 233–4
enhancement, 79–80, 221
enhancement consequences, 221
gamut, 103–4, 167–9, 173–4
Inter-observer differences, 69
International Color Consortium *see* ICC
Interpolation, 81–82, 102, 189–190
Intra-observer differences, 69
IPT, 116

Just noticeable difference (JND), 27

Kolmogorov–Smirnov test, 56

L^*, 28
LABHNU, 118
LabPQR, 236
Lateral geniculate nucleus (LGN), 17, 21
LCh, 194
LCLIP, 188
LGB *see* Line, gamut boundary
Light, 22
source, 22
Lightness, 14
Line, 125
gamut boundary (LGB), 143
intersection with line, 127
intersection with plane, 128
segment, 128
Linear basis, 235
LLIN, 200–3
for expansion, 221–2, 223–4
Look-up table (LUT), 81

LSH, 239
Luminous efficiency function, 26
LUT *see* Look-up table

Mahalanobis distance, 218
Mapping
  along a path, 98–100
  along connections between corresponding distance points, 100
  along curves, 100, 211–2
  along electrostatic field lines, 190
  along fixed-slope line, 99, 189
  based on corresponding area, 213
  color space, 106
  color-categorical, 98, 216, 218–9
  combination of separate mappings, 214–6
  for multiple regions in hue planes, 208–210
  in hue planes, 204–216
  in nonappearance spaces, 120
  in three dimensions, 216–9
  inverse-gamma-inverse *see* Mapping, inverse-power-inverse
  inverse-power-inverse, 197, 227–9
  knee function, 101, 199–200
  linear, 100, 194–5
  manual, 98, 189–190
  non-linear, 199–200
  of chroma, 99, 188, 199–200, 203
  of hue, 80, 203–4
  of lightness, 99, 194–9
  of lightness (chroma dependent), 198–9
  of skin tones, 205
  onion-peel, 217–8
  path types, 99–100
  piece-wise linear, 100
  sigmoidal, 101, 195–7, 224–5
  spring-primary, 216, 242
  towards center of lightness axis, 99, 187, 206
  towards cusp lightness on lightness axis, 99, 206
  towards lightness axis and cusp-connecting line, 99, 206
  towards point off lightness axis, 99, 206–8
  towards point on chroma axis, 99
  towards single focal point in hue plane, 206–8
  towards source-color dependent lightness axis point, 99, 189
  type, 100
Mean filtering, 159
Measurement
  geometry, 25
  spectral power, 24
Memory colors, 35
Metameric black, 236
Metamerism, 20
MinCD, 245
MINDE *see* Minimum color difference clipping
Minimum color difference clipping, 97, 176–8
  consequences of, 177–8
  effect of mapping color space on, 183
  preserving hue, 188
  redundancy due to concavity, 178–9
  weighted, 98, 183–5, 242, 252
  weighted plus Black Point Compensation, 185–7
  weights for visual similarity, 185–6
Mixed illumination, 34
Monochromator, 24
Morphing *see* Interpolation
Mountain range, 155

Normal vector, 129
  of line, 130
  of plane, 130
Notation, 194

Object
  emissive, 22
  reflective, 23
  self-luminous, 22
  transmissive, 23
Object Color Solid, 141, 160–1
Observer
  indirectness of judgments, 59
  instructions, 62–3
Opponent color signals, 21
Optic nerve, 20
Original, 46, 171–5
Overcompensation, 138–9

Pair comparison, 61–2
  data analysis, 64–7
Paradigm shift, 263
Perfect diffuser, 24
PCS *see* Profile Connection Space
Photodetector, 24
Photoreceptor, 18–9
Plane, 126

Point, 123
  focal, 206
  inside line segment, 128
  inside triangle, 128
Polytope, 144
Precision, 131
Principal component analysis, 235
Print, 39, 162–3, 172–3, 175
Profile Connection Space (PCS), 39, 85
  encoding efficiency of, 75
Projected motion picture film, 163
Projection, 39, 164–5
Psychometrics, 58
Psychophysics, 58

Radiance ratio, 232
Radiometry, 24
Ranking, 61
  data analysis, 63
Re-purposing, 50
Re-rendering, 92
Re-targeting, 50
Reference medium gamut (RMG), 87, 244
Reference white, 30
Rendering intent, 49
  ICC-absolute colorimetric, 49, 87
  media-relative colorimetric, 49, 86
  perceptual, 50, 87
  saturation, 50, 87
Reproduction, 46
  colorimetric, 48
  compromise, 52
  corresponding, 48
  creative, 52
  equivalent, 48
  exact (Field's), 52
  exact (Hunt's), 48
  Hunt's objectives for, 48–9
  optimum, 52
  preferred (Field's), 52
  preferred (Hunt's), 48
  spectral, 48
  Yule's properties of, 47
Retinex, 34
  multi-scale, 232
RMG *see* Reference Medium Gamut
Rods, 19
ROMM, 255

Sampling, 54–5
  for multi-primary devices, 141
  inputs to input devices, 141–3
  inputs to output devices, 140–1
Saturation, 16
  in gamut mapping, 115, 215–6
SCACLIP, 231
SCACOMP, 231
Scale, 61
  interval, 61
  nominal, 61
  ordinal, 61
  ratio, 61
Scanner, 37–8, 165
SCLIP, 187
  spectral, 237
Segment maxima, 150–5
  GBD matrix, 151
  piece-wise bilinear surface, 153
  triangulated surface, 153
SGCK, 198, 206, 212, 243, 245, 253–4
SGDA, 213
SLIN, 206
  for expansion, 223
SOCS, 161–2
Soft-clipping, 199
Solid angle, 24
Source, 46
Space
  cylindrical, 124
  orthogonal, 124
  spherical, 125
Spatial mapping *see* Gamut mapping, spatial
  frequency-sequential, 226–231
  Retinex-based, 232–3
SPD *see* Spectral, power distribution
Spectral
  mapping *see* Mapping, spectral
  power distribution (SPD), 22
  reflectance, 23
  reflectance (dimensionality of), 24
  transmittance, 23
Spectrophotometer, 24
Specular highlight, 85–6
Sphere packing, 158
Spherical segment, 150
Spreading, 32
sRGB, 40, 118
  color management, 83–5
Standard Colorimetric Observer, 25–6

Statistics
  choice of, 56–7
Steradian, 24
Stimulus, 25
  presentation, 60
Subjective accuracy, 49
Surface
  glossy, 23
  matte, 23
  splines, 155
SWOP, 118

Telespectroradiometer (TSR), 24
Tessellation, 131–2
  Delaunay, 131–2, 148
Thurstone's law of comparative judgment, 64
TOPO, 211, 213
  for expansion, 223
Torgerson's law of categorical judgment, 67
TRIA, 98, 210–211, 245
Triangulation *see* Tessellation
Tristimulus values, 26
  absolute, 93

UltraColor, 101, 216, 242
Unwanted absorptions, 37
Unwanted stimulations, 36

Viewing conditions, 46
  effect of changes on color, 3
  effect on gamut, 165–7
  ISO 3664, 85, 171
  differences necessitating gamut
    mapping, 109
Viewing mode, 34–5
  effect on gamut, 166–7
Vision
  mesopic, 19
  photopic, 19
  scotopic, 19
  trichromacy of, 20
Visual
  acuity, 19
  cortex, "17, 21"
  pathway, 18
  similarity, 29

Watt, 24
Wavelength, 22
Wilcoxon rank-sum test, 69–70
Windows Color System (WCS), 40, 88–89
WUV, 118

$X_LY_LZ_L$, 93
xy chromaticity, 26
XYZ, 26
  non-uniformity, 28

YIQ, 239
YSP, 238

z-score, 64

Printed and bound by CPI Group (UK) Ltd, Croydon, CR0 4YY